D0894210

THE FOSSIL HUNTERS

In Search of Ancient Plants

ALSO BY HENRY N. ANDREWS:

Ancient Plants and the World They Lived In
Index of Generic Names of Fossil Plants, 1820–1965
Studies in Paleobotany

THE FOSSIL HUNTERS

In Search of Ancient Plants

Henry N. Andrews

University of Connecticut

Cornell University Press

ITHACA AND LONDON

First published 1980 by Cornell University Press.
Published in the United Kingdom by Cornell University Press Ltd.,
2–4 Brook Street, London W1Y 1AA.

International Standard Book Number 0-8014-1248-X
Library of Congress Catalog Card Number 79–24101
Printed in the United States of America
Librarians: Library of Congress cataloging information appears on the last page of the book

To my wife, Elisabeth, for encouragement at all times, and to our many colleagues of the past whose story this is

Contents

Acknowledgments

A GREAT many people have contributed in one way or another to the preparation of this book. Many are acknowledged in the text itself. Others are cited in my "Sources of Information," and I am most grateful to the authors of the numerous biographies from which much of my account has been drawn.

For direct financial aid I wish to acknowledge a research grant from the History and Philosophy of Science Section of the National Science Foundation. Also, several grants, which helped in many ways, were received from the University of Connecticut Research Foundation.

A large portion of the library research was carried on in the University of Connecticut Library and several libraries in Cambridge, England. In Cambridge I spent many months in the University Library as well as in the Botany, Geology, History of Science, and Scientific Periodicals libraries. This was a delightful experience and I thank all of the staff members who helped me for their cordial and very efficient service.

Dealing with the various languages from which my source material comes has presented a major problem, and I am very grateful to the several scholars who have assisted me by translating Russian, Swedish, German, Polish, and Czech literature.

I am grateful also to all those who provided me with photographs or gave permission to reproduce them in this book. In some cases it has proved impossible to trace the precise origins of a picture, but I have made every effort to give proper credit in the captions.

HENRY N. ANDREWS

Laconia, New Hampshire

THE FOSSIL HUNTERS

In Search of Ancient Plants

1

Introduction

IN the mid-forties I wrote a book called *Ancient Plants and the World They Lived In*. I believe my enthusiasm at that time was somewhat greater than my knowledge but writing it was a pleasant task, carried out in the evenings partly as a relief from teaching mathematics to innumerable classes of civilians and soldiers. The book includes a chapter entitled "The Fossil Hunters," in which I tried to tell just a little about the lives of the paleobotanists themselves. I have thought a good deal about this aspect of our science and, with the encouragement of some of my colleagues, I began seriously to gather biographical material about ten years ago. It was not possible for me to do very much with biographical history along with my studies of Devonian plants, so I turned over the latter to my younger colleagues.

This is, therefore, a story about paleobotanists, about men and women who have, during the past three hundred years, probed about in the rocks of the Earth looking for vestiges of the life of ages long gone. It is not a textbook, nor is it intended to be complete or encyclopedic. There are now textbooks on fossil plants in several languages, there are many journals in which reports of research are published, there are bibliographic reports and summary accounts of varying kinds, and there have been numerous historical sketches, although most are brief and of a regional nature. These will be referred to specifically in due course. This great mass of literature, some of which is excellent and very useful, deals chiefly with the fossil plants themselves and their presumed living relatives. But we tend to learn rather little about the men and women who have done the digging, worked in the laboratory, and written their reports. I think it helps us to understand the scientific works of an author if we know something about the other facets of his life, and in many cases these are interesting, and varied, and worth recording.

We have had, and still have, in our coterie of fossil botanists some unique and fascinating people, and some of them have been very versatile, making diverse and important contributions in other fields as well. A few examples may reveal in some degree the flavor of the chapters that follow: Hugh Hamshaw Thomas, of Mesozoic pteridosperm fame and a great morphologist with a strong interest in angiosperm origins, was one of the early authorities on the military use of aerial photography. Marie Stopes, one of the most noteworthy students of coal microscopy in her day, also a seeker of early angiosperm fossils who spent a year and a half traveling alone in Japan to this objective, became the pioneer in birth-control education in Britain. Roland W. Brown, an important contributor to our knowledge of western American Tertiary floras and an acknowledged authority on the Cretaceous-Tertiary boundary, was a master lexicographer and is widely appreciated for his book *Composition of Scientific Words*. W. C. Williamson, whose monumental and basic studies on the structure of Carboniferous plants were written in his "spare time," had a full-time and very busy medical pratice and held the Professorship of Natural History at Owen's College, in Manchester, where he taught botany, zoology, and geology. And Lester Ward, one of the most learned of them all, known for his studies of Mesozoic floras and his bibliographic and historical works in paleobotany, left it eventually to found the science of sociology in the United States. These were great people by any standards, and there are many more.

The study of fossil plants, although young by some historical yardsticks, is now represented by a vast literature produced by many people. It is therefore imperative to state the approximate limitations of the field I shall try to encompass. It was my original intention to begin with the nomenclatorial starting point of "scientific" paleobotany, namely, the great publications of Kaspar Sternberg and Adolphe Brongniart in the early 1800's. But as a matter of personal interest I spent some time delving into the lives and studies of several of the naturalists of the late seventeenth and early eighteenth centuries—men such as Martin Lister, Edward Lhwyd, John Woodward, John Ray, and Robert Hooke. It is my impression that their contributions are more important than most writers of paleobotanical history have acknowledged. They began to clear a rough path through a thick forest of superstition and theological tradition, a task that had to be done and one that made possible the respected work of the early nineteenth century. They struggled, knowingly or otherwise, with two great stumbling blocks in paleontology: the medieval concept that fossils were meaningless rock formations ("formed stones"), in no way related to formerly existing plants and animals; and the theory that the Deluge, the Great Flood, had stirred up the lands of the Earth and left plant and animal remains entombed in the resultant sediments.

Numerous writers have dealt with this aspect of the history of science, and I have tried to select, from the original sources insofar as possible, passages that are especially pertinent to paleobotany. Some of these early workers were brilliant men, and some were not, but in the context of the intellectual atmosphere of their time it is more appropriate to try to understand their interpretations than to ridicule them. Although it may not have been entirely clear and straight, they started us on the right track.

Records of man's interest in fossils, of course, date back much earlier than the seventeenth century. In his "Fossil Lore in Greek and Latin Literature," Eugene S. McCartney (1924) notes:

> Strange to say, the earliest speculation about fossils that I can find evinces a fairly clear understanding of their geological significance. By appealing to the evidence provided by shells and the petrifactions of marine life discovered on mountains and in quarries, Xenophanes, a philosopher of the sixth century B.C., endeavoured to prove that the surface of the earth had once been beneath the waters of the sea. In the quarries of Syracuse he noticed the impression of a fish and of seaweed (?); in Paros, that of a small fish called *aphua*; in Malta, impressions of many kinds of sea life. He held that these changes took place when there was a stream of mud which later dried and kept the stamp or the imprint, so to speak, of the animals it overwhelmed. All humankind was destroyed when the land was thus carried into the sea; after this calamity man had to begin life anew. [P. 37]

And Lester Ward (1885) offers the following about the "ancients":

> It may surprise some to learn that the conclusions reached by the ancients were far more correct than those drawn twelve to sixteen centuries later, from much more ample data. Strabo, Xenophanes, Xanthus, Eratosthenes, and even Herodotus believed that the fossil shells they had seen once contained living animals, and that in process of time they had been turned into stone. They further concluded that the mountains in which they were found imbedded were once under the sea. [Pp. 385–86]

While recognizing the validity of these very early observations I think it is fair to say that the truths they recorded went unnoticed for many centuries. In Ward's words "the flicker that Pliny kindled upon the dying embers of Grecian learning was allowed to go entirely out" (p. 388). It is really not until we come to the "new" observations of certain Italian scholars in the fifteenth century and the British and Germans of the late seventeenth century that we find a continuing interest in fossils, and more especially in fossil plants. There was a lull in the latter half of the eighteenth century that I do not entirely understand, but the sparks were kept alive to burst into flame in the early 1800's.

I have thus chosen to start my account with a fairly detailed consideration of the British workers of the late 1600's, when developments were representative of what was going on generally in Europe at the time. It was fortunately a period when most workers in natural history were beginning to write in their national languages rather than in Latin. Some of the more important early continental naturalists will be introduced but, in keeping with my basic pattern in this book, I will offer what seem to be representative accounts rather than attempt a complete survey which could be little more than statistical in the space available.

Most of those who concerned themselves with fossils up until the start of the nineteenth century were either medical men or theologians by training. A few emerge in the early 1800's who were primarily botanists, but most were naturalists who were interested in, and contributed to, a wide range of subjects even into the latter part of the last century. Universities did not hire men to teach *paleobotany*, a word that Lester Ward coined in 1885. Lucrative academic posts were rare and the duties broad. The study of fossils was closely tied to geological investigation—or geognosy, as it was called by the German mineralogists—and the developing understanding of sedimentary rocks was accompanied by an interest in their entombed fossils.

Two threads woven into the paleobotanical fabric are especially significant through the work of the past three centuries. The first is an interest in the fossils as such, with the objective of correlating them with living plants and an understanding of the origins of the latter—which we now call evolutionary studies. The second is the use of the fossils in practical stratigraphical studies. We shall be concerned with the first throughout. The second is particularly important in the earlier years; more recently, like other aspects of the subject, it has become a science in itself.

A few specific areas in fossil botany I consider to be outside my central theme for one reason or another. I have not tried to deal with palynology, which has expanded so tremendously in the last few decades. The reader may refer to Traverse's recent excellent account (1974) for a brief summary of the past quarter century. Spores, however, both isolated and in the organs that produced them, have been studied for a long time and they will occasionally enter into the story. I shall deal with megafossils, but even here there are several branches of the subject that will be slighted. In particular many of the angiosperm (flowering-plant) floristic studies prior to the start of the present century quite clearly require very extensive revision.

There is a strong tendency in all of the historical-paleontological accounts to emphasize or deal exclusively with the great names as represented by their impressive discoveries and major publications. This practice is to some degree justified, but it is by no means easy always to point to one man as deserving the bulk of the credit for an important advance in knowledge. I will try to show how the many inconspicuous contributors, working in the background, have made possible some of the great advances. In my own career I have been aided by coal miners, sheep herders, museum curators, "rock hounds," students, and many other nonprofessionals whose services are indispensable.

The problem of where to end the story has proved more difficult than selecting a starting point. I had thought of excluding my colleagues who are still alive, but this seems a very artificial and unsatisfactory kind of cut-off—and all with whom I have discussed this project have objected to such a termination. Too much has transpired in the past few decades to deal with all of it critically here, and it is hardly appropriate for me to judge the work of the many younger workers. As a compromise I have set a period of about twenty-five years ago as an approximate stopping point. But I leave the door partly open; many significant developments in paleobotany show a continuity of progress up to the present day, and some of these are included. I hope that they will be accepted as examples of significant and continuing lines of research.

A few notes on my sources of information are in order. My story is derived in part from personal recollections of my colleagues over the past forty years; aside from that, I can make no claim to originality other than the way in which the facts I have sought out are put together. This is a story about many people, and I will let them speak in their own words whenever possible; but it is in large part gleaned from biographical sketches written, presumably, by those who knew best the character and achievements of the fossil botanists themselves.

I have combed through private and institutional reprint collections and thumbed through many bibliographies for biographical publications. National biography encyclopedias, although often brief, are helpful and usually lead one to more detailed accounts.

Some paleobotanists with a special bent for history have left excellent regional and specialized reviews. H. R. Goeppert, for example, prefaced his major writings with long and valuable historical accounts. Reid and Chandler's great work *The London Clay Flora* contains an exemplary historical introduction as well as a consideration of other topics that are basic to understanding the work itself. A great many such technical publications include historical data and I will refer to them in their proper places.

I must make a special acknowledgment of two great historical records prepared by Lester Ward: his "Sketch of Paleobotany," which appeared in 1885, and "The Geographical Distribution of Fossil Plants" of 1889. The "Sketch" is a chronological account of the history of fossil-plant studies up to 1885; it is readable and informative and reveals Ward's vast knowledge of languages and his incredibly retentive mind. The "Geographical Distribution" is a very detailed account of fossil-plant studies based on their geographical origin. It is a vast storehouse of information on the literature up to the time of its publication.

A few words are required on the matter of documentation of the information that is presented here. In the many places where statements are drawn, or quoted, from specific sources I have tried to cite the reference as precisely as possible yet without interfering with the readability of the text. Since my story relates to some forty years of experience with both people and literature, my sources of information in some instances may be less precise than a reader would desire. In this connection it may be noted that only in the present century has the practice of giving literature citations at the end of a research paper become in any way formalized and really helpful. I suppose that prior to that time most naturalists who published regularly knew the important people in their field and what they were doing, and expected that others were correspondingly informed. It thus can be very time-consuming, and sometimes impossible, to dig out the origins of some of their statements. Among the D. H. Scott letters there is one that he received from F. O. Bower (undated but evidently written about 1900) in which he makes it clear that he is not enthusiastic about the new system: "Do you not think these lists of literature at the end of papers, now so fashionable, are a trifle absurd. What does Miss——— want with telling us she has referred to Sach's Textbook & Goebel's outlines & Hooker's Genera & Species Filicum. It is really absurd waste of print. She might as well tell us what spelling books she used. Thank heaven, I never indulged that harmless form of vanity."

Finally, I would like to emphasize that my primary objective has been to record impressions of the people who are the subject of this book. Of the many hundreds of biographical sketches that I have read this approach to the kind of historical record that I have tried to develop is best expressed by B. L. Robinson in a short memorial that he wrote for the botanist Hermann, Count Solms-Laubach:

> In estimating a great scholar, inspiring teacher, or productive scientist, statistical matter takes secondary rank. The bulk of his published output, size of his classes, length of service, extent of his organizing and administrative activities—all these quantitative matters, which seem to have made the sum of his achievements, are much less significant, certainly far less interesting, than those human traits which made him what he was. His energy, sincerity, breadth of mind, his clarity of diction, skill of hand, quickness of apprehension, fairness in controversy, kindly interest, even his sense of humor—just these things which cannot be measured are what

make a man a vital force in his epoch. It is through them that he catches the attention of his contemporaries, influences his colleagues, guides and stimulates his students, and thus leaves his impress upon a branch of knowledge. [1925, p. 651]

2

Of Mineral Leaves and the Great Flood

IN the pages that follow we will look into the lives of a few men who were among those responsible for what seem to be the "significant beginnings" of fossil plant studies—studies that have been more or less continuous, and expanding, from their time to our own. We shall see a little of their characters, their interests, and their problems.

The men are Robert Plot, first curator of the Ashmolean Museum at Oxford University; his successor Edward Lhwyd, a great Welsh antiquarian and traveler; Martin Lister, a distinguished naturalist and physician to some of Britain's highest ranking royalty; John Woodward, of Cambridge, perhaps the most important paleobotanical collector of his time; and Johann J. Scheuchzer, a great Swiss naturalist who fits into the story as I wish to present it. These are our leading characters, but their work was in a general fashion supported, criticized, and developed by two subsidiary contributors: John Ray, one of the greatest botanists of all time, and Robert Hooke, a sort of universal genius, far ahead of his time, who probed into almost every aspect of science and unfortunately never followed any one to a full fruition.

The problems that these men dealt with were numerous but the most important were the real nature of fossil plants and animals, how they came to be present in the rocks of the Earth, how the rocks themselves originated, the concept of time as it relates to paleontological matters, and the nature of species, whether changing with time or not.

Lhwyd and Lister were rather strongly committed to the belief that fossils were, in large part, "sports" or freaks of nature and did not represent real plants or animals that once lived on the Earth. They were, however, uncertain in some cases; both of them were intelligent men who reflect some of the better, if somewhat imperfect, thinking among naturalists of the late seventeenth century. They thought a good deal about the matter and communicated freely with John Ray; they were hard workers and their scientific contributions were numerous and diverse. Lhwyd was an especially significant and interesting man whose work and thoughts are worth understanding.

It may be noted that the word *fossil* at this time and for a considerable period thereafter was applied to almost anything that was dug out of the earth. Unless otherwise indicated I am confining its meaning to specimens that reveal the former existence of plant or animal life in the modern context.

Edward Lhwyd (1660–1709) was born at Llanvorda near Oswestry in Wales; he was a patriotic Welshman who loved and served his native land all of his life. He differs from most of the students of natural history of his era in having been devoted in this direction from an early age without the bias of either clerical or medical studies. He entered Jesus College at Oxford in 1682, where he came under the influence of the botanist Professor Jacob Bobart and of Dr. Robert Plot. He also made the acquaintance of Martin Lister, and John Ray with whom he maintained a lasting correspondence.

In 1684, Lhwyd was appointed under-keeper of the recently established Ashmolean Museum in Oxford, and when the keeper, Dr. Plot, resigned in 1690, Lhwyd was appointed to the vacant post. Lhwyd was a man of broad antiquarian interests; among his published works are studies of the Celtic language and an Irish-English dictionary. But one of his most significant works is his *Lithophylacii Britannici Ichnographia* (A Picture Book of British Preserved Rock), a catalogue of the fossils ("formed stones") and minerals in the museum's collections which was printed in an edition of 120 copies in 1699. That Lhwyd was well regarded by his contemporaries is evidenced by the fact that this book was brought to life by financial aid from several of his friends, including Isaac Newton and Hans Sloane, the former best known in connection with his mathematical and gravity studies, and the latter as the maker of the collection that eventually led to the formation of the British Museum. Samuel Pepys also saw merit in Lhwyd's work: on July 6, 1698, Sloane wrote to Dr. Charlett, the Master of University College at Oxford: "Mr. Pepys and I drank yr health this day, after and before a great deale of serious discours. . . . Wee have in ¼ of an hour order'd Mr. Floid's [Lhwyd's] book of form'd stones to be printed" (North, 1931, p. 45). It may be noted as a matter of clarification that Lhwyd's name has been spelled in a variety of ways.

Pages 11 to 14 of the *Lithophylacii* are devoted to fossil plants and there are two quite respectable plates in which one can identify such Carboniferous genera as *Neuropteris*, *Alethopteris*, and *Annularia*.

Schneer (1954) has made a fitting tribute to Lhwyd's paleontological contributions:

> his real interest was in figured stones. He saw the need for a handbook of fossils that could be carried into the field, and proposed to catalogue his fossils of Oxfordshire. This project grew into a catalogue of the fossils of Britain, the *Lithophylacii Britannici Ichnographia*. Upon its publication, Hearne noted in his diary that Lhwyd was now esteemed the foremost naturalist in Europe. Plot's laboratory boy, whose father had dispensed with the formality of marriage to his mother, had earned the encomiums of men so widely distinct in their interests as Dean Hickes and Hans Sloane. [P. 260]

One's admiration of Lhwyd's achievements can only be increased with an understanding of the struggles he engaged in. His income at the Ashmolean was at best at a subsistence level, and not until 1697 was Lhwyd able in some measure to fulfill his ambition to travel. Then for a period of about four years he was mostly away from Oxford, traveling extensively in England, Wales, Scotland, and Ireland, and briefly in France. His objectives may best be described as a general cultural survey of the areas he visited. He wanted to preserve anything of real interest, whether in the category of natural history, archaeology, or languages.

Travel three hundred years ago was precarious, especially for one who was trying to collect the kind of information that concerned Lhwyd. He and his companions frequently met with suspicion and opposition approaching violent treatment—even in his native land; from Tenby in South Wales he wrote on February 28, 1698: "I find many of my Letters this last year have miscarried, intercepted I suppose by the Country people who were very jealous of us & suspected us to be employed by the Parliament in order to some further Taxes, & in some places for Jacobit spies" (Ellis, in Gunther, 1945, p. 32).

It was in Brittany, in February, 1700, however, that he encountered his most severe difficulties and his French travels were terminated abruptly. Lhwyd and his companion Parry were reported as being spies to the authorities in Brest. When they arrived the Intendant promptly and without examination ordered them to the Castle, where they were imprisoned for eighteen days. An examination of their papers finally convinced the authorities that they were not spies and they were released, but they were refused a pass to Paris and ordered to leave the kingdom.

Robert Plot (1640–1696), was a rather strong proponent of the formed-stone concept, although not completely set in his opinion. He was born to a family of property in Kent and received his B.A. and M.A. degrees at Oxford. He was elected to the Royal Society in 1677, was appointed secretary to that organization in 1682, and edited the *Philosophical Transactions* from No. 143 to No. 166. He was the first custodian of the Tradescant collections,

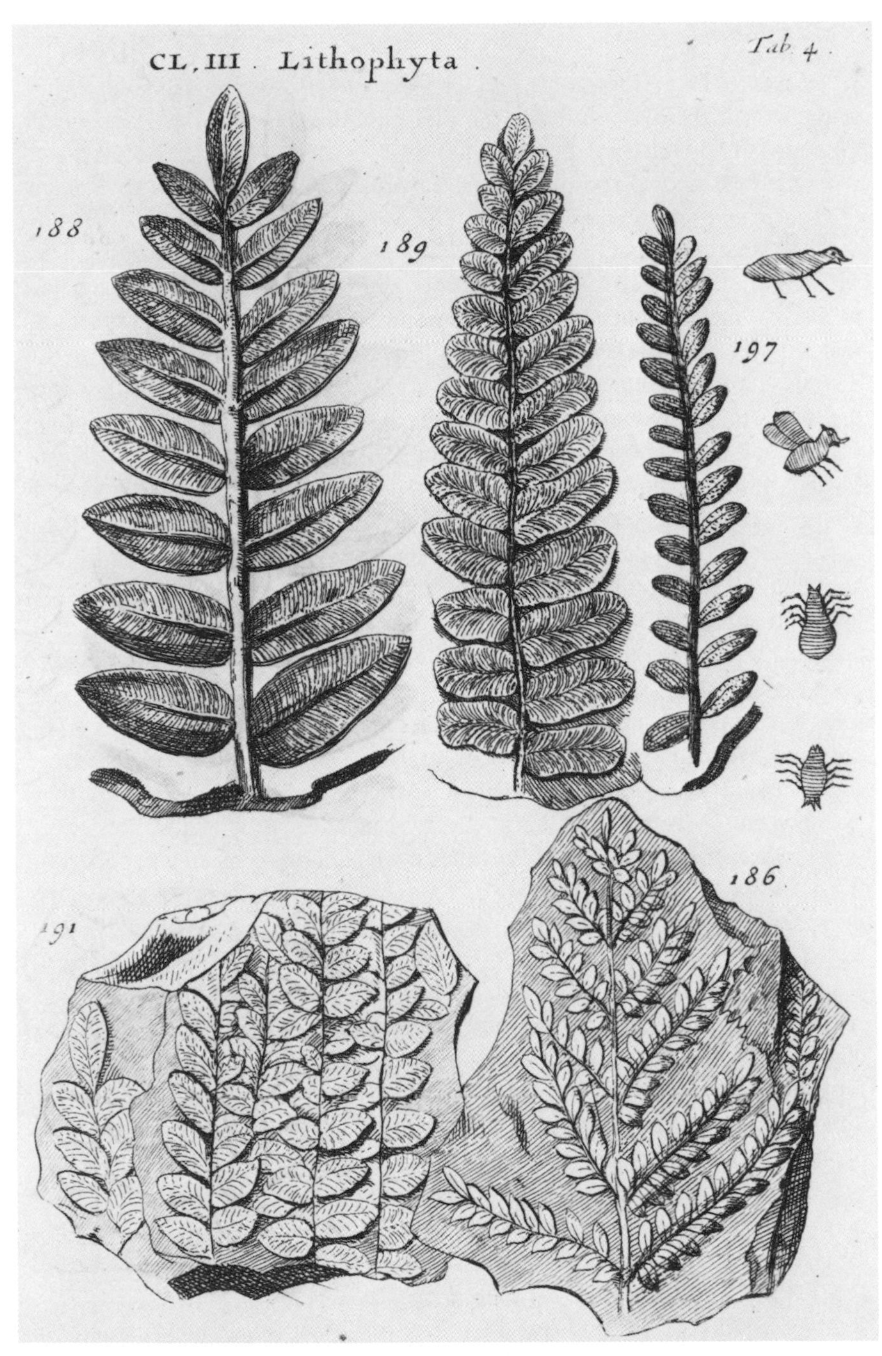

A plate from Edward Lhwyd's *Lithophylacii Britannici Ichnographia*, 1760.
Photo courtesy of the Cambridge University Library.

which arrived in Oxford in 1683 to form the nucleus for the Ashmolean Museum, of which Plot was made the first custodian; he was also appointed Professor of Chemistry in that year.

Plot is best known for his regional natural histories: *The Natural History of Oxfordshire* appeared in 1677 and *The Natural History of Staffordshire* in 1686. In the Oxfordshire volume Plot states that he accepts the opinion of Mr. Lister that fossils are *lapides sui generis* ("stones generated spontaneously"), and he touches on several points in the controversy of the time that are significant. He notes that the Flood might have been expected to distribute animal and plant remains rather indiscriminately, "whereas we find the stones that resemble them many times at the tops of hills, and but in few valleys; and those not scattered neither indifferently one amongst another, but for the most part those of a kind together" (1677, p. 112). He also regarded the forty days assigned for the Flood as having been "too small a time for so many shell-fish, so dispersed, as they must be presumed to be by so violent a motion, to get together and sequester them selves from all other company, and set them down, each sort, in a convenient station" (p. 113).

It was a particular concern of the scholars of the time to identify fossils, plants and animals, with living species. When this could not be done the most logical argument was that many parts of the Earth and the oceans were not known and that eventually these strange fossils would be matched by living counterparts. Plot adds a unique twist of his own: "If it be said, that possibly these Species may be now lost, I shall leave it to the Reader to judge, whether it be likely that Providence which took so much care to secure the works of the Creation in Noah's Flood, should either then, or since have been so unmindful as to suffer any one Species to be lost" (1677, pp. 114–15).

In answer to the argument that Nature does nothing without some good reason, and in reference to the intricacy of form of many of the fossils, so close to that of living organisms, Plot offers a charmingly naive answer based on aesthetics: "Nature herein acts neither contrary to her prudence, human raciocination, or in vain, it being the wisdom and goodness of the Supreme Nature, by the School-men called Naturans, that governs and directs the Natura naturata here below, to beautifie the World with these varieties; which I take to be the end of such productions as well as of most Flowers, such as Tulips, Anemones, etc. of which we know as little use of as formed stones" (1677, p. 121).

In his history of Staffordshire Plot discusses plant remains buried in peat and concludes that the tree trunks found there are truly such and were growing in the immediate vicinity. This would seem to be a case of dealing with "easy" fossils whose origin was quite obvious. With older plant remains found in the rocks of the earth he tended to adhere to the "formed-stone" concept, but even here he did leave the door open, and in the Oxfordshire history he sums up as follows:

> And thus I have given the grounds of my present opinion, which has not been taken up out of humor or contradiction, with intent only to affront other worthy Authors modest conjectures, but rather friendly to excite them, or in others, to endeavour collections of shell-fish, and parts of other Animals, that may answer such formed stones as are here already, or may hereafter be produced: Which when ever I find one, and the reasons alleged solidly answered, I shall be ready with acknowledgment to retract my opinion, which I am not so in love with, but for sake of Truth I can chearfully cast off without the least reluctancy. [P. 121]

In his short biographical sketch of Plot, Seccombe (1909) says: "Plot, who is said to have been a bon vivant, was a witty man and he knew how to render his stores of learning attractive to a wide circle of readers. . . . His acquisitiveness was such as to disgust some of his fellow-antiquaries, and Edward Lhuyd . . . credits him with as 'bad morals as ever' characterised a master of arts" (p. 1311). Several writers who have dealt with Plot remark on his presumed credulity, referring to an oft-quoted comment that the Staffordshire squires who supplied him with some of his information took delight in having "humbugged old Plot," which, as Challinor (1945) notes, is only to the discredit of the squires.

In weighing such criticisms it is well to remember that Lhwyd was anxious to take over Plot's position in the museum. A good many antiquaries (using the term in a broad sense) tend to be acquisitive, and I have been subjected myself to the opinions, witty and otherwise, of "country squires." I think that Plot's county histories were better than most such works of the time, at least as far as the natural history in them is concerned, and Challinor notes that the Staffordshire work stood for 150 years as the only natural history of the county.

Both Lhwyd and Martin Lister carried on a considerable correspondence concerning fossils and other aspects of natural history with John Ray (1627–1705). Few men have accomplished so much working with so little beyond their own innate abilities; there are numerous biographical sketches about him, the most detailed being Charles Raven's informative and delightful book (1942).

John Ray, the son of a blacksmith, was born in the small Essex village of Black Notley. He was admitted at Katherine Hall, Cambridge, in 1644 and transferred to Trinity in 1646, where "the polite arts and sciences were principally minded and cultivated" (Seward, 1937, p. 7). Although of lowly social origins by the standards of the time he was well received at Cambridge; he had a good knowledge of Latin and Greek as well as several modern languages. He was made a Fellow of Trinity in 1649 and his career there seemed promising and secure. This did not prove to be the case, and since it has considerable bearing on his later life and scientific productivity, a few lines of explanation are in order.

In 1643 the Parlimentarian party concluded a Solemn League and Covenant with the Scots Presbyterians to impose the reformed religion on all of Great Britain. Many Fellows refused to accept it and were expelled, but Ray was a Puritan and his position seemed secure. He was of a strongly religious turn of mind from an early age, and although he had once stated that he did not intend to take orders he was ordained in 1660, the year the monarchy returned with Charles II on the throne. Two years later the Act of Uniformity was given Royal assent and the Solemn League and Covenant declared unlawful. The ordained Fellows teaching in the colleges were obliged to concur, but Ray refused, saying: "I have already taken so many oathes & subscriptions as have taught me to disgust such pills" (Eyles, 1955, p. 104). He resigned his fellowship and left Cambridge in 1662. After several years of travel Ray and his wife settled at the home he had built for his mother in Black Notley, where he lived for the rest of his life, supported chiefly by a small stipend of seventy-two pounds a year that was bequeathed to him by a former student, Francis Willughby.

Although isolated in his small Essex village Ray's scientific productivity was enormous. His greatest study in botany was the *Historia Plantarum*, dealing with some twenty thousand plants and published in three folio volumes. Seward (1937) says of his biological work: "Ray's classification of plants, more than that of any other botanist of the seventeenth century, was a definite attempt to recognize and express natural affinities; he was a pioneer in the systematic treatment of animals as well as plants; he prepared the way for Linnaeus, who was born two years after Ray died" (pp. 23–24). But his scientific studies were not all; he produced *A Complete Collection of English Proverbs . . .*, and in 1678 he wrote to a friend: "Divinity is my Profession. The study of plants I never lookt upon as my businesse more than I doe now, but my diversion only" (Seward, 1937, p. 21).

After his departure from Cambridge in 1662 Ray was able to satisfy his desire to travel, his most extensive venture being an extended tour of several European countries from Belgium to Malta, which started in April of 1663. Fossil plants and animals were not a primary interest with him, but he has recorded his observations on them on several occasions. His general inclination, based on his own observations and to some extent the influence of Robert Hooke and Nicolaus Steno, was to accept fossils at their face value, as the remains of once-living plants and animals. He says: "The first and to me most probable Opinion is that they were originally the Shells or Bones of living Fishes and other Animals bred in the Sea" (1673, p. 120). In some of his writings he definitely rejects the idea that a "plastic virtue" in the earth was responsible for the formation of "formed stones," or fossil plants and animals as we know them today, although it is true that in some of his later writings and letters he was unable to quite make up his mind. He did waver in his opinion but I think for rather good reasons. He was a keen observer and a sharp thinker; moreover his strong theological makeup caused him to reject the explanation of fossils as meaningless mineral formations: "Nature doth nothing in vain".

Lhwyd, although not completely convinced in the matter tended to uphold the concept of fossils as being *lusus naturae*—that is, sports or freaks of nature. However, like Ray he strongly rejected the Flood concept to explain their presence in the rocks of the Earth. The element of doubt that he felt about the nature of fossils is expressed in a letter to Ray of November 25, 1690, relative to a collection of invertebrates: "Whether they were ever the tegumenta of animals or are only primary productions of nature in imitation of them, I am constrained to leave in medio, and to confess I find in myself no sufficient ability or confidence to maintain either opinion, though I incline much to the latter" (Lankester, 1858, pp. 227–228).

He was, however, concerned about the strong resemblance of some of the "formed stones" to living marine animals. He was also concerned with what he did *not* find: "On the other hand, it seems as remarkable that we seldom or never find any resemblance of horns, teeth, or bones of land animals, or of birds, which might be apt to petrify, if we respect their consistence" (Gunther, 1945, pp. 110–111).

Lhwyd was worried about how the "plastic power of salts" could have produced such a great variety of fossils if the latter were only accidental. In referring to certain apparent invertebrates he does offer a suggestion: "I have in short imagin'd they might be partly owing to fish-spawn, received into chincks and other meatus's of ye earth in the water of the Deluge, and so be deriv'd (as the water would make way) amongst the shelvs or layers of stone, earth, &c" (Gunther, p. 389). And as to plant fossils: "I imagined farther that the like origin might be ascrib'd to the mineral leavs and branches, seeing we find that they are for the most part the leavs of ferns and other capillaries, & of mosses & such like plants as are call'd less perfect: whose seeds may be easily allow'd to be washed down by the rain into the depth here requir'd, seeing they are so minute, as not at all to be distinguished by the naked eye" (Gunther, p. 390).

Ray gives a detailed discussion of the whole problem, including a review of the work of others, in his *Miscellaneous Discourses Concerning the Dissolution and Changes of the World* (1692). This appeared in a later edition in 1713 under the title *Three Physico-Theological Discourses*, etc. It is a little difficult at points to know whether Lhwyd was influencing Ray or vice-versa. In reference to some Carboniferous plants that Lhwyd had obtained and sent to Ray the latter inclined toward Lhwyd's view that they were not the real thing! "Yet I must not dissemble, that there is a *Phenomenon* in Nature, which doth somewhat puzzle me to reconcile with the Prudence observable in all its Works, and seems strongly to prove, that Nature doth sometimes *ludere* [amuse itself], and delineate Figures, for no other End, but for the Ornament of some Stones, and to entertain and gratify our Curiosity, or exercise our Wits" (1713 ed., p. 125). I do not think that this is Ray at his best, but he clearly was a bit uncertain.

Two points that appear in Lhwyd's writings seem to me to be particularly important and reveal the depth of his perception: In describing some Coal Measure plants from Wales he comments on their distinctive morphological features, noting that they are distinguishable into *Species*, as in the case with living plants. It would seem here that he is entertaining a close relationship between the living and the dead (see Challinor, 1953, p. 141). The second is an interesting comment on geologic time. In a letter to Ray dated February 30, 1692, he described the disintegration of some extensive cliffs in Wales:

> From hence I gather, that all the other vast stones that lie in our mountainous valleys, have, by such accidents as this fallen down; unless perhaps we may do better to refer the greatest part of them to the universal deluge. For, considering there are some thousands of them in these two valleys, . . . whereof (for what I can learn) there are but two or three that have fallen in the memory of any man now living, in the ordinary course of nature we shall be compelled to allow the rest many thousands of years more than the age of the world. [Gunther, 1945. pp. 158–59]

Gaining a better understanding of geologic time was one of the real focal points about which paleontological progress was centered. To do this, two requirements had first to be met: The need for more careful and detailed field studies of sedimentary rocks, and minds capable of correctly interpreting what they saw (Lhwyd's observation quoted above is representative of what was needed). And overcoming or circumventing the religious influence stemming from James Ussher's calculations.

In his "Epistle to the Reader" in *The Anals of the World* Ussher (1581–1656) had said: "I incline to this opinion. that from the evening ushering in the first day of the World, to that midnight which began the first day of the Christian era. there was 4003 years, seventy dayes, and six a temporarie howers. . . . " It did not seem profitable to me to try to read the entire 907 large pages of small print that compose this work of Ussher's. As scholarship standards of his day were measured, one may admire the incredible attention to detail and organization that went into it; whether it was time well spent is another matter. Some measure of the respect that Ussher commanded at the time may be judged from these brief abstracts from Gordon's (1909) biographical sketch: He was one of the first students at Trinity College, Dublin, and "before graduating B.A. (probably in July 1597) he had drawn up in Latin a biblical chronology (to the end of the Hebrew monarchy), which formed the basis of his "Annales." . . . His Augustinian theology commended him to the puritans, his veneration for antiquity to the high churchmen; no royalist surpassed him in his deference to the divine right of kings. All parties had confidence in his character, and marvelled at his learning" (pp. 64, 70).

Before considering in general the developing struggle with the nature of fossils, Martin Lister (1638–1712) should be mentioned. Lister's contributions to paleobotany were minimal but he was a good naturalist, known for his studies in conchology, living and fossil, and he was a distinguished physician. He was born at Radclive, Buckinghamshire, in 1638, the grandnephew of Sir Matthew Lister, physician in ordinary to Charles I. Martin received some of his medical education from Sir Matthew and he studied at St. John's College, Cambridge, where he received his B.A. in 1659. He developed a lucrative medical practice in York and transferred to London in 1684. He seems to have been successful there as well, for in the latter years of his life he served as physician to Queen Anne.

He held, generally, to the view that fossils were freaks of nature and not the remains of formerly living organisms. But this opinion was tempered with reason and he realized that there might be exceptions. In a sort of open letter, addressed to Steno, that appeared in the *Philosophical Transactions* in 1671, he says that sea shells may well be found in the rocks along the shores of the Mediterranean but "for our English-inland *Quarries*, which also abound with infinite number and great varieties of shells, I am apt to think, there is no such matter, as Petrifying of Shells in the business . . . but that these Cockle-like stones ever were, as they are at present, *Lapides sui generis* [spontaneously produced stones], and never any part of an Animal" (p. 2282).

Lister is also remembered for his contribution of 1684 in the *Philosophical Transactions* in which he suggests a kind of stratigraphic map. Of this Sheppard says: "Dr. Lister's scheme for a map of England, distinguishing the soils and their boundaries by colours, has certainly the merit of priority. Sir

Charles Lyell acknowledged that Lister was the first who was aware of the continuity over large districts of the principal groups of strata in the British series, and who proposed the construction of regular geological maps. The scheme, however, was never carried out in his time" (1922, p. 93).

These leaders of the time influenced the thinking, at least in some degree, of others of less prominent stature. As an example, in a communication to the Royal Society in 1683 one Griff. Hatley reports finding fossil shells, and, in accordance with Lister's views, he gives reasons for "their never having been the spoils of Animals." But he does, in a curious way, correlate fossils with living organisms:

> But by what means they receive this likeness to shells, is hard to determine, your own conjecture satisfies me best. . . . And there can be no convincing argument given, why the salts of Plants, or animal Bodies, washed down with rains, and lodged under ground; should not there be disposed into such like Figures, as well as above it: probably in some cases much better, as in a colder place; and where therefore, the work not being done in a hurry, but more slowly, may be so much the more regular. [Hatley, 1683, p. 465]

Hatley's reference to "the salts of Plants" indicates a contemporary lack of knowledge of plant reproductive mechanisms. Pichi-Sermolli (1959) states that the nature of fern sporangia became known only in the 1680's, although some understanding of spore germination had been evidence by Cordus a century before. Apparently a few more decades were to elapse before the real significance of the "brown dots" on the underside of fern leaves was to be at all widely recognized. I have encountered many people, intelligent but lacking botanical training, who have assumed that the sori (aggregates of sporangia) of ferns were some sort of disease; thus it is easy to understand the mistakes of the seventeenth century!

It seems to me that men like Lhwyd, Lister, and Ray represent the pattern of thought concerning the nature and origin of fossils as it was developing among the more intelligent and informed naturalists of their time. In some cases they leaned strongly toward accepting fossils at their face value—as the remains of formerly living organisms—but in large part they rejected them as such. It is easy for those who have some knowledge of fossils at the present time to ridicule their ideas, but even today there are relatively few people who do possess this knowledge, and actually the problem has not changed very much in the past three centuries. Two points seem especially pertinent.

In his admirable book *The Meaning of Fossils*, Martin Rudwick (1972) draws a distinction between "easy" and "difficult" fossils. The "easy" ones are well-preserved specimens revealing structures so obviously of a plant or animal nature or so closely comparable to living plants or animals that no intelligent person can doubt their origin. The "difficult" ones are those that are poorly preserved or present structures that do not clearly reveal whether they are of organic origin or not. Complicating the problem for the layman today are the difficult concepts of geologic time, the formation of rocks, and the interpretation of plant structures, whether living or fossil. Even professional paleontologists occasionally have trouble in agreeing on the relationships of their fossils, and sometimes, indeed, in agreeing on whether a particular specimen (a "very difficult" one!) is of organic origin or not.

A few present-day, homely examples that bear on the matter of "easy" vs. "difficult" fossils may help to put the matter into perspective. I am sure that any of my professional colleagues could offer comparable experiences.

In my early days at the University of Connecticut a man struggled into my office one morning with a huge slab of Connecticut River (Triassic) sandstone. It contained a medley of mud cracks and I was asked to determine, on the basis of the "venation," what kind of leaf it was. The owner of the specimen had not come to ask me whether or not it was a fossil leaf; this he had decided for himself. My task was to tell him what *kind* of a leaf it was. My efforts to explain why this quite irregular mud-crack pattern could not represent the veins of a (very huge) leaf were not successful, and he left in something short of a pleasant frame of mind.

One summer some twenty-five years ago I was searching along the conglomerate cliffs of southern Quebec for specimens of *Prototaxites*, a Devonian plant known from petrified trunks up to nearly three feet in diameter. Well-preserved specimens reveal two series of tubes which compose a structure that is most closely comparable with the stipes of large brown seaweeds, but paleobotanists are still uncertain about its affinities. (A more detailed discussion of this curious fossil is given elsewhere in this book.) A very helpful old man took me to a place along the beach where several large specimens were exposed, for which I was most grateful. He explained that they were pieces of maple trees that had fallen from the top of the cliff and become petrified. I could see that he was quite content with this explanation and quite certain about it. It was a unique and highly erroneous interpretation, but I could not see that much would be gained from trying to disillusion him, especially since I could not be sure that I knew what *Prototaxites* was myself!

One of the better coal-ball localities in the central United States occurs in a coal seam that outcrops along the streambank of a forested piece of Illinois farmland. The owner allowed us to collect there for a modest fee. He sat down one day to watch us, and I am sure he wondered why we spent so much time digging out the petrifactions, although I had given him a couple of informal lectures on the subject. He stated quite simply that the "fossils" were just pieces of plant debris that had washed down the creek and become lodged in the coal. In view of the richness of the deposit and our hope to continue digging there I could see no point in contradicting or possibly antagonizing him.

These are examples of specimens that a paleontologist would have little trouble identifying as being organic or inorganic in origin ("easy" fossils). One of the most intriguing in the "difficult" category that I have seen are the filamentous, plantlike pseudofossils found in spherulitic geodes, which have been described by Roland Brown (1957). On first examination they resemble filaments of certain blue-green algae, although a careful study reveals a branching pattern that is definitely not plantlike. Roland Brown had a flair for collecting curious and problematical fossils and pseudofossils, as well as examples of mistakes that his colleagues had made. He had several drawers of these in his office in the National Museum in Washington, and it was a "hall of fame" that I hoped never to be elected to.

Aside from the nature of fossils. whether representing former living organisms or not, the naturalists of the late seventeenth and early eighteenth centuries were concerned with a second great problem: how they came to be encased in the rocks of the Earth, sometimes at sea level and sometimes at the tops of high mountains. A solution to the problem had to wait for geology to catch up—for a better understanding of the origins of what we now call sedimentary and igneous rocks. Until that was accomplished the Great Flood or Deluge was to many naturalists the only explanation. Although some, such as Ray and Lhwyd, rejected it, others were fanatic in supporting it. Certainly one of the most enthusiastic was John Woodward. It may be useful to remind the reader of the origin of this phenomenon as it is recorded in the book of Genesis: "And the raine was upon the earth fortie dayes and fortie nights. . . . The waters prevailed so exceedingly upon the earth, that all the high mountains that were under the whole heaven were covered. . . . And the waters prevailed upon the earth an hundred and fiftie dayes" (Gen. 7:12, 19, 24; quoted from the 1723 London edition of Robert Barker).

Acceptance of the Flood in paleontological literature continued into the early nineteenth century, and it is not entirely a dead issue today, as I will point out on a later page. It is easy to ridicule Woodward, as his contemporaries were ridiculed for their ideas about "formed stones," but I think the justification is even less valid if one keeps in mind the intellectual dominance of religious matters and the fact that the formal education that many of the naturalists of the day had received was chiefly theological. And there are still aspects of geology, in this context, that modern scholars may accept but still find a little perplexing. In his journal of a trip across the Andes, E. W. Berry (1921) writes: "The usual association of granitic cores with high mountains makes the Paleozoic slates of the Unduavi Pass seem strange, and although it is a commonplace of geology, I could never get quite used to finding sedimentary rocks with marine fossils 16,000 or 17,000 feet above sea level" (p. 503).

John Woodward (1665–1728), an arch defender and interpreter of the Mosaic Deluge, wrote in *An Essay toward a Natural History of the Earth . . .* in 1695:

> That during the time of the Deluge, whilst the Water was out upon, and covered the Terrestrial Globe, All the Stone and Marble of the Antediluvian Earth: all the Metals of it: all Mineral Concretions: and, in a word, all Fossils whatever that had before obtained Solidity, were totally dissolved, and their constitutent Corpuscles all disjoyned, their Cohaesion perfectly ceasing. That the said Corpuscles of those solid Fossils, together with the Corpuscles of those which were not before solid, such as Sand, Earth, and the like: and also all Animal Bodies, and parts of Animals. Bones, Teeth, Shells: Vegetables, and parts of Vegetables, Trees, Shrubs, Herbs: and, to be short, all Bodies whatsoever that were either upon the Earth, or that constituted the Mass of it, if not quite down to the Abyss, yet at least to the greatest depth we ever dig; I say all these were assumed up promiscuously into the Water, and sustained in it, in such a manner that the Water, and Bodies in it, together made up one common confused Mass.
>
> That at length all the Mass that was borne in the Water, was again precipitated and subsided towards the bottom. That this Subsidence happened generally, and as near as possibly could be expected in so great a Confusion, according to the Laws of Gravity; that Matter, Body, or Bodies, which had the greatest quantity or degree of Gravity, subsiding first in order, and falling lowest: . . . That the Matter, subsiding thus, formed the Strata of Stone, of Marble, of Cole, of Earth, and the rest; of which Strata, lying one upon another, the Terrestrial Globe, or at least as much of it as is ever displayed to view, doth mainly consist. [Pp. 74–75]

John Woodward was born in Derbyshire in May, 1665, and educated at a country school; sometime before he was sixteen he was sent to London, to be apprenticed to a linen draper according to some accounts. He became acquainted with a prominent physician, Dr. Peter Barwick, who took Woodward into his home, where he presumably received such medical training as he acquired. Woodward apparently studied diligently and he seems to have been clever at using such education as he had to impress the right people. Dr. Barwick himself had been educated at St. John's College, Cambridge, and was a devoted churchman as well as a prominent physician. It is not clear from the biographies I have read how Woodward moved ahead as fast as he did, but in 1692 we find him with a Professorship of Physic in Gresham College, London. Gresham seems to have been a sort of adult-education center and his lecturing there contributed to his social and academic status. He was elected a Fellow of the Royal Society in 1693; two years later he was made a Doctor of Medicine by Archbishop Tenison, granted the same degree by Cambridge University, and made a member of what is now Pembroke College at Cambridge.

Woodward apparently became attracted to fossils while visiting with Barwick's son-in-law Sir Ralph Dutton, in Gloucestershire. The interest certainly expanded and he traveled extensively in England and on the continent, making many local contacts which enabled him to bring together a large and excellent collection of geological specimens that was especially rich in fossil plants and animals. His collections were described in the Catalogue of his book, *An Attempt towards a Natural History of the Fossils of England* (1729). He deserves to be remembered, if for nothing else, as one of the most enthusiastic and precise collectors of his time. His many specimens were accompanied by data as exact as he could obtain and are nicely preserved today in a special cubicle of the Sedgwick Museum in Cambridge. This is one of the finest and oldest fossil-plant collections still in existence.

Woodward fully realized the advantages of obtaining the cooperation of local collectors and prepared special instructions which he distributed in his travels. His *Fossils of All Kinds . . .* (1728) includes a chapter entitled "Brief Directions for Making Observations and Collections and for Composing a Travelling Register of all Sorts of Fossils." With just a little modernization of the English it would still serve rather well as field instructions for the uninitiated.

As a man Woodward seems to have been something less than a delightful companion. Charles Raven, in his biography of Ray (1942), says "he was violent and bitter of speech, and regarded every comment as an insult: as such he invited criticism and provoked enmity." He quarreled with Sir Hans Sloane in 1710 and was expelled from the Council of the Royal Society. And a controversy over the new treatment of small pox led to an armed encounter with one Dr. Mead; in retrospect this seems more amusing than serious and is perhaps the only "duel" recorded in paleobotanical history:

> On the 10th instant, about eight in the evening, passing on foot, without a servant, by the Royal Exchange, I there saw Dr. Mead's chariot, with him in it, and heard him bid his footman open the door. But Dr. Mead made no sign to speak to me, nor did I in the least suspect that he would

follow me. I walked so gently, that, had he intended to have come up with me, he might have done that in less than 20 paces. When I came to the College-gate, which stood wide open, just as I turned to enter it, I received a blow, grazing on the side of my head (which was then uncovered), and lighting upon my shoulder. As soon as I felt the blow, I looked back, and

John Woodward. Portrait by an unknown artist, now in the Geology Department, Cambridge University.

saw Dr. Mead, who made a second blow at me, and said, I had abused him. I told him that was false, stepped back, and drew my sword at the instant; but offered to make no pass at him until he had drawn; in doing which he was very slow. At the moment I saw he was ready, I made a pass at him; upon which he retreated back about four feet. I immediately made a second, and he retired as before. I still pressed on, making two or three more passes; he constantly retiring, and keeping out of reach of my sword; nor did he ever attempt to make so much as one single pass at me. I had by this time drove him from the street quite through the gateway, almost to the middle of the College-yard; when, making another pass, my right foot was stopped by some accident, so that I fell down flat on my breast. In an instant I felt Dr. Mead, with his whole weight upon me. It was then easy for him to wrest my sword out of my hand, as he did; and after that, gave me very abusive language, and bid me ask my life. I told him, I scorned to ask it of one who, through this whole affair, had acted so like a coward and a scoundrel; and at the same time, endeavoured to lay hold of his sword, but could not reach it. He again bid me ask my life. I replied, as before, I scorned to do that, adding terms of reproach suitable to his behavior. By this time some persons coming in interposed, and parted us. As I was getting up, I heard Dr. Mead, amidst a crowd of people, now got together, exclaiming loudly against me for refusing to ask my life. I told him, in answer, he had shewn himself a coward, and it was owing wholly to chance, and not to any act of his, that I happened to be in his power. I added, that had he been to have given me any of his physick, I would, rather than take it, have asked my life of him, but for his sword, it was very harmless; and I was far from being in any the least apprehension of it.

J. Woodward, Gresham College, June 13, 1719
[Nichols' Literary Anecdotes, pp. 641–42]

One of the problems concerned with the Flood was the amount of water needed to cover the highest mountains. Woodward assumed that it came from the "Abyss" deep within the earth. A somewhat more novel explanation was supplied by a contemporary, William Whiston (1667–1752) who, in his *A New Theory of the Earth* (1696) suggested that it came from the condensation of vapor from the tail of a comet passing close to the Earth. Whiston was at Clare College, Cambridge, for a time, having fallen heir to Newton's professorship when the latter retired. Charles Lyell (1872) notes that Whiston was one of the first to venture the opinion that the text of Genesis might be interpreted so as to allow a longer time period for the age of the Earth prior to the creation of man.

Woodward carried his defense a bit far in trying to pinpoint the time of year of the Flood; in his *Essay* he says: "of all the various *Leaves*, which I have yet seen, thus lodg'd in *Stone*, I have observed none in any other State, nor *Fruits* further advanc'd in *Growth*, and towards *Maturity*, than they were wont to be at the latter End of the *Spring* Season" (p. 81). He ventures a more precise time in the *An Attempt towards a Natural History*: "The Hazle Nuts digg'd up in Ergland, are rarely such as appear to be ripen'd. The Pine Cone are in their vernal State; as are all the Vegetables, and the young shells. The Deluge came on, and a stop was put to their further Growth, at the End of May" (p. 59).

James Parsons (1705–1770), basing his opinion on the famous Eocene-age fruits and seeds of Sheppey, chose a somewhat later season:

> If these fruits, which I have the honour to lay before you, are antediluvian, one would be apt to imagine they, in some measure, point out, with Dr. Woodward, the time of year in which the deluge began; which he thinks was in May: and yet this very opinion is liable to some objections; because altho' fruits capable of being petrified, from their green state, may be pretty well formed in May here, as well as in the same latitude elsewhere, in favour of this opinion; yet there are the stones of fruits, found fossil, so perfect, as to make one imagine they were very ripe, when deposited in the places where they are discovered; which would induce one to think the deluge happened nearer Autumn, unless we could think them the productions of more southern latitudes, where perhaps their fruits are brought to perfection before ours are well formed. [1757, p. 402]

This is actually more than an amusing comment, for it is one of the earliest suggestions about the previous distribution of plants that are represented as fossils, and speculations on the climatic conditions under which they lived—a topic which greatly concerned many later paleobotanists, including Brongniart. Parsons was born at Barnstaple, Devonshire; he spent several years studying medicine in Paris and took a medical degree in Rheims in 1736. He spent much of his life practicing medicine in London, and his antiquarian interests led him to dabble a bit with fossils.

Woodward was not generous with the use of his fossils; a physician, Dr. Erndl, who visited him recorded: "It is wonderful how chary and churlish he is in showing his cabinet of curiosities. If you do get a peep at it, mind you do not touch the smallest object with so much as the tip of your finger" (North, 1931, p. 64).

But let us give him credit for assuring the preservation of at least one fine eighteenth-century collection. In a bequest to Cambridge University he included one hundred pounds for a lecturer who would also have the responsibility of caring for his fossils. There was to be no nonsense about this; the lecturer must be a bachelor; "and in case of the marriage of any of the said lecturers afterwards, his election shall be thereby immediately made void, lest the care of a wife and children should take the Lecturer too much from study, and the care of the Lecture" (Clark and Hughes, 1890, p. 182).

I have chosen to include in this unit of my story one continental naturalist, Johann Jakob Scheuchzer of Zurich, Switzerland. He was a diluvialist, a strong supporter of Woodward in this respect, and maintained a close correspondence with both Woodward and Lhwyd. There is thus a real link with the British work related above, but Scheuchzer carried it to a more advanced level, and if one wishes to point to any one man as "the first paleobotanist," he seems deserving of the title.

Scheuchzer's *Herbarium Diluvianum*, which first appeared in 1709, stands as the first really comprehensive and well-illustrated book on fossil plants. The second edition (1723) includes fourteen large plates, about two-thirds of which are devoted to illustrations of fossil plants. One can readily recognize many Carboniferous genera such as *Asterophyllites*, *Alethopteris*, *Sphenopteris*, *Annularia*, several dicot leaves, and others. In some cases the figures are certainly better than those of other authors published at later dates.

A recent biography of Scheuchzer (Fischer, 1973) gives us an informative account of this interesting and influential naturalist who received little recognition in his own country during his lifetime. He was born in Zurich on August 2, 1672; both his father and grandfather had strong botanical interests, and much of his early education was received from his father, who had a good knowledge of natural science and mathematics and was himself a collector of "figured stones."

A small stipend from the Zurich town council enabled him to attend the University of Altorf (dissolved in 1809), where he received his first medical training. During his two-year stay, his most important contact was with Johann Christoph Sturm (1635–1703). Sturm emphasized experimental science rather than the older "philosophical" approach, a path that Scheuchzer followed his entire life. He devoted considerable time to botanical studies at Altorf, making good use of the botanical garden. Sturm, like Scheuchzer's father, collected fossils, and both regarded them as mineral accidents ("figured stones") in the rocks. Since Scheuchzer was also familiar with Lister's work it is understandable that he would have accepted this idea.

In August of 1693 he went to Utrecht, where for four months he devoted himself to intensive studies of medicine, attending lectures in anatomy, surgery, and botany. He noted that the students there were very hard-working, and he was especially glad for the opportunity to conduct numerous dissections that had not been possible with the more limited resources at Altorf. His latter medical practice does not seem to have been very lucrative, undoubtedly due to his many other interests. He did acquire a reputation for his medical knowledge, and in about 1710 he was offered a position as private physician to Peter the Great, on the recommendation of Gottfried Leiniz.

Johann J. Scheuchzer. From the frontispiece of his *Herbarium Diluvianum*, 1723.

After careful thought Scheuchzer rejected the offer, preferring to remain in his own country where he knew there was much to be done.

He returned to Switzerland after his stay in Utrecht, seeking some kind of regular and reasonably remunerative employment, which, however, seems to have been elusive for much of his life. Eventually in 1710 he was appointed to a long-promised professorship in mathematics in Zurich. He was a good mathematician and probably familiar with Newton's and Leibnitz' calculus. In addition to Latin and Greek, which he learned in his earlier years, he also mastered English, French, and Italian. He engaged in considerable mountaineering in Switzerland, and numerous reports of his alpine journeys were published.

In Zurich he joined a small scientific club, the Collegium Insulanum, which gave him an opportunity to develop his own ideas and probably share and discuss them with sympathetic scholars. In a lecture to this group in September, 1694, he considered in some detail the nature of fossils. He was familiar with Lister's work in conchology and apparently had wondered about the close resemblance between certain fossil and living shellfish. In the course of this lecture he seems to be clearly leaning toward an interpretation of fossils as the remains of formerly living organisms: "When I consider this remarkable circumstance, that the same stone mussel corresponds everywhere in the special parts of its inner cavity to those of the sea mussel, then I must simply doubt whether these classes of mussels did not really originate in the sea or some other bodies of water, unless some other possible genesis of these particular parts can be discovered" (Fischer, 1973, p. 20, translation). Fischer notes that Scheuchzer's change of mind from the "figured stone" concept to the Deluge hypothesis is evident in his book *Specimen lithographiae Helveticae Curiosae* (*Treatise on Curious Swiss Stone-pictures*), which appeared in 1702. This includes eighty-eight illustrations of animal fossils and in it he refers to the universal flooding of the earth.

Discarding the idea of fossils as mineral accidents and recognizing them as the remains of former organisms that lived on the earth was a great advance. It is unfortunate that a strong religious bias committed Scheuchzer, as it did Woodward, so deeply to the Flood hypothesis.

Scheuchzer was acquainted with and drew from the work of Lhwyd; many of the latter's specimens are described in the *Specimen Lithographiae*, and Coal Measure plants from Glamorgan are illustrated in the *Herbarium Diluvianum*. Scheuchzer, like Woodward, was confident that his fossil plants could be related to living ones and the appendix of the 1723 edition of the *Herbarium* records several hundred fossil plants according to Tournefort's system as drawn up in his *Éléments de botanique* of 1694.

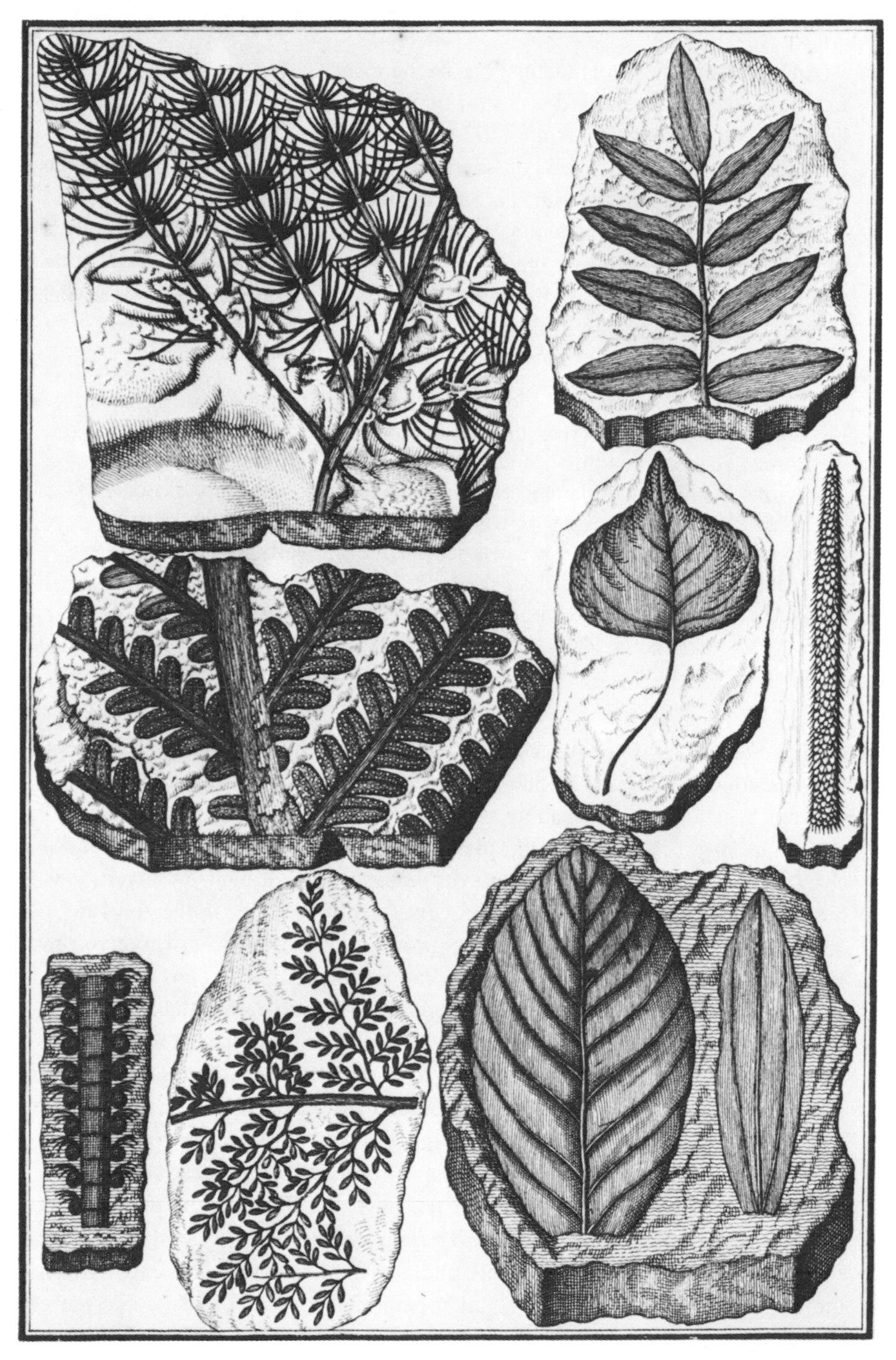

Plate II from the *Herbarium Diluvianum* of Johann J. Scheuchzer, 1723.

An interest in the Great Deluge has by no means come to an end. A recent, rational, and enlightening discussion of it is to be found in Hirsch Cohen's book *The Drunkenness of Noah* (1974). In a chapter entitled "The Aegean—Source of the Flood," Cohen considers in some detail a volcanic eruption on the island of Thera in the Aegean Sea, at about 1450 B.C. It is described as "the most devastating explosion, natural or man-made, ever witnessed by a human being" (p. 100) and may have destroyed the Minoan civilization on Crete some seventy miles to the south. He also cites evidence for a somewhat comparable flood in the region of the Tigris and Euphrates rivers:

> To account for the biblical Flood story incorporating such Mesopotamian themes as the hero-survivor, the ark, and the gathering of wildlife species, it must be understood first of all that the biblical narrator was engaged in writing the history of the universe in monotheistic terms. In the course of his work he had to come across the widely circulated story of a great flood. To have ignored it because its polytheism was theologically unacceptable probably would have rendered his history of the world incomplete, or even erroneous, in the eyes of his less intellectual contemporaries. Accordingly, he had to rewrite the early flood story to conform to his monotheistic faith. [P. 115]

I have greatly enjoyed discussing this subject with Hirsch Cohen, rabbi of the synagogue at Storrs, Connecticut. The chapter in his book that I have referred to is well worth reading for additional information. Lest the reader think that Hirsch Cohen and the present writer, like Woodward and Scheuchzer two and a half centuries ago, have been carried away by a religious bias it may be well to add a note on recent scientific studies in the Aegean. A quite detailed account of the Thera explosion, or explosions, was given in 1969 by the Dutch geologist Rein W. Van Bemmelen, who considered possible correlations of the great ash cloud and resultant rains, the column of smoke and fire, the terrible roar of the eruption (possibly heard as far away as central Africa), and the sea waves with such Biblical events as the ten Egyptian plagues and the momentary retreat of waters in the Sea of Reeds that allowed Moses' group to pass through while the pursuing Egyptians were caught by the wave crest. It is not easy to sift out the later literary embellishments from the natural phenomena of the time, but there can be no doubt as to the widespread devastation of the Thera phenomena.

It seems appropriate to close this phase of the story with the entrance on the scene of a man who "put it all together." There are few people more intriguing and of greater significance than Robert Hooke (1635–1703); we can hardly claim him as a paleobotanist but we cannot overlook the importance of his astute and accurate observations and deductions. No one of his time seems to have been so capable of sorting out sense from the intertwined maze of superstition, dogma, and truths and half-truths of the time. As a sort of summary of the best contemporary thinking I have selected a few of Hooke's observations that bear directly on paleobotanical progress. (Unless otherwise indicated, the following quotations are taken from *The Posthumous Works of Robert Hooke*, 1705.)

Hooke was among the first to recognize that a large part of the world of living things could never reach the fossil stage—this is the "imperfection" of the fossil record that is so often cast into the face of the paleontologist: "we find that most things, especially Animal and Vegetable Substances, after they have left off to vegetate, do soon decay, and by divers ways of Putrefaction and Rotting, loose their Forms and return into Dust" (p. 293). He understood and explained the natural erosive forces at work on the earth:

> Another Cause there is which has been also a very great Instrument in the promoting the alterations on the Surface of the Earth, and that is the motion of the Water; whether caus'd 1st. By its Descent from some higher place, such as Rivers and Streams, caus'd by the immediate falls of Rain, or Snow, or by the melting of Snow from the sides of Hills. Or, 2dly. By the natural Motions of the Sea, such as are the Tides and Currents. Or, 3dly. By the accidental motions of it caus'd by Winds and Storms. . . . The former Principle seems to be that which generates Hills, and Holes, Cliffs, and Caverns, and all manner of Asperity and irregularity in the Surface of the Earth; and this is that which endeavours to reduce them back again to their pristine Regularity, by washing down the tops of Hills, and filling up the bottoms of Pits, which is indeed consonant to all the other methods of Nature. [P. 312]

He was also concerned with the time element: "Nor are these Changes now only, but they have in all probability been of as long standing as the World. So 'tis probable there may have been several vicissitudes of changes wrought upon the same part of the Earth" (p. 313). And he takes up the rather delicate matter of the apparent loss of some species or the creation of new ones: "Thirdly, That there may have been divers Species of things wholly destroyed and annihilated, and divers others changed and varied, for since we find that there are some kinds of Animals and Vegetables peculiar to certain places, and not to be found elsewhere; if such a place have been swallowed up, 'tis not improbable but that those Animal Beings may have been destroyed with them; and this may be true both of aerial and aquatick Animals" (p. 327). On another page he words this a little differently: "That there have been many other Species of Creatures in former Ages, of which we can find none at present; and that 'tis not unlikely also but that there may be divers new kinds now, which have not been from the beginning" (p. 291).

And in another passage he deals with the matter of time and gives his opinion concerning the Flood:

> I think it will be evident, that it could not be from the Flood of Noah, since the duration of that which was about two hundred Natural Days, or half an Year could not afford time enough for the production and perfection of so many and so great and full grown Shells, as these which are so found do testify; besides the quantity and thickness of the Beds of Sand with which they are many times found mixed, do argue that there must needs be a much longer time of the Seas Residence above the same, than so short a space can afford. [P. 341]

In his *Micrographia* of 1665 Hooke expresses his views on fossils in general and presents what is, to the best of my knowledge, the first illustration of petrified wood. He shows a clear understanding of the origin of the specimen at hand:

> That this *petrify'd* Wood having lain in some place where it was well soak'd with *petrifying* water (that is, such a water as is well *impregnated* with stony and earthy particles) did by degrees separate, either by straining and *filtration*, or perhaps, by *precipitation*, *cohesion* or *coagulation*, abundance of stony particles from the permeating water, which stony particles, being by means of the fluid *vehicle* convey'd, not onely into the *Microscopical* pores, and so perfectly stoping them up, but also into the pores or *interstitia*. [P. 109]

Although he did not write very much specifically on fossils it is clear that his understanding of them was far more accurate than that of most of his predecessors or contemporaries:

> From all which, and several other particulars which I observ'd, I cannot but think, that all these, and most other kinds of stony bodies which are found thus strangely figured, do owe their formation and figuration, not to any kind of *Plastick virtue* inherent in the earth, but to the Shells of certain Shel-fishes, which, either by some Deluge, Inundation, Earthquake, or some such other means, came to be thrown to that place, and there to be fill'd with some kind of Mudd or Clay, or *petrifying* Water. [1665, p. 111]

Hooke was born on July 18, 1635, in the town of Freshwater on the Isle of Wight. He was not strong physically and as a child was left much to himself. He showed genius for contriving mechanical devices at an early age, which at that time was diverted into making toys. His father died when he was thirteen, leaving him a hundred pounds; he went to London and studied for a time at Westminster School, where he mastered the first six books of Euclid in a week. In 1653 he went to Christ Church, Oxford, where he soon attracted the attention of leading people such as Robert Boyle. While at Oxford he invented the balance spring for watches and carried on experiments with the pendulum that led him into studies concerned with the determination of longitude. He claimed to have devised a successful method to this end, but for reasons that are not entirely clear this work was not brought to a satisfactory conclusion.

Hooke was one of the leading lights in the early development of the Royal Society, founded in 1660; two years later Hooke was appointed to the position of Curator with the task of delivering lectures and devising "experiments" to be presented at the meetings. This position occupied him more or less for the next forty years of his life and diverted his great abilities and energy into innumerable channels. There seems to have been little that he did not concern himself with, and he did most of it extraordinarily well: he established the standard for the thermometer from the freezing point; he contrived a way to make the barometer more sensitive; following the great London fire of 1666 he developed a plan for rebuilding the city, and although this was not followed he was appointed City Surveyor and was one of the dominant figures in charge of reconstruction; he developed instruments to aid in hearing; and according to a biographer, "Hooke divined before Newton the true doctrine of universal gravitation, but wanted the mathematical ability to demonstrate it" (Clerge, 1908, p. 1179).

In her very readable biography Margaret 'Espinasse (1956) describes him as "irascible, generous, brilliant, simple; honest through and through, a person of complete integrity" (p. 154). He was a mechanic and an experimental scientist equaled by few in history. One cannot help but wonder what he would have done if he had marshaled his efforts into fewer channels and followed them through.

3

The Eighteenth Century

THERE was a rather meager period in productive fossil-plant studies from the time of Woodward and Scheuchzer in the 1720's to the great surge of activity that came in the early years of the nineteenth century. The knowledge of living plants and geological processes was increasing notably, and the naturalists of the time seem to have been concentrating on such basic investigations, preparing the way for later studies of the plants of the past. But there are a few works of real significance that form an integral part of our history, although I have had some difficulty in aligning them in a meaningful fashion.

The interpretation of fossils as "sports of nature" faded noticeably from the philosophy of paleontology after the time of Scheuchzer, although adherents of the Flood concept have never entirely given up. Yet one of the strangest contributions to the study of fossils, one of the great curios in natural history literature, appeared in 1726: Johann Beringer's *Lithographiae Wirceburgensis*. In reference to this work Edwards (1967) notes that the "sports of nature" concept was killed by ridicule.

Johann Bartholomew Adam Beringer (1667–1740), a Doctor of Philosophy and Medicine, was Professor and Dean of the Faculty of Medicine at the University of Würzburg. His book is illustrated with some twenty-three plates of presumed fossils which consist of crudely carved "art" works placed in an area where he was in the habit of digging. The "fossils" include plants in full flower, spiders with cobwebs, shooting stars, and Hebrew letters. These were long thought to be the pranks of some of Beringer's students, and indeed I continued in this error in my book *Ancient Plants* (1947). Beringer's book is a very rare item; I have seen copies in the British Museum and in the Paleobotanical Library of the U.S. Geological Survey in Washington, and I am still puzzled by it. Rather recently Jahn and Woolf (1963) have given us

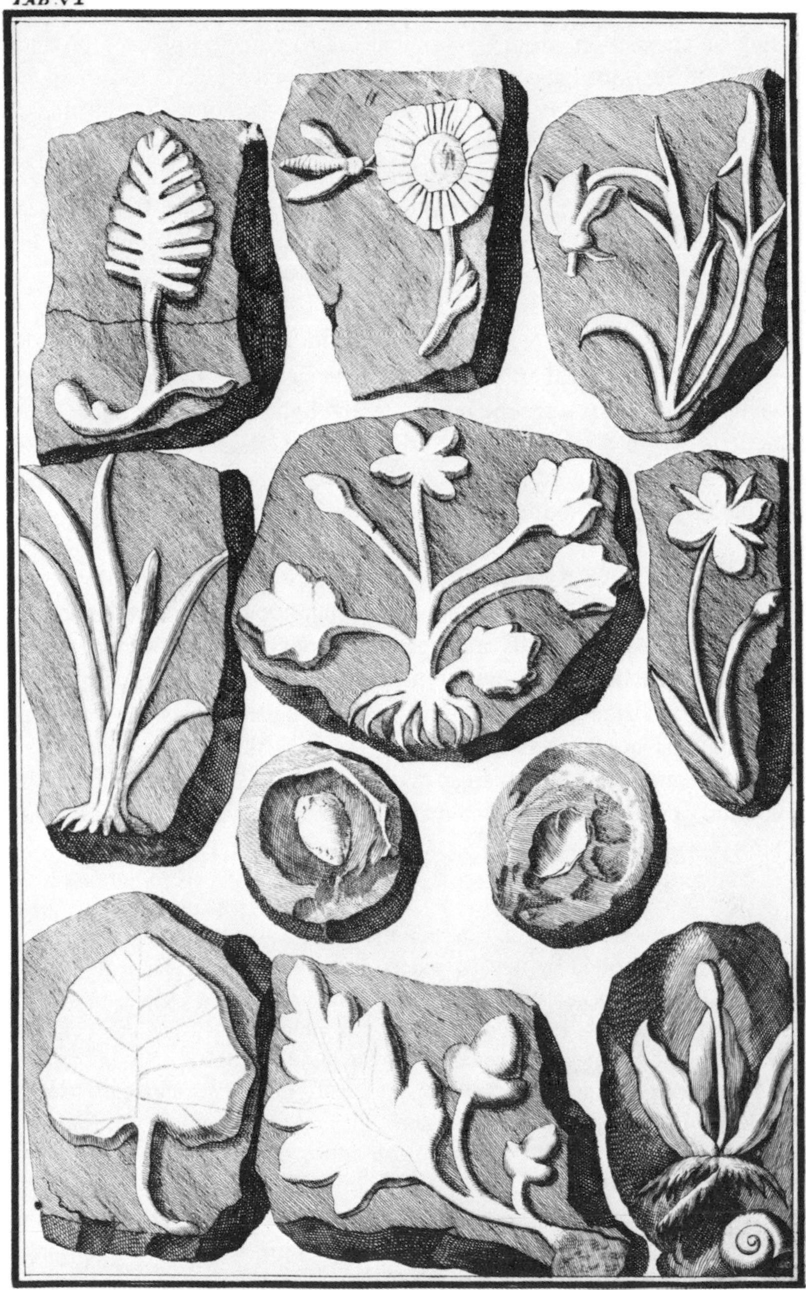

A plate from Johann Beringer's *Lithographiae Wirceburgensis*, 1726. From a copy in the Paleobotanical Library, U.S. Geological Survey

an English translation (from the original Latin) with the plates reproduced, and most of the information here is taken from their study.

The truth apparently is that this famous hoax was not a student prank but a rather insidious maneuver on the part of two of Beringer's university colleagues, Professor J. Ignatz Roderick and Georg von Eckhart, Privy Councillor and Librarian. The ill feeling between these two and Beringer seems to have been very real. Beringer had a favorite collecting locality in a mountain a mile from Würzburg and hired several boys to help him with the digging. Presumably one or more of these boys were brought into the plot and buried the "fossils" that had been prepared by the two conspirators. These pieces of paleontological art work in Beringer's collections and the *Lithographiae Wirceburgensis* were the results. In the light of what we know about it today it is still a strange affair. Beringer was certainly credulous; he was acquainted with the more important naturalists of the time, and as a general principle he accepted the Flood concept to explain the presence of fossils in rocks.

He seems to have been apprised rather early in his explorations of the possibly fraudulent nature of his specimens, and he discusses at some length the possibility that his fossils are pagan art works of earlier German peoples. He also distinguishes between what he regarded as ordinary fossils (probably real fossils) and the fabrications of his two colleagues. There is a strong note throughout his book relative to (in his opinion) the unique nature of the mountain from which he obtained the specimens. Beringer seems to have been quite confident that his interpretation of his fossils would prevail. He says:

> No intelligent person could entertain the slightest suspicion that our iconoliths were petrified by the sea waters—whether in the general flood, or drawn from the ocean by some other channel. However, by reason of the contrast, which we have intimated in this chapter, between our controverted figured stones, and genuine diluvial and maritime specimens found on the same mountain, certain dabblers and other mere amateurs in the field of Lithography have underhandedly and falsely taken occasion to insinuate that, while the others are true relics of the Flood, ours are the supposititious product of recent artificiality. Thanks to their vicious raillery, their false rumors and gossip, they have forced me—though I shrink from the task—to refute and confound them all in good time. [Jahn and Woolf, 1963, p. 71]

Beringer does express doubt about the origin of some of his "fossils":

> As I have frequently intimated in the course of this dissertation, the answer to this controversial question is still a matter of dispute. Weighty arguments favor art; no less impressive are those which argue for Nature. In this chapter I shall present both contentions, withholding my own opinion, though I shall not hesitate to declare my stand when, in due time, the diggers will have penetrated more deeply into the mountain, and will have uncovered more lucid evidence to resolve the doubt in one direction of the other. [P. 83]

And on a later page he does give a little to the possibility of fraud: "Let us grant that among our stones there are some which are spurious, and that they are found and foisted off on the Lithophiles. Does this mean that every iconolith I have uncovered during the past six months smacks of imposture?" (p. 95).

In connection with the ideas of Woodward and others concerning the time of year of the Flood, Beringer has some ideas of his own: "According to the more probable opinion of both theologians and lithographers as related by Scheuchzer in his *Herbarium Diluvianum*, the Flood occurred in the spring, probably May, at which time flowers and leaves (rather than ripe fruits) flourish. How, then, did the diluvial tempest miraculously deposit on the shores of Franconia a ripe and integral apricot, complete with pit, meat, and skin, and a mature acorn appended to a small branch?" (p. 70).

We shall never know exactly what went on in Beringer's mind as he pondered the true nature of his specimens. It seems to me that, once he had accepted them as true fossils—and had invested a considerable sum of money in the project in the bargain—although a shadow lingered in the back of his mind, he could not quite bring himself to admit that it was all a hoax.

Some good may have come from all of this; in 1804 James Parkinson had the following rather helpful comment to offer about Beringer's book: "it appeared that the censure and ridicule, to which its author was exposed, served, not only to render his contemporaries less liable to imposition; but also more cautious in indulging in unsupported hypotheses" (p. 26). Beringer was not the first or the last to commit foolish errors of interpretation of fossils, but I think there is nothing to be gained from exposing more recent ones here.

Among the few who were contributing to fossil botany in Britain in the mid-eighteenth century probably the most interesting and significant is Emanuel Mendes Da Costa (1717–1791). Arber (1921b) credits him with being the first to publish a memoir as such on Carboniferous plants. This appeared in 1758 in the *Philosophical Transactions* under the title "An Account of the Impressions of Plants on the Slates of Coals." His figures show specimens of

Lepidodendron, *Sigillaria*, and *Stigmaria*, although those generic names are not used. His specimens are described quite carefully as "impressions of ferns, grasses, etc."; although he knew little of the affinities of the plants his morphological interpretations were more astute. In reference to a *Lepidodendron* branch compression from Yorkshire he says: "The impression is much like what might be made by the branches of the common fir, after the leaves are fallen or stript off" (1758, p. 230).

Da Costa's life is described by his biographer (Goodwin) as "a continual struggle with adversity." It would seem that either he was persecuted for his Portuguese Jewish background or else he was something of a rascal! He was born in London on June 5, 1717, and in his early days received some legal training. He became an accomplished naturalist, apparently by his own effort, with a special interest in conchology, in which field he published several articles and books. In 1747 he was elected a Fellow of the Royal Society and was later appointed to a clerkship. But in 1754 he was imprisoned for debt and his collections held in bond. Five years after his appointment to the clerkship he was "detected in various acts of dishonesty" and was dismissed. The Society apparently felt strongly about the matter, for Da Costa was arrested and committed to the King's Bench prison, his library and collections being seized and sold at auction; he remained a prisoner until 1772. In 1774 he asked to be allowed to give a course of lectures on "fossilology" at Oxford, but his reputation prevented him from obtaining the appointment.

Da Costa recognized the difference between the living and the long dead: "the remains of those plants and animals, we know, are . . . the inhabitants of the most remote climes from those, where they now lie buried" (1758, p. 234). It is not easy to determine where the "exotic theory" originated. Ward (1885) says: "The earliest case of this kind on record is that of Leibnitz, who in 1700 furnished a note on the occurrence of impressions of supposed Indian plants in Germany, a conclusion which he arrived at from a comparison of fossils with living species from India, and believed them to agree" (p. 396).

Somewhat more notable is the 1718 account of Antoine de Jussieu of plants from the Carboniferous of St.-Chaumont in France. In his two plates of illustrations he includes quite good figures of *Neuropteris* and *Alethopteris*, which he suggests may be tropical ferns with affinities in the East Indies and "les Iles de l'Amerique," which I judge to mean the West Indies. And in 1757 James Parsons sent to the Royal Society a "most curious parcel of fossil fruits" from the Eocene clays of Sheppey, England. Some of them he called "absolutely exotic" and in his figure explanations he describes the various fruits and seeds as, or compares them to, figs, coffee berries, mangoes, sand boxes and others.

Arber (1921b) gives David Ure credit for producing the first illustrations of fossil plants from Scotland in his *History of Rutherglen and East-Kilbride* in 1793. This book deserves only brief mention; fossil plants were not Ure's primary interest and he seems to have understood his own lack of exact knowledge in this respect. His book includes four reasonably good plates of Carboniferous plants including a calamite pith cast, which he compared with bamboo; a specimen that probably represents calamite roots, which he compared with *Equisetum*; a *Lepidodendron* stem impression, which he tentatively called a pine; and several others. He notes that some of the fossil species are exotic, either not found in Europe or unknown in any part of the world.

Leading more significantly to the important events of the early 1800's, a comprehensive German work on fossils appeared in the latter part of the eighteenth century, by G. W. Knorr and Johann Walch, the latter's name being commemorated in the well-known Permian conifer genus *Walchia*. Their work is usually found in the form of four bound volumes, which, like so many comprehensive endeavours of the period, were published separately over a period of several years. I have examined the set in the library of the British Museum. The first volume, by George Wolfgang Knorr (1705–1761), appeared in 1755 under the title *Sammlung von Merkwürdigkeiten der Natur und Alterthumer des Erdbodens*. The project was continued after Knorr's death by J. E. I. Walch (1725–1778) as *Die Naturgeschichte der Versteinerungen zur Erläuterung der Knorrischen Sammlung von Merkwürdigkeiten der Natur heraugsgegehen*. The first volume of this work, as I have seen it, contains several plates devoted to fossil plants, including Carboniferous plants and dicot leaves. Zittel (1901) notes that "the third volume begins with a dissertation about fossil wood, followed by the description of a number of Carboniferous plants. . . . The masterly text of Walch sets forth his own original observations, and displays a knowledge of the older literature unsurpassed for its completeness and accuracy" (p. 22).

Ward (1885) comments as follows on this work: "Walch was the first to offer anything like a nomenclature of fossil plants, and although most of his names have now disappeared from the text books, they still served a useful purpose during a long embryonic period in the history of the science. He called petrified trunks by the terms *Lithodendron* and *Dendrolithus*; pieces of petrified wood *Lithoxylon*, and also *Stelechites*; petrified roots, *Rhizolithus*" (p. 426).

G. W. Knorr, an art dealer and well-known engraver with a special interest in natural-history materials, was born in Nuremberg on December 30, 1705. There is some continuity here as Knorr derived his interest in natural history at least in part from Scheuchzer, having assisted in the preparation of the engravings for the latter's book *Physica sacre*, which dealt with Scheuchzer's biblical studies relative to the Flood. Knorr's biographer (*All. deutsch. Bio.*) says of his work with Walch: "Knorr's description of the completely unsystematically presented objects was brief and unscientific." And he goes on to say that Walch tried to correct the deficiency in the other volumes.

Walch was born in Jena on August 29, 1725, and apparently to please his father, who was a theology professor, he started his academic career with similar studies. He traveled extensively with his brother through western and southern Europe and in 1755 was appointed Professor of Logic and Metaphysics at the University of Jena. He wrote and lectured on religious topics but his primary interest seems to have been classical philology. He was a conservative "solid citizen" who held the respect of his colleagues and friends; as an "academician," I find it interesting to read that he was "opposed to all controversies."

Several writers refer to the works of Ernst Friedrich von Scholotheim (1764–1821) as a starting point in strictly scientific paleobotany. I have noted elsewhere that it is not always easy to cite a "first" in the matter of scientific contributions, but the quality of Schlotheim's work is clear, and two recent appraisals of the man and his work (Langer, 1966; Daber, 1970) make it easier to assess his contributions.

In 1804 Schlotheim produced a book entitled *Beschreibung merkwürdiger Kräuter-Abdrücke und Pflanzen-Versteinerungen. Ein Beiträge zur Flora der Vorwelt*. The subtitle (*Ein Beitrage . . .*) is one that several authors have used. And in 1820 he published his great work, *Die Petrefactenkunde*.

Ward (1885) says: "These works, though few in number, were systematic and conscientious, and constituted by far the most important contribution yet made to the knowledge of the primordial vegetation of the globe. They form the earliest strictly scientific record we have in Paleobotany" (p. 371). And according to Langer, Brongniart, who was critical of the 1804 study, is said to have stated that had Schlotheim used the binomial system in 1804 his *Flora der Vorwelt* would have been the starting point for all later paleobotanical work.

The 1804 treatise is accompanied by fourteen excellent plates illustrating chiefly Carboniferous fernlike foliage; it is my understanding that these were prepared by Schlotheim himself. Langer points out that this work was actually developed from an article that Schlotheim wrote in 1801 entitled "Abhandlung über die Kräuterabdrücke im Schlieferthon und Sandstein der Steinkohlenformationen." In 1813 he came out with a large but less well known study, *Beiträge zur Naturgeschichte der Versteinerungen in geognostischer*

Hinsicht. As the title indicates this was concerned more with the geological importance of fossils than his other accounts.

Although Henry Steinhauer (1818) is usually given credit for being the first to use binomials in describing fossil plants it seems a rather insignificant matter in view of the distinct superiority of Schlotheim's work; he did use binomials in 1820, and all but two of the twenty-nine plates are devoted to fossil plants.

Schlotheim's collections came from the Lower New Red Sandstone of Thuringia, from the Carboniferous of the Saar, and from France. He made extensive collections himself and studied them carefully, without, it would seem, being unduly encumbered in his interpretations and thinking by the foggy notions of the past. He regarded the Carboniferous plants that he described as representing a warmer climate than prevailed in the same areas in the early nineteenth century and recognized that they could not be compared with living species, and must therefore represent extinct ones. His 1813 book shows some understanding of the stratigraphical importance of fossil plants.

Schlotheim was born on April 2, 1764, in Allmenhausen, Thuringia. In 1781 he began the study of public affairs in Göttingen and also came into contract with the zoologist J. Fr. Blumenbach. From June, 1792, until February, 1793, he studied with A. G. Werner at Freiburg, Saxony. He was a friend of Alexander von Humboldt and was aided in his fossil-plant studies by Samuel von Bridel. His studies of the metallurgy of iron and silver apparently advanced his career; in 1828 he became Correspondent (Member) of the Ducal Ministry, and in 1822 he became supervisor of the library and of the art and natural-history collections of the Duke of Saxony.

Schlotheim's contributions were rather numerous and his work of a caliber considerably more precise and of more lasting value than that of any of his predecessors. There is reason for Daber's (1970) complaint that although Schlotheim's work was accepted as classic and genuinely scientific, he was passed over in the matter of establishing a starting date for paleobotanical nomenclature. Daber also notes that the collections that form the basis of the 1820 book were given to the museum of Berlin University, where they are presently preserved.

The works of two men in Britain in the early years of the nineteenth century also deserve more consideration than they have received. James Parkinson (d. 1824) is known for his monumental three-volume *Organic Remains of a Former World*; the first volume of 461 pages is devoted to fossil plants and the other two to fossil animals. Volume 1 first came out in 1804 and was reprinted at least once, as my own copy is dated 1820, and volumes 2 and 3

Ernst Friedrich von Schlotheim. From W. Langer, *Argumenta Palaeobotanica*, 1966; after B. von Freyberg, 1932.

are dated 1808 and 1811. A few decades ago this work was readily available in the second-hand book shops and was probably quite influential in its day. Volume 1 is written in the form of a series of letters to an unenlightened, imaginary friend concerning the nature and significance of fossils. It is a detailed summary of the contemporary state of knowledge regarding many aspects of fossil plants and is the first comprehensive attempt to "popularize" the science. Parkinson adhered firmly to the great Flood concept and his basic knowledge of plant classification was not extensive, but he was an intelligent man and apparently had a good deal of field experience.

He gives the following classification for plant fossils: "Vegetable fossils, I shall divide into the following orders: –1. Fossil trees. –2. Fossil plants. –3. Fossil roots. –4. Fossil stalks. –5. Fossil leaves. –6. Fossil fruits and seed vessels" (1820, p. 53). Several of the letters are devoted to discussions of a variety of peaty materials which reveal a considerable knowledge of the modes of deposition of plant materials and processes of preservation. Of peat he says:

> The best and most perfect peat has very little, if any, earth in it; but is a composition of wood, branches, twigs, leaves, and roots of trees, with grass, straw, plants, and weeds; which lying continually in water becomes soft, . . . and has no gritty matter in it. It is, indeed, of a different consistence, in different places; some being softer and some firmer and harder; which may, perhaps, arise from the different sorts of trees it is composed of. [1820, p. 94]

Parkinson understood that coal was derived from plant materials but could not quite separate the Flood from an explanation of its origin. Other letters are devoted to a discussion of many different kinds of petrifactions. He was acquainted with a great variety of fossil-plant materials from many localities: Tertiary wood from Ireland, the Permian "Star stones" of Chemnitz, the Cretaceous palms and other woods of Egypt, as well as jet, amber, and various other kinds of fossil plants; and he knew something of the minerals that were involved in the processes of petrifaction. His identification of fossil plants leaves something to be desired; he tended to make comparisons with living plants but had to conclude that this was not always possible.

In an article that appeared in 1811 Parkinson notes the importance of fossils in stratigraphical studies:

> To derive any information of consequence from them, on these subjects, it is necessary that their examination should be connected with that of the several strata, in which they are found: . . . That exactly similar fossils are found in distant parts of the same stratum, not only where it traverses this island, but where it appears again on the opposite coast: that, in strata of considerable comparative depth, fossils are found, which are not discovered in any of the superincumbent beds. [1811, p. 325]

I suspect that this was based largely on his own observations but he was familiar with the work of William Smith, and that of Cuvier and Brongniart in France.

Like most others of the time Parkinson dealt with fossils only as an avocation. We do not have a birth date for him but it is known that he was a practicing physician, apparently in London, in 1785. He also seems to have been something of a political activist. His biographer says:

> In October 1794 Parkinson was examined on oath before the privy council in connection with the so-called "Pop-gun Plot" to assassinate George III in the theatre by means of a poisoned dart. He admitted being a member of the Committee of Correspondence of the London Corresponding Society, and of the Constitutional Society, and also that he was the author of "Revolution without Bloodshed: or Reformation preferable to Revolt," a penny pamphlet published "for the benefit of the wives and children of the persons imprisoned on charges of High Treason." [Boulger, 1909, p. 314]

Parkinson apparently lived a bit dangerously in his political reform activities!

He published a smaller book in 1822 under the general title *Outlines of Oryctology* (an old word for paleontology), which was intended "to aid the student in his enquiries respecting the nature of fossils."

In summary, his *Organic Remains* is a considerable mine of information about various aspects of paleontology and parts of it are quite readable, if somewhat verbose. It probably did much to disseminate knowledge about fossil plants and animals among the general public interested in natural history at the time. As to its botanical importance it cannot be regarded as great.

There is surely no greater contrast in the history of fossil botany between a man's scientific interests and his pursuit of a living than in the case of our second neglected Briton, William Martin (1767–1810). Martin is known almost exclusively for his *Petrificata Derbiensia*, which appeared in 1809, although parts of it were published in 1793 under the title *Figures and Descriptions of Petrifactions Collected in Derbyshire*. Martin says the object was "to give coloured figures of extraneous fossils." It is based, according to Edwards (1967), chiefly on fossils collected by White Watson, who proposed to collaborate with Martin in a work on the fossils of Derbyshire; Watson is not mentioned in the 1809 book, which caused him some anguish. There are numerous reasonably good plates illustrating Carboniferous plants such as *Calamites*, *Alethopteris*, *Stigmaria*, *Lepidodendron*, and others. The naming is admittedly peculiar, the plants all being assigned the generic (?) name

Phytolithus. For example, the fossil now known as *Alethopteris lonchitica* is called "PHYTOLITHUS Filicites (striatus)," followed by a few Latin words of description. This work is not especially commendable and both Edwards (1967) and Arber (1921b), in their historical sketches, give it very faint praise.

However, it puzzles me as it did J. Challinor (1948), on whose article I have relied for much of my information about Martin, why a very different work of Martin's which also appeared in 1809 is not mentioned in such historical reviews. I refer to his *Outlines of an Attempt to Establish a Knowledge of Extraneous Fossils, on Scientific Principles.* The term "extraneous fossils" was used at this time to refer to specimens that clearly represented former plant or animal life, in contrast to simple rock or mineral specimens. To the best of my knowledge, John Hill, in his three-volume treatise *A General Natural History* (1748), was the first to use the phrase; much the greater part of this work is concerned with minerals as such, but Hill has a short appendix devoted to "the extraneous Fossils; which are bodies of the animal or vegetable kingdom accidentally bury'd in the earth, [and] belong properly to the histories of plants and animals." He figured some Carboniferous plants and, recognizing that they did not compare closely with modern British species, attributed them to an American origin.

Martin's scientific work is paradoxical. It seems evident that he did not have a good basic knowledge of the structure and classification of living plants. He was a good observer, did considerable field work, and was especially interested in basic principles rather than producing monographs or memoirs. Challinor gives a concise description of the *Outlines*: "Published in the same year as the *Petrificata*, it treats its subject comprehensively and philosophically and at the same time adopts a minutely analytical method. It will be found to contain a remarkably thorough and precise exposition of all those numerous detailed considerations that must be borne in mind in the study of fossils and fossilization in general" (1948, p. 49).

As an introduction to his *Outlines* it is appropriate to note his classification. He divides all "Natural Bodies" as follows:

A.1. ANIMALS are natural bodies, organized, living, and sentient.
B.2. VEGETABLES are natural bodies, organized, and living, but not sentient.
C.3. FOSSILS are natural bodies, unorganized, and neither living, nor sentient.

His "Fossils" are treated as follows:

A.4. NATIVE FOSSILS, or MINERALS
. . . fossils destitute of an organic form: exhibiting such a structure only, as arises from the apposition of the particles, of which they are composed. Minerals arrange under 4 classes—*Earths*, *Inflammables*, *Metals*, and *Salts*.

B.5. EXTRANEOUS FOSSILS, or RELICS
. . . fossils, which have the form or structure of animal or vegetable bodies.

The "Extraneous fossils" are fossils in the modern concept and Martin separates these on the basis of preservation:

A.1. CONSERVATA are the remains of animals or vegetables preserved by various operations of nature amongst *minerals* [petrifactions in the modern concept].

B.2. PETRIFICATA are *mineral bodies* which have, mediately or immediately received their form from *animals* or *vegetables* [impression and compression type fossils].

The *Outlines* is essentially a very detailed text on the nature of fossils, what they are, how they are formed, and the geological processes involved. It is not easy reading and suffers from long, distracting footnotes, but it contains a great deal of precise information. There is some similarity with the first volume of Parkinson's *Organic Remains* as to subject material. If the precision of Martin's writing could have been blended with the more readable form of Parkinson's (and the text reduced in volume) the result would be quite good.

In the preface to his *Outlines* Martin says:

> The study of *extraneous fossils* is confessedly useful to the *Geologist*—it enables him to distinguish the relative ages of the various strata, which compose the surface of our globe; and to explain, in some degree, the processes of nature, in the formation of the mineral world—To the *Botanist* and *Zoologist*, an investigation, which leads to the knowledge of organic forms no longer found in a recent state, must always prove interesting—And the causes, that have operated to produce the distinctions existing between plants and animals of the present day, and those of former unknown ages, offer, to every contemplative mind, an inexhaustable course of rational enquiry.

Martin was born in 1767 at Marsfield in Nottinghamshire, the son of a hosier who abandoned his wife and son when William was a year and a half old. His mother was an actress, and his father went on the stage for a time, but with indifferent success. William followed: when he was five he sang on the stage to the accompaniment of a German flute. When he was twelve he started taking drawing lessons from a James Bolton, who also awakened a taste for natural history. At about this time he began to learn the Latin language, as well as something of the art of engraving. An interest in the stage and theater business seems to have occupied him much of his life, but his stage abilities were something short of spectacular; one of his biographers notes: "As he never possessed a good voice, he did not deem it advisable to devote much time to music; but he excelled in singing humorous and ludicrous songs" (anon., 1811, p. 561). He was not sufficiently successful in this pursuit to support his own family when it started to grow, and he left the stage to take a position as a drawing master, in which, for a time, he fared somewhat better.

It is remarkable that Martin achieved all that he was able to, with little formal education, living most of his life at a subsistence level and working long hours as time was available for his scientific work. Fossil plants were not his only interest; he started a project on the river fishes of Great Britain and also accumulated a considerable fossil shell collection. All of this was sufficient to bring about his election to the Linnean Society (London) in 1796, but, constantly driving himself to do far more than he could, he contracted tuberculosis. He struggled on, devoting the last two years of his life to the completion of the *Outlines*, and shortly before he died he wrote to his physician saying that it might be a hopeful source of income for his wife and six children.

The era merging into that of Brongniart and Sternberg produced several naturalists who are of some importance as peripheral contributors to our story and who reflect the growing interest among botanists and geologists in the plant life of the past. Two factors that were especially important in the correct interpretation of fossil plants were a better understanding of geological phenomena and the allowance of adequate time for them to have taken place. It is hardly possible to leave these matters without mentioning James Hutton (1726–1797) and his work. Hutton received his basic education in Edinburgh and continued his studies in Paris and Leyden, where he received his medical degree in 1749. He did not practice medicine; having inherited land from his father in Berwickshire, and apparently having a natural interest in both farming and geology, he devoted his life to these pursuits.

Hutton was not a voluminous writer, but with some urging on the part of his friends he was induced to set some of his ideas down on paper; the great work for which he is best known, his "Theory of the Earth," was read before the Royal Society of Edinburgh on March 7 and April 4. 1785. This paper has a reputation of being rather difficult reading, which is true to some extent, but much of it seems quite remarkable for the time and is worth a little patience on the part of a modern reader.

Hutton had a good understanding of the composition of coal and he discusses the petrifying process of wood, although his interests were not especially with fossil plants. His knowledge of the processes that are responsible for changing the face of the earth is well summed up in part by the following passage:

> The raising up on a continent of land from the bottom of the sea, is an idea that is too great to be conceived easily in all the parts of its operation, many of which are perhaps unknown to us; and without being properly understood, so great an idea may appear like a thing that is imaginary. In like manner, the co-relative, or corresponding operation, the destruction of the land, is an idea that does not easily enter into the mind of man in its totality, although he is daily witness to parts of the operation. We never see a river in a flood, but we acknowledge the carrying away of part of our land, to be sunk at the bottom of the sea; we never see a storm upon the coast, but we are informed of a hostile attack of the sea upon our country; attacks which must, in time wear away the bulwarks of our soil, and sap the foundations of our dwellings. Thus, great things are not understood without the analyzing of many operations, and the combination of time with many events happening in succession. [1788, p. 217]

As scientists developed a clearer understanding of the way in which sedimentary rocks came into existence, they needed also to recognize that time was needed for all of these processes to take place—much more time than had generally been allowed. We have seen that men like Ray and Lhwyd realized that something more than a few thousands of years was necessary for the formation of the earth's crust but I think that up to his time Hutton gives us the clearest concept, and he was generous with time! He says: "Time, which measures every thing in our idea, and is often deficient to our schemes, is to nature endless and as nothing" (p. 215). "The result, therefore, of our present enquiry is, that we find no vestige of a beginning,—no prospect of an end" (p. 304).

Karl Friedrich Philipp von Martius was born in 1794 at Erlangen, Germany, and is probably best known for his explorations in Brazil during the years 1817–1820, on which he based his three-volume *Historia Naturalis Palmarum* (1823–1850); this work includes a short section, "De Palmis Fossilibus," by Franz Unger, which is illustrated with three plates. This fossil-plant portion is not especially important but Martius' work as a whole is quite magnificent; and the great color plates of living palms include superb landscape illustrations that are a glory to behold.

Martius gave some thought to fossil plants and in 1825 he wrote a treatise entitled *On Certain Antediluvian Plants Susceptible of Being Illustrated by Means of Species Now Living within the Tropics*. He was familiar with the work being done by Schlotheim and Sternberg, and notes: "I directed my attention, on a journey made by me through Brazil, toward the investigation of those forms of plants, which might be considered as prototypes of the antediluvian vegetables discovered in our own countries; nor have my efforts been altogether without success, as some things occurred to me capable of throwing light upon the nature of antediluvian plants, and which I now proceed to announce" (p. 49). His "announcements" actually consist of rather confused comparisons of Carboniferous plants with modern tree ferns, palms, and cacti, but it is of some importance that a noted botanist was concerned with the fossils; he refers to such geological studies as "among the most elevated pursuits, and the most becoming in which men can engage" (p. 47).

Johann Gottlieb Rhode (1762–1827) produced a work of some note on the Carboniferous plants of Silesia in 1820. It is devoted very largely to the arborescent lycopods, especially *Lepidodendron*; the plates are large and impressive, certainly some of the best that had appeared on the genus up to that time. Rhode was quite a versatile writer, being chiefly concerned with the theater and literary pursuits. He served in a directive capacity at the Breslau Theater, published the *General Theater Magazine* and was editor of the *Silesian Newspaper for the Upper Classes*. He was appointed Professor of Geography and German in 1809 at the military academy in Breslau; his paleobotanical contribution, which seems in contrast to his other interests, appeared rather late in his life.

In the early 1800's in France several minor contributions to paleobotany were made by Faujas de Saint-Fond (1741–1819); he was a prominent figure in Paris and a vigorous field worker, and he probably encouraged others in the study of fossil plants. He was among the first to describe fossil dicotyledonous leaves; his first paper in this category is one on plants from the Ardèche province in southeastern France which appeared in 1803, and in 1819 he described fossil plants from Mt. Bolca in Italy. In his *Essai de geologie* of 1803, he devoted about thirty-five pages to a consideration of various kinds of petrified woods. He was in correspondence with Sternberg, who in March, 1818, informed him of his collections of fossil plants from Bohemia and suggested that Faujas encourage French naturalists to engage in similar explorations.

Faujas de Saint-Fond was born in 1741 at Montélimart in the valley of the

Rhone. He studied law and at twenty-four became president of the Senéchal court, but geological pursuits were clearly what interested him most. Through the aid of Georges Buffon he obtained an appointment as assistant naturalist in the Natural History Museum in Paris, and later became Royal Commissioner of Mines and Professor of Geology. He seems to have been deft at handling people and in carrying on his work through the years of revolution. He had some contact with Adolphe Brongniart, but the latter gives him only passing mention in his *Histoire*. Faujas traveled widely in Europe and in Britain, one of his better known works being his *Journey through England and Scotland to the Hebrides in 1784*. He suffered the misfortune, as did several later paleobotanists, of losing much of his collections by shipwreck on their way back to France.

As a geologist Faujas is probably best known for his study of the volcanics of central France, but he was a man of broad interests, one of which was the subject of aerial navigation, and he published a two-volume treatise on the art of making balloons and navigating them. His biographer recounts an amusing incident that took place during Faujas's studies of the volcanics that reflects the not uncommon viewpoint that the uninformed hold regarding fossils. In those days (as sometime even now!) a man with a hammer and sack slung over his shoulders was regarded with suspicion.

> Faujas tells us, that, accoutred in this manner, he attracted the notice of two distinct types of onlookers. In the small towns, his critics of superior discernment used to ask him of what use those pursuits of his could be; what good come of gathering together such quantities of stones which after all might be collected anywhere. But the peasants in the valleys and among the puys had more common sense than to put such questions. They felt sure that no one would take the trouble to pick up and preserve what was really worthless. When they saw a man so engaged, they were certain that he must be in search of some valuable mine. They would, therefore, watch attentively his movements, and as soon as he left the place, they would repair to it and procure an abundant store of what they had seen him collecting. They would then carefully guard it until an opportunity offered of visiting the nearest little town, where they would carry the treasures mysteriously to some jeweller who, to their surprise and disgust, would only laugh at them. [Geikie, 1907, pp. xix-xx]

It seems appropriate at this point to try to sort out the important factors involved in the progress of fossil-plant investigations from the late seventeenth century up to the early 1800's. I think we can identify two partially distinct sets of ideas or events which, if my understanding is correct, were stated rather well by my teacher Hamshaw Thomas in 1947:

> The great revolution in thought about man's place in the universe, about his relations to less fortunate dwellers on the earth and to other living creatures, came about in two stages. First, the recognition of the features which indicated something of the history of the earth's surface and the probable causes of the events which seem to have occurred. Secondly, the attempt to integrate the historical facts of organic life on the earth with the results of investigations on the plants and animals now living. This second phase resulted in the hypothesis of the origin of species by evolution. [1947, p. 325]

During the period of about 1675 to 1800 I do not see any strikingly specific developments that might be called breakthroughs, but rather a gradual chain of progress in which two stages may be recognized, although they certainly overlap. *First*, beginning in the late 1600's: suggestions that the earth may be much older than formerly supposed; acceptance of fossils at their face value as the remains of previously living organisms; a start toward systematic searching for fossils as they relate to living organisms rather than as mere curiosities; a consideration of the nature of species and the possibility of their change through time; and the beginnings of a liberation from theological influence. *Second*, also beginning in the late 1600's but becoming more clearly defined in the middle to late eighteenth century: a much clearer understanding of the nature and origin of rocks, especially sedimentaries; the development of botanical gardens and herbaria based on worldwide explorations, rendering available vastly more information about living floras; better communications and the growth of scientific societies; and, finally, the education of men specifically for botanical and geological pursuits rather than medicine and theology.

4

Brongniart and Sternberg

Two great studies mark the beginning of what has often been called "the scientific period" in paleobotany: Adolphe Brongnairt's *Histoire des végétaux fossiles* and Kaspar Sternberg's *Flora der Vorwelt*. I have tried to place in a proper context the work of some of the many others who preceded these two men, and especially that of Schlotheim; the important contributions of Brongniart and Sternberg were made possible in part by their own genius and industry and in part by the general level of knowledge that had been reached at the beginning of the nineteenth century. Both men, while dependent on their many predecessors, took full advantage of the possibilities then available; these they distilled and crystallized into the two works noted above. They came at a critical time and started the surge of progress in fossil botany that has continued ever since. If my discussion tends to place Brongniart in the brighter light it is because I think that he clearly stands as the foremost figure of his time in our science.

Adolphe Brongniart's life was unique in many ways and it is easier to understand his accomplishments if we look into the life first. Stafleu (1966) has summed it up rather well: "Adolphe's life went undoubtedly before the wind: a peaceful Europe, an unprecedented expansion of technology, sympathetic and tradition-conscious surroundings, a name with status; but he also had a logical and methodical mind, with a great leaning towards recognition of fundamental principles, completely free from the irrational myths of the past" (p. 320).

He was born on January 14, 1801, the son of a great paleontologist, Alexandre Brongniart. His family came from Arras but had been settled in Paris for a century; it included naturalists and architects—a background conducive to encouraging his natural interests. There were no financial problems and he was able to travel quite widely in his early years. His family life seems to have been a tranquil one; he preferred the calm of his laboratory, where he

Adolphe Brongniart. From the frontispiece of his *Recherches sur les graines fossiles silicifées*, G. Masson, Paris, 1881.

worked hard and long hours, to "involvement" in outside affairs. Although he loved life in Paris he usually spent August and September at a country place near Gisors, some twenty-five miles northeast of the city, that had been established by his father. His earliest education was received from his highly educated and liberal-minded father and mother and he showed a serious interest in scientific studies when he was ten. He became a doctor of medicine at the age of twenty-five, his thesis being a monograph on the family Rhamnaceae, although he did not intend to practice medicine and was actually deeply immersed in his paleobotanical studies at this time.

Before he was twenty he accompanied his father on excursions to Switzerland and also visited Italy; in 1824 he traveled through Scandinavia on a paleontological excursion with his father and met Nilsson, Agardh, and others. In 1825 he went with his grandfather to Great Britain, where he met Robert Brown; and on this trip or in later years, he either met or corresponded with W. C. Williamson, Charles Lyell, William Buckland, and other leaders in botany and geology. In 1835 his travels in Europe brought him into contact with de Jussieu in Brussels and Martius in Munich. Thus he developed a considerable acquaintance with people and places at an early age, and his brilliant and retentive mind made the best of it. He was probably the first person to devote his entire life almost exclusively to the study of fossil plants and not simply as a side line to medicine, the clergy, or a business. In 1831 he was appointed aid to Desfontaines in the Natural History Museum in Paris and replaced him in the Chair of Botany in 1834. The most detailed biographies of Brongniart are those of Saporta (1876) and de Launay (1940, which also includes a good bibliography), and Stafleu (1966) gives a very good analysis of his *Histoire*; I have drawn freely from these sources.

Adolphe was a paleobotanist from the start but his success in dealing with fossil plants was based in large part on a great knowledge of, and an active research interest in, living plants. His bibliography is liberally sprinkled with items that bear this out—articles on the flowers of *Zea mays*, the stem structure of the cycads, the structure and function of leaves, pollination studies, the fruit of *Lemna*, as well as floristic and monographic studies.

His first paleobotanical contribution was a lengthy one, *Sur la classification et la distribution des végétaux fossiles en général* which appeared in 1822 and which de Launay (1940) refers to with some reason as a "vaste sujet un peu prématuré." Nevertheless it is a remarkable document for a man of twenty-one and might be worth a perusal by young paleobotanists starting out today. It runs to more than eighty pages; it draws heavily from the work of his predecessors and is in fact quite a good historical sketch, but it is more than that: he gets into a number of important biological matters, some of which were to be of great concern to him in later years. For example, he makes comparisons between *Equisetum* and the calamites, and the living and fossil lycopods, noting that the size factor is not a problem in drawing natural relationships. And he recognizes that the huge genus *Filicites* of previous writers includes a diverse assemblage and divides it into *Sphenopteris, Pecopteris, Glossopteris, Odontopteris,* and *Neuropteris*. This is not done formally, the names being introduced in his plate explanations as *Filicites (Sphenopteris) elegans*, etc. And along with such specific matters Brongniart established three basic factors or principles in his earliest work that were to be guidelines for much that he did thereafter: the development of a classification, which of necessity was somewhat artificial in considerable part; the recognition of a distinct succession of major floras through time; and the application of fossil plants to stratigraphical studies in geology—that is, the recognition of distinct plant assemblages at different horizons. Since it is not possible or necessary to consider all of Brongniart's contributions, I shall deal in some detail with two that from a historical point of view are most important.

The *Prodrome d'une histoire des végétaux fossiles* is a small volume of 223 pages, which, according to Stafleu (1966), appeared either in the last few days of 1828 or the first days of 1829. As the name implies it is a summary or condensation of his classification of fossil plants that was to be presented in detail in the *Histoire*. As a basis for discussion it will be helpful for the more serious student of paleobotany to include here an outline of Brongniart's classification.

Classe I. Agames
- Famille: Conferves
- Famille: Algues

Classe II. Cryptogames Celluleuses
- Famille Mousses: *Muscites*

Classe III. Cryptogames Vasculaires
- Famille: Equistétacées: *Equisetum, Calamites*
- Famille: Fougères
  - Frondes: *Pachyteris, Sphenopteris, Cyclopteris, Nevropteris, Glossopteris, Pecopteris, Lonchopteris, Odontopteris, Anomopteris, Taeniopteris, Clathropteris, Schizopteris*
  - Tiges: *Sigillaria*
- Famille: Marsiléacées: *Sphenophyllum*
- Famille: Characées: *Chara*
- Famille: Lycopodiacées: *Lycopodites, Selaginites, Lepidodendron, Lepidophyllum, Lepidostrobus, Cardiocarpon, Stigmaria*

Classe IV. Phanérogames Gymnospermes
- Famille: Cycadées
  - Frondes: *Cycadites, Zamia, Pterophyllum, Nilsonia*
  - Tiges: *Mantellia*
- Famille: Conifères: *Pinus, Abies, Taxites, Voltzia, Juniperites, Cupressites, Thuya, Thuytes*
  - Conifères douteuse: *Brachyphyllum*

Classe V. Phanerogames Monocotyledons
- Famille: Nayades: *Potamophyllites, Zosterites, Caulinites*

Famille: Palmiers
Tiges: *Palmacites*
Feuilles: *Flabellaria, Phoenicites, Noeggerathia, Zeugophyllites*
Fructifications: *Cocos*
Famille: Liliacées
Tiges: *Bucklandia, Clathraria*
Feuilles: *Smilacites, Convallarites*
Fleurs: *Antholithes*
Famille: Cannées: *Cannophyllites*
Monocotylédons dont la famille n'est déterminée:
Tiges: *Endogenites, Culmites, Sternbergia*
Feuilles: *Poacites*
Inflorescences: *Palaeoxyris, Echinostachys, Aethophyllum*
Fruits: *Trigonocarpum, Amomocarpum, Musocarpum, Pandanocarpum*
Classe VI. Phanérogames Dicotylédons
Famille: Amentacées, Juglandées, Acerinées, Nympheacées
Végétaux dicotyledons dont la famille ne peut être déterminée.
Végétaux dont la classe est incertaine: *Phillotheca, Annularia, Asterophyllites, Volkmannia, Carpolithes*

The first part of the great *Histoire* also appeared in 1828; it continued in a series of fifteen parts or fascicles, the last of which was published in 1838. (One usually finds this work today bound in two volumes.) The specific dates of the separate parts have been given by Stafleu (1966) and Andrews (1970); in several cases the dates vary slightly, and I should say that the information I published in 1955 was taken from the Compendium Index of Fossil Plant Names of the U.S. Geological Survey. Stafleu's treatment gives the year, the month, and in some cases the day of publication, and it should probably be accepted as the most accurate dating that we have.

The *Histoire* was never finished; Brongniart had planned to issue twenty-four parts but it came to an end in the middle of a sentence in the fifteenth. I have encountered rumors that other parts exist in manuscript, but I have not been able to verify them. One can only guess as to why it stopped, and in the middle of a sentence. Brongniart was still a young man with forty-eight years of his life ahead of him, and he produced scores of works in those years. Since others have guessed at the reason for the abrupt termination I will take the liberty to add my own. There are times when one finds oneself involved in a task so overwhelming that it is impossible to go on, or other things prove to be of greater importance; I conjecture that this may have happened to Brongniart. The *Histoire* goes pretty well through the vascular cryptograms—the Carboniferous plants that many others had been concerned with and Brongniart knew so well. He was approaching the seed plants, which were less well known as fossils; the task of dealing with them must have seemed formidable and one that could not have been done nearly as precisely. He probably intended to go on, but put the whole thing aside for a time, and then could never bring himself back to it.

The order of presentation of the plant materials in the *Histoire* follows closely that of the *Prodrome*, with the ferns, or rather, fernlike foilage in Volume 1 and the lycopods in Volume 2. A few interesting points immediately attract attention in the light of present knowledge. *Sigillaria* gave him some trouble; he did point out that these plants were not cacti, as Martius had suggested, or euphorbs, as Artis thought; Brongniart's assignment to the ferns was due to the superficial comparison of the leaf scars with comparable structures in some tree ferns. This mistake would be corrected when something was learned about the vascular anatomy of the sigillarias. He was doubtful about *Annularia* and *Asterophyllites*, and had real difficulty with *Sphenophyllum*, placing it in the Marsileaceae. But for the most part his classification is a tremendous advance over anything that had been proposed previously for fossil plants. An especially fine feature of the *Histoire* is the inclusion of pertinent comparative material on living plants, which adds unity and meaning to the work and reveals the excellent informational background that Brongniart possessed.

His general concept of the Carboniferous floras seems especially important since this was the age he knew best. In an article written in 1838 ("Reflections on the Nature of the Vegetables Which Have Covered the Surface of the Earth, at the Different Periods of Its Formation") he notes:

> Of these several distinct associations of vegetables which have successively flourished upon our globe, none is so worthy of our attention, as that which appears to have been the first developed [Carboniferous], and which, during a long space of time, seems to have covered with thick forests, all those parts of the earth which were above the level of the waters, and the remains of which, piled one upon another, have formed those frequently thick and numerous beds of coal, which are the preserved relics of the primitive forests that preceded, by many ages, the existence of man. [1838, p. 3]

As to climate he followed the prevailing notion that the Carboniferous floras existed in a warm temperate or tropical area: "We are consequently led, by the study of the vegetables which accompany the beds of coal, to infer that at this early period the surface of the earth, in the countries producing the best known of these coal deposits, namely Europe and North America, possessed a state of climate similar to that now existing in the Archipelagoes of the equinoctial regions, and, probably, a geographical configuration very little different" (1938, p. 8). In reference to the "geographical configuration" he states that "Geology and botany, therefore, appear to us to agree in announcing that, at this epoch, the parts of the earth which rose above the waters formed only islands of small extent, disposed in archipelagoes in the midst of vast seas" (1829, p. 363).

A perplexing aspect of Brongniart's overall philosophy is, to me, his treatment of the succession of floras that have existed on the earth from Paleozoic times to the present. It seems clear from his writings that he did not accept (or perhaps devote much thought to) evolution in the Darwinian sense. J. B. Dumas *fils*, in a preface to Brongniart's great work on fossil seeds, notes that he never accepted evolution. And Stafleu says: "It should be realized that Brongniart was not a Darwinian *avant-la-lettre*. The changes in the floras of previous ages were described by him as transformations of life: evolutionary thinking in the modern sense should not be attributed to him" (1966, p. 321).

Brongniart postulated four major periods of geologic time, beginning with the Carboniferous.

> *During* each of these periods, the vegetation has only presented gradual and unlimited changes, which have not influenced the essential characters of the vegetation. *From one period to that following it*, there is, on the contrary, an abrupt transition, a sudden difference in the most important characters of the vegetation.
>
> This supposition of a complete or almost complete interruption of vegetation at the surface of the globe, between two of the periods of vegetation which we have admitted, is so much the more probable, that there exists no species common to two successive periods. Everything is different in them, and the idea cannot be rejected, that a new vegetation, arising under influences different from those which previously existed, has occupied the place of the old vegetation. [1829, pp. 353–54; my italics]

The idea of a series of successive creations was expressed as late as 1857 in one of Brongniart's writings, but I have found no indication as to what he thought might have been responsible for their origins.

Brongniart may be compared, I believe, with a few other paleobotanists, such as Scott, Seward, Kryshtofovich, and Němejc, for his great ability to bring the accumulating information together, making an understandable whole out of it, and in a readable fashion. He wrote several papers on the general floristic composition of the major geologic periods which have appeared in both French and English. Two of the earlier ones are his "Observations on Some Fossil Vegetables of the Coal Formation" (1826) and "General Considerations on the Nature of the Vegetation Which Has Covered the Surface of the Earth, at the Different Epochs" (1829). And a later and more detailed study is more useful than others published in the middle of the century, *Chronological Exposition of the Periods of Vegetation and the Different Floras Which Have Successively Occupied the Surface of the Earth* (1850).

His interest in the stratigraphic use of fossils is evidenced by a rather lengthy footnote in an 1857 paper in which he defends the priority of the work of Georgeŝ Cuvier and his own father; in a study published in 1811 (*Essai sur la géographie minéralogique des environs de Paris*) they had noted that specific fossils distinguished the several horizons of the area in which they were working.

National pride enters in here, the chief "competitor" being the great English geologist William Smith, whose *Strata Identified by Organized Fossils* was published in 1815. We cannot claim Smith as a paleobotanist, but his work does fit into our story as a whole. He was born in 1769 at Churchill in Oxfordshire. At the age of eighteen he became associated with a Mr. Webb, from whom he learned the business of land surveying. As a result, Smith traveled extensively through parts of England. His biographer, John Phillips (1839), notes that as a young man he was quite unacquainted with the natural-history literature and had no teacher other than a "habit of observation." He apparently had extraordinary powers of observation and made very extensive collections of fossils, becoming adept in using them to identify rock formations. He was not the first to do this but he carried this kind of study to a great point of refinement. He took up residence in London in 1804 and his studies became well known. He was also concerned, like others before and after him, with drainage problems in the fenlands of Suffolk and Norfolk; his first publication was a *Treatise on Irrigation*, which appeared in 1806. The *Strata* of 1815 was followed in 1817 by his *Stratigraphical System of Organized Fossils*. Thus although his work was certainly known before 1815, that of Cuvier and Alexandre Brongniart appeared in print first.

W. C. Williamson knew William Smith and his wife quite well as they stayed at the Williamson home for extended periods when Williamson was a boy. In his *Reminiscences of a Yorkshire Naturalist* he gives a short character sketch of Smith that is representative of some of the good reading that the little book contains:

> One of the grandest figures that ever frequented Eastern Yorkshire was William Smith, the distinguished Father of English Geology. My boyish reminiscence of the old engineer, as he sketched a triangle on the flags of our yard, and taught me how to measure it, is very vivid. The drab knee-breeches and grey worsted stockings, the deep waistcoat, with its pockets well furnished with snuff—of which ample quantities continually disappeared within the finely chiselled nostril—and the dark coat with its rounded outline and somewhat quakerish cut, are all clearly present to my memory.

> Spending the greater portion of his morning in writing, towards noon he would slowly wend his way to the museum, where he always found my father a friend with whom to gossip about the rocks of the Cotswolds, the clays of Kimmeredge, or the drainage of the Eastern Fens. He would expound in a Coleridgean fashion his ideas of their relation to the strata of Yorkshire and of the other parts of England. His walking pace never varied; it was slow and dignified; he was usually followed a few yards in the rear by his rose-cheeked partner in life. We have a thousand times contemplated the fine old man, who, amid his favourite haunts, thus laid the foundations of geological Science. [1896, pp. 12–13]

Brongniart enters into our story on other pages of this book, but I cannot leave him here without a brief mention of his *Recherches sur les graines fossiles silicifiées* (Researches on Fossil Silicified Seeds), which was published posthumously in 1881. (He died on February 18, 1876.) This is a quarto volume with twenty-one magnificent colored plates describing the silicified seeds from Autun and St.-Étienne. It is certainly one of the great landmarks in paleobotanical literature, a volume to be treasured for its scientific and aesthetic values. Bernard Renault and Cyrille Grand'Eury had long had a close working relationship with Brongniart and in a broad sense both may be regarded as his students. The important works of Renault and Grand'Eury will be considered on later pages but it may be noted here that the volume on the silicified seeds was in part the work of Renault, although just how much credit goes to him I cannot be sure.

The *Versuch einer geognostisch-botanischen Darstellung der Flora der Vorwelt* (An Attempt at a Geographical-Botanical Description of the Flora of Prehistoric Times) by Kaspar Maria, Graf von Sternberg, is cited in the Botanical Nomenclature Code as the official starting point for the naming of fossil plants. I have noted elsewhere there is some feeling that Schlotheim's work has not received adequate recognition; there is certainly great merit in his work and, indeed, Schlotheim, Sternberg, and Brongniart stand as a most distinguished trio whose studies ushered in a new era in paleobotanical research.

Sternberg was actually born three years before Schlotheim, but the latter's works were published before Sternberg's *Flora der Vorwelt.*

The *Flora der Vorwelt* was published in separate numbers from 1820 to 1838 (the dates are given in the *Generic Index*; Andrews, 1970, p. 335). Like Brongniart's *Histoire*, the parts are usually found in libraries bound into two volumes. It was translated into French by the Comte de Bray (*Essai d'un exposé géognostico-botanique de la flore du monde primitif*), but the translation apparently was issued in a very limited edition; there is a copy in the Paleobotanical Library of the U.S. Geological Survey in Washington.

This basic account is well illustrated and is more comprehensive than Brongniart's *Histoire*. It displays a good knowledge on Sternberg's part of the status of fossil botany and draws upon the best work of his predecessors and contemporaries. One finds in it the starting point for many genera that have stood the test of time and are still in use, such as *Alethopteris, Annularia, Equisetites, Lepidodendron*, and *Walchia*, to mention but a few. Others have fallen by the wayside, and one rather amusing example may be given: It was the custom then, I think, even more than now, to commemorate one's colleagues with generic names; for specimens which later proved to be cordaitean pith remains Sternberg created the genus *Artisia* for Artis, in return for Artis's *Sternbergia*!

Ward (1885) says of the *Flora der Vorwelt*:

> The most important departure effected in this work was in establishing vegetable paleontology for the first time upon a geognostic basis. He assumed three periods of vegetation: (1) an insular period characterized by the great coal plants; (2) a period characterized by the predominance of cycadean types; and (3) a period . . . characterized by dicotyledonous forms. It will be at once perceived that these three periods correspond substantially with the Paleozoic, Mesozoic, and Cenozoic ages of modern geology. [P. 428]

In view of the unfinished nature of the *Histoire* it is difficult to compare the works of the two men. Brongniart reveals a greater knowledge of plants and a keener sense of organization. Sternberg introduced much that is new and incorporated the best of the past, drawing on the studies of Brongniart, Corda, Schlotheim, Lindley and Hutton, and others.

The *Flora der Vorwelt* includes a final section written by Corda, whose life and work are mentioned elsewhere, and Sternberg apparently also received considerable aid from the botanist Karel B. Presl. Presl was born in Prague in 1794 and received his doctorate in medicine in 1818. His dissertation was a monograph on Sicilian grasses which he had collected, at least in part, on a journey to Italy and Sicily in 1817. Presl apparently found botany more to his liking than medicine; he was quite a prolific researcher and among his more important works are the *Reliquiae Haenkeanae* based on large collections made in South America by Theodor Haenke, and his *Symbolae botanicae*, which includes descriptions of numerous new genera. He also produced notable works on the Lobeliaceae and the ferns, the latter being both anatomical and taxonomic. Presl was a curator in the National Museum in Prague for many years, and in 1832 he was appointed to the Chair of Natural History and Technology in the Philosophical Faculty.

The scientific differences between Sternberg's *Flora der Vorwelt* and Brongniart's *Histoire* are reflected by the very different kinds of lives that their authors experienced. Sternberg was also well-off financially, but that is about the only point in common; much of his life extended through a war-torn and tumultuous time in Europe.

Kaspar Sternberg was born on January 6, 1761, in Prague. His mother, the Countess Anna Josepha von Kolowrat, is described as a lovely woman who spoke and wrote fluent German, French, Italian, and English; thus the son was exposed to a liberal linguistic education, adding Latin to it in school.

In 1789, when Sternberg was twenty-eight and nearing the start of his scientific career, the French Revolution began; Louis XVI was executed in 1793 and the French Republic developed a rather enormous army, whose military power under Napoleon dominated Europe for more than twenty years. The military turmoil was accompanied by internal squabbles among the Austrian, Czech, Bavarian, Prussian, and Polish interests. During this turbulent period in Europe much of Sternberg's botanical life was lived.

Since two of his brothers had entered military service his parents thought it best to enroll him in a holy order. Thus, following earlier studies in Freising, Regensburg, and Prague he journeyed to Vienna, where he busied himself with theological studies and enjoyed an audience with the Empress. One gains the impression from Palacky's biography (1868) that Sternberg was quite adept at meeting the right people and at avoiding personal disaster, and that he was never totally devoted to a religious life. His uncle, Minister Count Leopold Kolowrat, arranged for him to visit Rome, where his studies at the Collegium Germanicum advanced well but were not all theological; he formed a small literary society there with five other students who met once a week to read scientific works. However, political developments in the spring of 1782 resulted in the closing of the Collegium and all Austrian subjects were recalled. He spent a few months in Naples and returned to Regensburg in January of 1783 and became attached to the regional cathedral, although he was then only twenty-two.

Along with his administrative duties at the cathedral he served as archivist of the history of the capitol, but political maneuvering does not seem to have appealed to him, and an acquaintance with Baron Gleichen, a man of broad knowledge, tended to steer him toward scientific matters. The retreating Prussian army, following battles with the French, marched through Regensburg in 1792. The next two years are not clear but in 1794 we find Sternberg trying to obtain shelter for some four hundred priests who had fled from the prospects of the guillotine in France. At about this time a combination of his distaste for religious-political intrigue and a chance meeting with Count Bray, President of the Botanical Society of Regensburg, seems to have led him to give up active church service and embark seriously on a botanical career. In 1804 Prince Primas placed him in charge of the development of a botanical garden in Regensburg, in connection with which he visited Italy and Paris; he became acquainted with Alexander von Humboldt, Laplace, Cuvier, De Candolle, and, perhaps of particular importance at this time, with Faujas de Saint-Fond, who showed him his fossil collections, and he also studied Scheuchzer's collections.

By 1808 he was giving regular lectures in the botanical garden and devoting his time to the study of living plants (including the Saxifragaceae) as well as fossil ones. This life, which he evidently enjoyed, was disturbed in 1809 when the garden, which was in the path of warring French and Austrian troops, was largely destroyed, although Sternberg was able to save his collections and library. In his later years he lived much of the time in Prague, where he was concerned with the development of a national museum. I do not know that he was the founder of international science meetings, but he was concerned about the need for more effective communication among scientists and was largely responsible for the Congress of Natural Scientists that was held in Breslau in 1833. He had a strong interest in what we call today economic geology, and in 1837, the year he died, he completed a two-volume work on the history of mining. He is also indirectly linked with later paleobotanical developments in Germany, for in 1835 Goeppert visited Prague to study Sternberg's collections. Sternberg died on December 31, 1837.

In summary, botany and geology owe much to the great works of Brongniart and Sternberg. It seems to me that it does not quite do justice to the many others who preceded them simply to refer to these two as the "fathers" or "founders" of scientific paleobotany. They depended on previous developments, and they added a great deal that was new; they gave the study of fossil plants a firm foundation for its development, in addition to their other services to botany and geology, and their successors were quick to expand the productive phase that they initiated. Sternberg's *Flora der Vorwelt* brought together a great mass of factual material, but it definitely lacks the superior classification found in Brongniart's *Histoire*. It is unfortunate that the latter was never finished but his many other published works mark Brongniart as a really great botanist; he had a fine "biological sense" and his work will probably be referred to as long as there is an interest in the plants of the past.

5

Paleobotany in Britain: Witham to Scott

ONE of the longest and most productive epochs in paleobotanical progress began in Britain in the early years of the nineteenth century with Edmund Artis, Henry Witham, John Lindley, and William Hutton and continued well into the present century with the tremendous contributions of numerous men and women such as Robert Kidston, D. H. Scott, A. C. Seward, and others. It is representative of some of the best work that has been done and it includes people who were fascinating characters and great botanists; having spent considerable time in Britain over the past four decades I have become especially well acquainted with the era through personal contacts, museum studies, and field excursions. If my emphasis is a bit strong in these chapters it is because I may be able to record some information that is not readily available elsewhere.

Improved methods of extracting information from fossil plants have resulted in great pulses of progress, and partly on this account I choose to start this period with Henry Witham (1779–1844) and some notes on the thin-section method. In 1831 Witham produced a small book entitled *Fossil Vegetables, Accompanied by Representations of Their Internal Structure as Seen through the Microscope*. In 1832 he brought out a short paper on petrified stem material that he described as *Lepidodendron Harcourtii*, named for the Reverend C. G. V. Harcourt, rector of Rothbury. The text is accompanied by two plates that show cellular structure in excellent fashion and he draws a comparison with the modern *Lycopodium*.

The second edition of his book came out in 1833 under the title *The Internal Structure of Fossil Vegetables found in the Carboniferous and Oolite Deposits of Great Britain* and is the work for which he is best known. It contains sixteen beautifully executed plates; the first two show basic cellular organization of the stems of modern conifers, dicots, and monocots. The others are chiefly devoted to fossil gymnosperms. This is not entirely Witham's work, for on page 4 he gives credit to "Mr Macgillivray, for his unremitted attention and assiduity in the difficult task of executing the beautiful drawings from which the engravings have been made, and for his kind assistance in the minute botanical descriptions of the various plants here presented to the public."

Witham's book is important for several reasons, but especially as the first major account of the internal structure of fossil plants based on the use of thin sections. In the "Concluding Remarks" of the 1831 edition Witham gave a detailed description of the thin-section method, and a number of people credit him with developing the technique. However, he stated clearly: "I have the pleasure of laying before my readers a full account of the process, for which I am indebted to Mr. Nicol" (p. 45). Some readers apparently did not notice this, and an annoyed William Nicol wrote in 1834:

> Not only my method of preparing thin slices, but that of examining the organic structure of fossil woods, by reducing them to thin translucent slices, has been ascribed by some unwise friends to Mr. Witham, although that gentleman has no more claim to the latter than I have. It was first employed in this quarter by Mr. Sanderson, the lapidary.
>
> When I first began to prepare such sections, I had recourse to a process I had practised upwards of fifteen years ago, in preparing thin slices of the most fragile substances, as calcareous spar, in order to examine their effects on polarized light. [1834, p. 157]

Nicol goes on to describe the technique in some detail. Actually Witham, in the appendix (p. 76) of his 1833 edition, referred the reader to "Mr. Sanderson, Lapidary, St. Andrew's Square, Edinburgh" if he should require his service in this time-consuming task! W. T. Gordon, whom we shall meet again on a later page, had this to say: "The advent of the new technique had more than one consequence. It divided the study of fossil plants into two sections, one biological and the other stratigraphical, and it also led indirectly to what was probably the greatest advance ever made in Petrology" (1935, p. 62).

It is fair to add that Nicol's interest in petrified woods was more than a casual one. He contributed papers to the *Edinburgh New Philosophical Journal* in 1831, 1833, and 1835 on fossil woods from New South Wales, the island of Mull, and Van Diemen's Land. Had these been better illustrated I am sure that he would have received much greater recognition as a student of fossil plants. In 1867 the British Museum of Natural History purchased a collection of nearly five hundred of his thin sections of woods, prepared from fossils from British and foreign localities.

Both D. H. Scott (1911) and Albert Long (1959) went to some trouble to obtain biographical data on Witham and to define the importance of his work. Long says:

> Concerning the interpretation of the fossil trees from Lennel Braes, he rightly regarded them as not being Vascular Cryptogams. This he emphasized as a remarkable fact, since Adolphe Brongniart—the "father of palaeobotany"—had stressed the overwhelming superiority in numbers of

Henry T. M. Witham of Lartington. Portrait by an unknown artist. From the *Makers of British Botany*, Cambridge University Press, 1913.

> the Vascular Cryptogams in the Carboniferous period, together with plants he called *Monocotyledons*—which were probably *Cordaitae*. . . . Witham regarded the fossil trees from Lennel Braes as *Dicotyledons*, in which group *Gymnosperms* were then included. . . . It was, therefore, Witham's greatest achievement to demonstrate that fossil Gymnosperms were prevalent in early Carboniferous times. [1959, pp. 255–56]

Witham was born in 1779, the second son of John Silvertop of Minster Acres, Northumberland, and as Henry Silvertop he inherited the Lartington property. He married Elisa Witham of Yorkshire and by Sign Manual took the name and arms of Witham; thus most of his writings appeared under the authorship of "Henry T. M. Witham of Lartington." His wife inherited valuable estates, and to a point Witham seems to have been quite astute in his personal affairs, but Walton (1959) notes that he did not confine his pursuits to paleobotany, and by injudicious management much of his property was lost. Lartington Hall, a magnificent house with a chapel attached, is still standing, and Albert Long, who visited it in July, 1977, says: "It was all very old-world and fascinated me. . . . The windows are shuttered. . . . The lawns were cut and evidently used for croquet but the Hall is not lived in" (personal letter).

Witham gave fossil botany a significant push with his studies of internal structure and the technique required to examine it, and I will leave him with a few words of his own:

> My pretensions to botanical knowledge are indeed very limited; nor do I presume to rank myself among the cultivators of a science to which so many eminent individuals have devoted themselves in this country. The only object I have always steadily kept in view, is to direct their attention to a department of botany which has hitherto been too much neglected; for, although the study of external forms of the stems, leaves, and fructifications, of recent vegetables, has elicited much knowledge respecting the nature of the former, little has been effected by an application to their internal composition, in which decided and characteristic differences are nevertheless to be found. [*Internal Structure of Fossil Vegetables*, 1833, pp. 1–2]

Shortly before and immediately after Witham's fine book came out there appeared two works of a somewhat controversial nature dealing with compression fossils that attracted considerable interest, namely, Artis's *Antediluvian Phytology* of 1825 and Lindley and Hutton's *Fossil Flora of Great Britain*, which was published between 1831 and 1837.

The *Antediluvian Phytology* of Edmund Tyrell Artis (1789–1847) is a smallish volume containing twenty-four plates of Coal Measure plants from Yorkshire. Brongniart, in the *Histoire*, dismisses it as being of little consequence, the specimens being small in number, imperfect, and referable to genera already established. National feelings may enter into the matter of making judgments in paleobotany, as in other facets of life. Lester Ward, who I think tried to seek out the best side of things, referred to it as a classic of

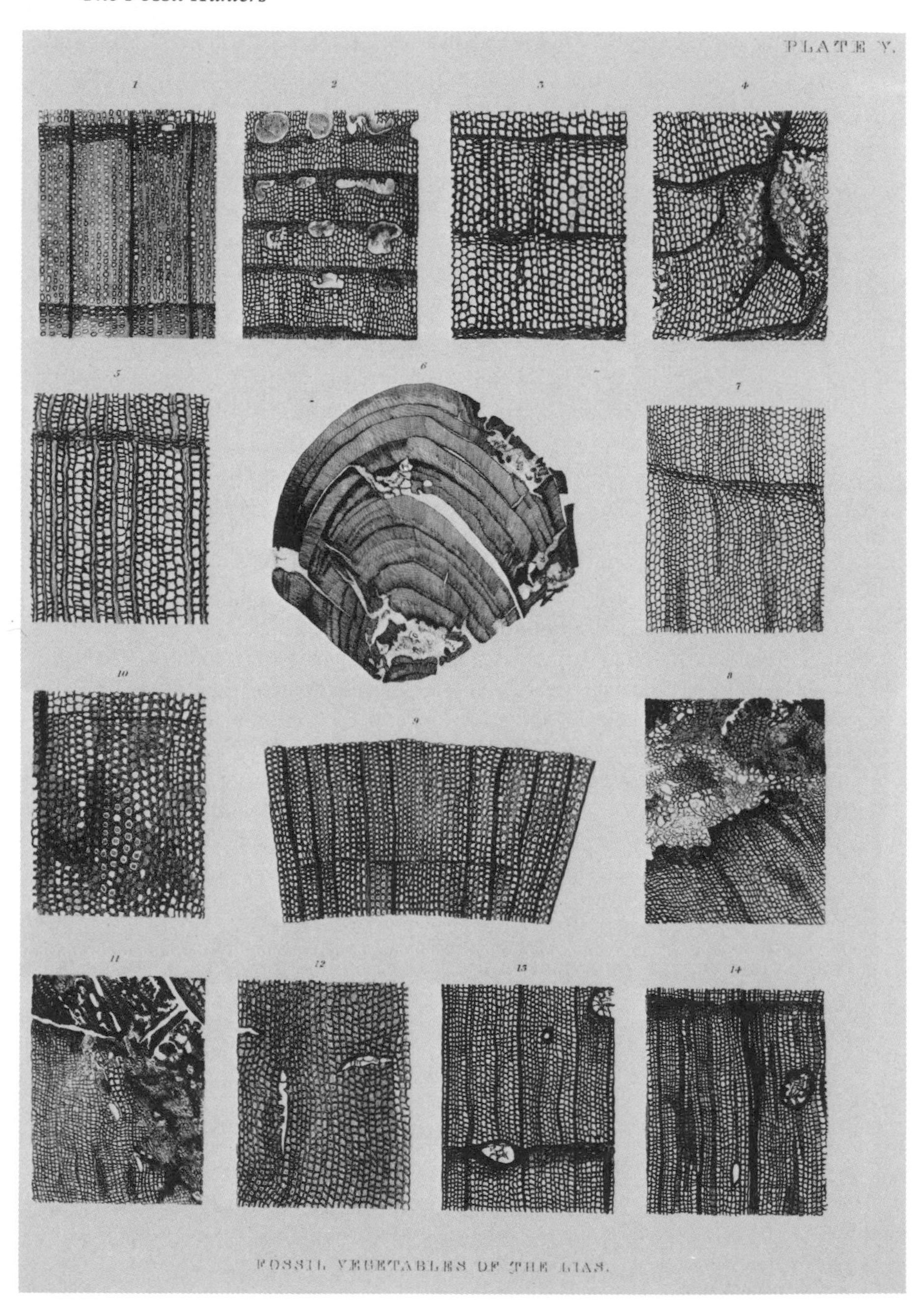

A plate from Witham's *Observations on Fossil Vegetables*, 1831.
Photo courtesy of the Cambridge University Library.

paleobotany, but he did qualify his appraisal in referring to it "rather as a work of art than science" (Ward, 1885, p. 405).

Arber (1921b) says that Artis's book was the first scientific work published in Britain on British fossils by a British subject. It deserves some credit for helping to stir up interest at a time when it was needed. There is very little biographical information available on Artis. He was born at Sweflin near Saxmundham in Suffolk in 1789 and exhibited a talent in art and an antiquarian interest early in life. He is said to have amassed a large collection of fossil plants from Yorkshire and Derbyshire which served as the basis of his book. He was very much interested in Roman antiquities and spent the last twenty-two years of his life near Peterborough in this pursuit. His enthusiasm must have been strong, for it is recorded that on one occasion, when digging at Sibson, "he bivouacked with his men in the depth of the winter of 1846–7 in a wood adjoining, until the weather caused his party to desert and leave him" (anon., 1849, p. xxiii).

Lindley and Hutton's *Fossil Flora of Great Britain* has been variously maligned by later workers, but I am inclined to think that, in the context of the times, it deserves somewhat better treatment. Arber, although he cites the inadequacies of the work, notes that "it unfortunately remains to this day [1921] the one illustrated British book containing a general account of our Coal Measure plants" (1921, p. 484).

The *Fossil Flora* appeared in several installments between 1831 and 1837, and these are usually found bound into three volumes. They contain 230 plates and the plants thus figured range from the Lower Carboniferous to the Tertiary. The fossils were collected for the most part by William Hutton (1797–1860), who lived in Newcastle from 1823 to 1839. He was an agent for an insurance company but had a strong interest, although apparently no training, in geology. He was also largely responsible for the formation of the Natural History Society of Northumberland, Durham, and Newcastle, and at least a part of his collections are now preserved in the Hancock Museum in Newcastle.

Hutton collected the specimens and had the necessary drawings prepared, the two being sent periodically to John Lindley (1799–1865), who was for many years Professor of Botany at University College, London. Lindley was one of the great botanists of the first half of the century and is perhaps best known for his studies of orchids. He devoted many years of service to the Horticultural Society, planning exhibitions and guiding it through a time of financial difficulty. An anonymous biographer says that "his greatest merit consists in having successfully established in this country the Natural System. He also had the happy knack of popularizing and making clear the labours of others, but his own ideas were often crotchety. He was a capital follower, but an indifferent leader" (1865, p. 388).

Lindley and Hutton recognized their own inadequacies (see vol. 1, p. xxv) but had the courage to go ahead with a task that they thought needed to be done. Lesser men often simply complain, or say nothing, about the lack of progress. And there are some significant and interesting aspects of the *Fossil Flora*. The first volume includes, in the preface, a rather detailed discussion of the ways and means of identifying fossil plants and the use of anatomical characters and external morphological features in distinguishing the major groups of plants. Lindley was concerned with the reliability of the fossil record in conveying an accurate account of the plant life of the past. He seems to have been convinced that the apparent abundance of "ferns" in the Carboniferous was questionable (perhaps the first thought that they might not all be ferns!), and he decided to do something about this. Accordingly, on March 21, 1833, he filled a large tank with water and placed 177 specimens of living plants in it, including thallophytes, bryophytes, ferns, lycopods, cycads, conifers, and flowering plants. He examined the remains about two years later with the following results:

> This experiment appears to me to lead to most important conclusions. These things seem clear: firstly, that Dicotyledonous plants, in general, are unable to remain for two years in water without being totally decomposed; and that the principal part of those which do possess the power, are *Coniferae* and *Cycadeae*, which are exactly what we find in a Fossil state; secondly, that Monocotyledones are more capable of resisting the action of water, in particular Palms and Scitamineous plants, which are what we principally find as Fossils, but that Grasses and Sedges perish; so that we have no right to say that the earth was not originally clothed with Grasses because we no longer find their remains. [Vol. 3, pp. 11–12]

He concludes that the proportion of different kinds of plants preserved as fossils sheds no light on former climatic conditions, but rather reflects the preservability of the plants. The inadequacies of his experiment and the conclusions drawn are evident, but he was one of the first to take the trouble to experiment rather than simply to speculate.

It was well established by this time that coal was largely of plant origin, but the mode of deposition was in dispute. This matter is considered in Volume 2 with conclusions that seem to conform rather well with present-day concepts. Some of the mistakes in identification, although understandable, are amusing and were not confined to the less learned paleobotanists. As an example, Lindley and Hutton had some trouble with *Sphenophyllum*. In his

Prodrome (1828), Brongniart classified it with *Marsilea*. While praising Brongniart's work in general they disagreed with this treatment and referred it to "the Pine tribe of modern Floras." Schlotheim had suggested that *Sphenophyllum* might be placed with the palms. *Stigmaria* was, as I have mentioned elsewhere, a particularly vexing fossil. Sternberg had suggested an affinity with the Euphorbiaceae or cacti, while Brongniart preferred the Aroideae; Lindley and Hutton tended to support Sternberg's assignment (see *Fossil Flora*, vol. 1, pp. 103–4).

Arber (1921b) notes numerous inaccuracies in the preparation of the figures for the *Fossil Flora* and a lack of exact locality data, errors which made this work less valuable to subsequent workers than it might have been. In an attempt to straighten out some of the errors Kidston prepared a rather lengthy "revision" in 1891.

Albert Long has appraised Lindley and Hutton's work in a way which I think is accurate and fair:

> Unfortunately, Lindley only knew extant plants and his knowledge was almost an encumbrance impeding the correct interpretation of the fossils. In other words, his light was a kind of darkness and the fossils almost had to speak for themselves—poor things! Hutton deserves credit for collecting and preserving the fossils which would otherwise have been lost and he did the right thing in seeking the help of Lindley—a competent botanist—and they made a commendable attempt to illustrate them without the aid of photography. [Personal letter]

People of lesser professional stature were beginning to take a serious interest in fossil plants. In 1823 one Thomas Allan described a *Lepidodendron* stem compression that came from the Craigleith quarry near Edinburgh. He discussed possible affinities with the palms and cacti, but his enthusiasm makes up somewhat for the inaccuracy of his identification:

> It is not my intention to touch upon the interesting speculations which the occurrence of these unknown species of vegetables irresistibly opens to the mind. They give us a glimpse of former periods, which sets conjecture at defiance, and smiles at the vain attempts of theory to unravel. I only wish to call attention of the naturalist, to the advantage which may be derived from a proper attention to this department of natural history; and when it is known how much this country, I may say our immediate neighborhood, abounds in fossils of this description, it is rather to be lamented that no extensive collection of them, has been systematically made. [1823, pp. 236–37]

Two men of strikingly divergent character bridge the small gap between the time of Witham, Lindley, and Hutton and the later monumental contributions of Williamson: Edward W. Binney and Joseph D. Hooker.

Binney (1812–1881) was a self-made geologist and paleobotanist of the first order. He was born at Morton, Nottinghamshire, in 1812; in 1828 he was apprenticed to a soliciter in Chesterfield, and at twenty-one he went to London, where he continued his law studies for another two years. During these years he spent much time reading in libraries and was said to be constantly seen at the British Museum.

After completing his law studies he settled in Manchester. One of his biographers has suggested that his geological interests were initiated on the assumption that some scientific knowledge of the coal fields would be of use in a business way in dealing with the coal operators. In view of the intensity and apparent sincerity of his geological studies this seems partially true at best. His business affairs did prosper, and in addition to his law practice he operated a paraffin factory at Bathgate near Edinburgh.

He soon became much interested in the Lancashire coal field and spent much time exploring the area on foot. He acquired a good deal of first-hand knowledge from the miners and other local people. One of his biographers notes:

> He was tall and powerful, and he had no wish to seek society. Indeed, he always spoke in a disparaging manner of the usual social intercourse in the middle and upper ranks, and delighted in rambling over the country and mixing with the men he chanced to meet, studying their ways and learning their observations. It was in this way that he came to take much interest in the scientifically inclined working man, and he had a particular pride in speaking more highly of him than of the more learned or elaborately trained. [Anon., 1881, p. 447]

Through the interests of Binney and several others the Manchester Geological Society was formed in 1838, and he also rendered important services to the British Geological Survey resulting from his long experience in the Lancashire and Cheshire coal fields. He wrote some 134 papers; many were brief but a few stand out as important contributions. He is probably best remembered by fossil botanists for his *Observations on the Structure of Fossil Plants Found in the Carboniferous Strata*, which was published by the Paleontological Society of London in several parts between 1868 and 1875. Although not entirely original it is an excellent work on the anatomy of the Carboniferous calamites and lycopods and served as a comprehensive foundation for later studies of these groups. In 1847 he wrote a rather long article,

"On the Origin of Coal," which reveals his extensive knowledge of the coal fields and the coal itself.

One contribution of Binney's that will always be memorable to paleobotanists working with Carboniferous petrifactions is the report he wrote with J. D. Hooker in 1855 on coal balls—the first one on these most important fossils that I am aware of. Coal-ball petrifactions have played a great role in the development of Paleozoic paleobotany, and the plants contained within them are mentioned in many places in this book. Francis Oliver gave me an interesting description of them in a letter of January 2, 1946:

> In 1910 (I think it was) Sutcliffe of Shore-Littleborough near Rochdale, Lancashire, a cotton spinner, reopened a deserted coal mine on the property for paleobotanical reasons. The then new fossil *Sutcliffia* was one of the products. The entrance was a hole in a hillside and you just walked in. The mine was more or less horizontal—a main passage—laterals extending for several hundreds of yards. About ⅓ to ½ of the seam for which coal had been worked for his mill consisted of dense accumulations of coal balls ranging in size from a walnut to a foot or more in diameter. It was an incredible spectacle and Sutcliffe rigged up a temporary electric light installation, which made it really spectacular. At places the entire seam appeared to consist of coal balls in almost continuous contact. I helped myself to what filled several crates, and most of them must still be in a cellar I had for storage at University College, London. It was customary with the lapidaries to place coal balls on the tops of walls, where in time the surface would weather, showing very distinctly which would be more profitable balls to slice.

In view of the historic importance of coal-ball petrifactions it is appropriate to quote a few lines from the paper by Binney and Hooker:

> The specimens of plants which we are about to describe were found embedded in nodules of limestone, enclosed in a thin seam of bituminous coal not above 6 inches thick, in the lower part of the Lancashire coalfield. . . . The origin of these nodules may probably be ascribed to the presence of mineral matter, held in solution in water and precipitated upon, or aggregated around certain centres, in the mass of vegetable matter now for the most part turned into coal. . . . The immediate cause of the calcification was no doubt due to the abundance of fossil shells in the shales immediately overlying the coal and nodules . . . they present no appearance of these remains having been brought together by any mechanical agency; they appear to be associated together just as they fell from the plants that produced them, and to be the rotting remains of a redundant and luxuriant vegetation. [Hooker and Binney, 1855, pp. 149–50]

I believe that this article is very largely the work of Binney, but I take advantage of it to bring in a few lines on the paleobotanical works of his collaborator. It is rather an honor for paleobotany to be able to include in its annals a few works of one of the great botanists of the nineteenth century, Joseph D. Hooker (1817–1911). At the age of thirty-eight he was appointed Assistant Director at Kew, where his father served as Director. Joseph became Director himself ten years later and held the post for the next twenty years. He had taken his M.D. degree at Glasgow and then entered the Navy as assistant surgeon and botanist on James C. Ross's expedition to the Antarctic on the exploring ships *Erebus* and *Terror*, sailing on September 29, 1839. This was the start of a career that made Hooker one of the great plant geographers. On this trip he visited Madeira, Kerguelen Island, Antarctica, and New Zealand and collected much of his material for *The Flora of New Zealand*. The expedition returned in 1843, and four years later he started his second great journey, spending three years in India, exploring especially the Himalayan areas of Sikkim and Nepal. This is still an exciting area to travel in, but whether one has been there or not Hooker's two volumes of *Himalayan Journals* make very good reading.

He published a short note on fossil wood from the Macquarie Plains, Tasmania, but probably his most important contribution to fossil botany is his 1848 description of *Lepidostrobus* cones, which was based on quite a large number of specimens and is a fine piece of work, if somewhat tedious to read. It is well illustrated and gives us an understanding of the cone structure, and he clearly recognized their affinities: "If, now, these cones be examined with reference to the known contemporaneous fossils which accompany them, it will appear impossible to deny their having [been] the reproductive organs of *Lepidodendron*, not only from their association with the fragments of that genus, [but] because the arrangement of the tissue in the axis of the cone entirely accords with that of the stem of *Lepidodendron* (1848b, p. 451).

Another great British botanist, Robert Brown (1773–1858), who enters but briefly on the stage of fossil botany, also contributed to this particular facet of our science. In 1847 he read a paper before the Linnean Society on a petrified Coal Age cone fragment that he called *Triplosporites*, from the presumed grouping of the spores in threes, and which he recognized as having lycopod affinities. A brief note on this appeared in 1848, but his illustrated account came out in 1851. The cone was a well-preserved terminal part of a *Lepidostrobus*, and it is unfortunate that Brown and Hooker could not have collaborated in the study. Brown was born at Montrose, Scotland, in 1773 and attended the universities of Aberdeen and Edinburgh. He engaged in a four-year voyage to the Australian region and is noted for many important botanical discoveries: he elucidated the true nature of the cyathium in *Euphorbia* and described the curious parasite *Rafflesia Arnoldi*; probably most important of all, certain of his studies of the seed cone in the cycads and conifers led directly to the recognition of the gymnosperm group (Green, 1914).

Another of Hooker's papers shows the primitive status of Tertiary seed studies at the time, a status that has changed for the better quite drastically in more recent decades. In 1855 Hooker described some "Seed-vessels" from the Eocene of Lewisham, and although he was cautious in drawing conclusions he reported spores in them and suggested an alliance with the ferns. These were called *Carpolithes ovulum* Brongn. and *Folliculites minutulus* Bronn. At a considerably later date Clement Reid found that the former was the seed of a water lily; *Folliculites* is also a seed, belonging to *Stratiotes* (Scott, 1912).

In the same volume of the Geological Survey's *Memoirs* (1848) in which his account of *Lepidostrobus* appeared he wrote a rather long article, "On the Vegetation of the Carboniferous Period, as Compared with That of the Present Day," which is well worth reading today. Hooker had certain misgivings about fossil botany and he points out some of the pitfalls that the unwary may fall into. He noted the need for much more information on living plants, and the difficulty of dealing with sterile fern foliage, ferns with dimorphic foliage, and so on. In concluding he says: "Too much has been expected from the botanist, who wants materials for those bold generalizations which the fossils of the animal kingdom so abundantly supply. Except to individuals who have great facilities for this study, the collection and examination of the waifs and strays of a by-gone Flora is a forbidding pursuit" (1848a, p. 428).

Although most of Hooker's paleobotany was done in his earlier years he kept in touch with the progress that was being made. Scott says:

> Perhaps his most interesting letter in this connection was one written on receiving the preliminary communication by Prof. F. W. Oliver and myself on the seed of *Lyginodendron*, which, it may be remembered, was identified in the first instance by the glands on the cupule. He wrote (June 13, 1903): "I must write to thank you for sending me the Proceedings R.S. with your and Oliver's paper on *Lyginodendron*, which has interested me more than I can express. What can be the meaning of the capitate glands? they would seem to indicate the contemporaneous insect-life which I think has been demonstrated to exist in the Coal Measures. Has any one accounted for the quantity of pollen-grains in the sac of the ovule of Cycadeae? so many more than the wind is likely to have brought. [Scott, 1912, p. 37]

Until the period of D. H. Scott and his contemporaries no one influenced the course of paleobotanical studies in the area of Paleozoic petrifactions as did W. C. Williamson. He was a prodigious worker and a prolific writer in a time when paleobotanical knowledge was advancing rapidly, and he has left us a fascinating and informative record of the life of his time in his *Reminiscences of a Yorkshire Naturalist* (1896). His best-known unit of work is a series of nineteen memoirs entitled *On the Organization of the Fossil Plants of the Coal-Measures*, which appeared in the *Philosophical Transactions* of the Royal Society of London between 1871 and 1893. A series of studies based chiefly on coal-ball petrifactions, it brings in a considerable majority of the important Carboniferous genera and includes articles on the anatomy of the calamites, *Lepidodendron, Lyginopteris*, and several of the coenopterid ferns, and important reproductive organs such as *Lepidocarpon*, calamite cones (homosporous and heterosporous), stigmarian rootlets, and others. Each memoir is illustrated with several plates of Williamson's own drawings and is based largely on thin sections of his own preparation. The latter are preserved today in the collections of the British Museum.

Williamson's work leads us into important discoveries and controversies of his time and this accounts for the amount of space that is devoted to him here.

These articles were initiated rather late in Williamson's career, when he was fifty-five, and were done in his "spare time" from various other duties. Some botanists of the time were impatient with Williamson, and it is true that he had not kept up with botanical progress, but for very good reasons. One encounters comments such as that of W. T. Thiseleton-Dyer in a letter to D. H. Scott of May 11, 1892: "Williamson's grasp of the subject is so imperfect that I can never understand in the least what he is driving at" (Scott letters).

A summary appraisal by Solms-Laubach, in his biography of Williamson, is much fairer, and a few of his comments are worth recording:

> in his fifty-fifth year, [Williamson] began the great series of memoirs which mark the culminating point of his scientific activity, and which will assure to him, for all time, in conjunction with Brongniart, the honourable title of a founder of modern Paleobotany. . . . Williamson's method of anatomical description, clear as it is, bears the stamp of the scholastic ideas of a past time. For this reason it is only understood with difficulty by the botanists of the present day. [Solms-Laubach, 1895, p. 442]

William Crawford Williamson (1816–1895) was born on November 24, 1816, in Huntress Row, Scarborough. One may say that his interest in natural history began at birth, for his father, John, was a knowledgeable and avid collector—indeed, a wing had to be added to their house to afford space for the growing collections of fossils, minerals, shells, and insects. His father was acquainted with some of the leading natural scientists of the day; in 1826, for example, William Smith, the great stratigrapher and "father of British geology," and his eccentric wife established themselves in the Williamson home, where they lived for a considerable time.

When he was fifteen William was sent to school in Bourbourg, France, a short and somewhat disastrous experience that was in some measure redeemed on his way home in March of 1832: "Instructions had . . . been sent to me to spend a few days in London, and especially to call upon my father's old friend, Sir Roderick Murchison, the distinguished geologist. This was to me a solemn business. Murchison, Sedgwick, Lyell, and Buckland were the deities of my geological Olympus. When, having been invited to breakfast with the great man, I stood upon his doorstep in Bryanston Place, I had scarcely courage to ring the door bell" (1896, p. 24).

Shortly thereafter he was placed as a "medical student" with a rising general practitioner, Mr. Thomas Weddell. Williamson writes:

> the great time for work began about six o'clock in the evening, when the physic-making for the day took place. At that time few patients, excepting such as lived in the country, and whose visits involved the hire of a horse, paid for anything beyond the cost of medicines which they received. . . . Such practitioners were paid for their services by the sale of drugs, which their patients must swallow. As my governor had a very large practice among the middle and lower classes of a maritime fishing town such as Scarborough then was, the number of draughts and mixtures to be compounded, pills to be rolled, ointments to be rubbed up, and blisters and plasters to be spread, made the two or three hours after six o'clock a busy time. [1896, p. 27].

He started producing scientific articles before he was twenty and, aided by his father's acquaintances, his reputation spread. Lindley and Hutton's *Fossil Flora* was initiated when Williamson was still apprenticed to Weddell; of his being asked to prepare illustrations of the Gristhorpe plants he says: "I did so, and contributed to the pages of that work almost as long as its quarterly parts continued to be issued. . . . So far as my own communications to it are concerned, some of the paleontologists familiar with its pages may be amused to learn that most of the drawings were prepared at one end of Mr. Weddell's kitchen-table, whilst the housekeeper was occupied at the other end with the several processes of providing the day's dinner" (1896, pp. 35–36).

Before completing his medical apprenticeship he accepted a position as

William C. Williamson. Courtesy of the Cambridge University Press.

curator at the Manchester Natural History Society, which he later resigned in order to continue his medical career, for a time giving lectures to raise funds. In September of 1840 he entered University College, London, where he was able to study under some of the better men of the time, including Professor Lindley, who had never actually met him.

Upon completing his studies in London he returned to Manchester on January 1, 1842, and started a medical practice which he continued through the greater part of his life, along with all of his other occupations. He relates that one day he and his wife were invited to tea where one of the other guests called his attention to Mantell's *Medals of Creation*. He borrowed the book and read it, paying particular attention to passages about the microscopy of chalk. He obtained an old microscope of his father's, made the acquaintance of Dr. Mantell, and started his microscopic studies, which led to some of his well-known early work with living and fossil foraminifera and his studies of the alga *Volvox*.

In January, 1851, he was appointed Professor of Natural History at the newly created Owen's College in Manchester, where he was to teach classes in zoology, botany, and geology. In connection with his medical work, and partly stemming from an ear ailment of his own, he spent some time in Paris studying and observing in the clinic of the surgeon M. Menière, and he then helped to establish a special clinic in Manchester to treat ear ailments. He was in Switzerland with his family in the summer of 1870 and visited with Brongniart in Paris; the two had several paleobotanical discussions, of which I shall recount more shortly.

Lecturing came easy to Williamson, and he was apparently always ready to impart his knowledge to others. During the British Association meeting in Glasgow in 1876 he participated in a field excursion to the Island of Arran, where he and several others obtained a large *Lepidodendron* stump that was brought on board the steamer they were traveling in. At the request of Archibald Geikie he gave an impromptu lecture on the Carboniferous flora. "Afterwards, as I was explaining the peculiar features of this tree to the President of the Association and some of his family who were on the steamer, I was interrupted by a violent oratorical attack from a strange, clerical looking Scotchman, who denounced me, not only for what I had just been saying to the President, but also for my little lecture from the bridge. He said I was preaching a doctrine of devils, I presume by referring to the great antiquity of my plants." The man proved to be the preacher of a nearby Presbyterian chapel, and a friend of Williamson's chanced to hear him the following Sunday "expounding to his people what had taken place, and with what a 'Doctrine of Devils' I had sought to lead the souls of pious Scotchmen away from the truth" (1896, p. 170).

Chapter 13 of his *Reminiscences* contains a good summary of the origin of his work with petrified plants of the Coal Measures and the long controversy with the French paleobotanists, especially Brongniart.

> At this time [early in the 1850's] I had no intention of entering upon the long series of studies of the carboniferous plants in which I subsequently became engaged. My friend Mr. Binney was then investigating these plants, and I had no desire to interfere with his researches, but unfortunately he was not a botanist, and so fell into serious mistakes.
>
> . . . Though the "Prodrome" inevitably abounded in errors, it gave us

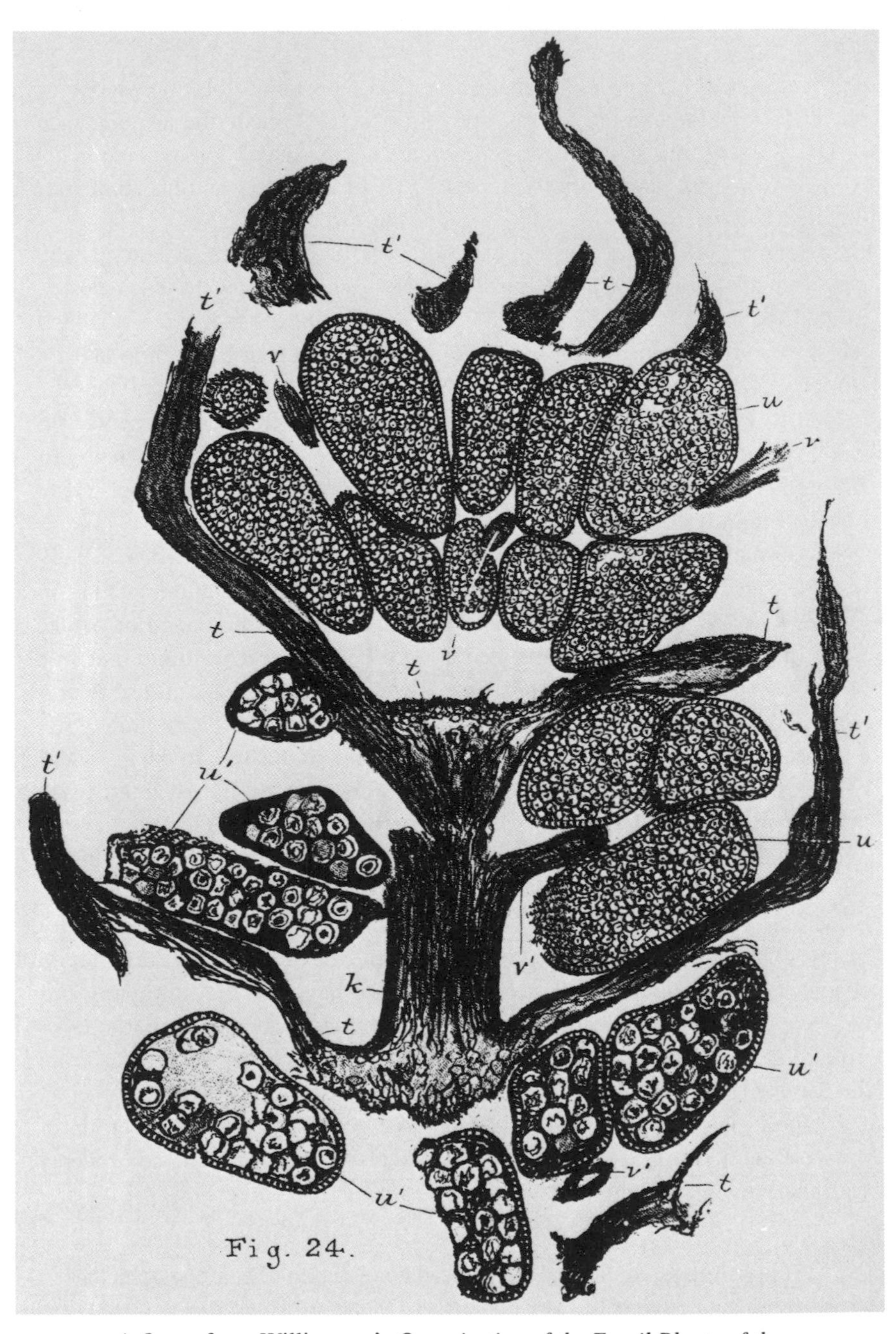

A figure from Williamson's *Organization of the Fossil Plants of the Coal-Measures*, Philosophical Transactions of the Royal Society, 1881. Photo courtesy of the Cambridge University Library.

> for the first time a philosophic arrangement of the classes, families, and genera of fossil plants then known, in four successive periods of the earth's past history. It would have been better had [Brongniart] adhered to this classification, because it was far more accurate than others which he substituted at later periods of his life. [1896, pp. 195, 197–98]

The most renowned dispute in paleobotany of the nineteenth century stemmed from the phenomenon of cambial activity in the arborescent cryptogams—the great calamites and lycopods of the Carboniferous forests. I think it is most interesting to let those who were present on the scene tell the story in so far as possible. To begin with, Brongniart had declared, on the basis of his studies of compression fossils, that a vast majority of the Carboniferous plants were cryptogams, belonging to the ferns, lycopods, and articulates (this being long before the pteridosperms were recognized). I do not think that anyone questioned this opinion until fossils with internal structure began to be studied; when such specimens were found the trouble began. Williamson says:

> Unhappily Brongniart was the first to strike a serious blow at his own philosophy. . . . Many years later two specimens fell into Brongniart's hands, one an equisetiform Calamite, and the other a lycopodiaceous Sigillaria, in each of which the internal structure was preserved. In one of these plants was found a central pith surrounded by an exogenously developed cylinder of wood. In the Sigillaria a similar condition existed. These woody zones were unquestionably formed by true cambium layers. Brongniart then believed that no living Cryptogam possessed a cambium; hence he concluded that both the Calamite and the Sigillaria must be removed from the positions in which he had originally placed them amongst the Cryptogams, and he classed them both with the Conifers and Cycads, believing them to be gymnospermous plants. This error has led to nearly thirty years of conflict amongst palaeo-botanists. [1896, pp. 198–199]

Two general sources of evidence eventually brought a recognition that the cryptogams included herbaceous as well as arborescent elements, the latter possessing cambial activity. These were the discovery of petrified specimens in which the external morphology could be correlated with comparable features of compression fossils, and the discovery of stem specimens with the spore-bearing cones intact.

Williamson has this to say concerning the origin of the fossil-plant material that formed the basis of his *Organization* series:

> Up to this time I did not know where Mr. Binney had got his specimens, but I soon had the advantage of an introduction to Mr. J. Butterworth of Shaw, near Oldham, who had not only collected plants from the localities which had supplied Mr. Binney with his, but had fitted up an excellent lapidary's lathe and prepared some sections of the fossil plants, of which he showed me a small but beautiful series. He also supplied me with a number of hard calcareous nodules dug out of the coal, from which these plants, the internal tissues of which were so beautifully preserved, were obtained. I at once provided myself with a jeweller's lathe, and commenced that series of practical researches which have continued until now. [1896, p. 200]

There is a considerable literature to be found on this great "controversy of the cambium," which is summarized here very briefly. For a long time Brongniart's position was strongly upheld by his disciples, especially such highly competent workers as Renault, Grand'Eury, and the Marquis de Saporta. Several of Williamson's papers in the *Organization* series deal with articulate cones that demonstrated the cryptogamic nature of the arborescent lycopods and articulates, and such evidence also eventually came from the French group. Bonnier (1917) goes to considerable length to give René Zeiller credit for determining that *Sigillaria* was a vascular cryptogam and not a seed plant. He refers to Zeiller's discovery of a Sigillarian cone with spores preserved that was attached to part of a stem with Sigillarian leaf scars; thus Zeiller definitely settled the issue, which had become something of a national one.

Solms-Laubach gives Williamson credit for establishing the fact that the cambium has appeared in various groups of plants and was by no means confined to phanerogams: "This is a general botanical result of the greatest importance and the widest bearing. In this conclusion Palaeontology has, for the first time, spoken the decisive word in a purely botanical question" (1895, p. 443).

It is fair to Brongniart and his followers to note that the cambium is a remarkably complex affair, little-known at the time, and much still remains to be learned about it—how a single layer of cells in the dicots, for instance, can give rise to such a remarkably diverse assemblage of mature cell types in the xylem and phloem. It gradually became evident that the cambium originated in several groups of vascular plants and that its method of functioning is variable. There is some evidence that it is unifacial in some of the lepidodendrids and calamites; for further information the reader may refer to contributions of Arnold (1960) and Eggert (1962).

There are many important discoveries scattered through Williamson's *Organization* series and his other papers. He described heterospory in the calamite cone known as *Calamostachys casheana*, one of his original drawings serving as the frontispiece of the second volume of Scott's *Studies in Fossil Botany*; he described pteridosperm stems such as *Lyginopteris* and *Heterangium* and suggested that in their anatomy they were intermediate between the ferns and cycads. Seward said in reference to the *Organization* papers: "It is impossible to give adequate expression of the importance of these memoirs in a short obituary notice. The labour and scientific skill embodied in this legacy to botanical science have only been at all fully realised in recent years" (1895, p. 142).

Williamson retired from his many official duties in 1892 and spent the last three years of his life near London. He had not completed his studies of the petrified plants of the Coal Measures but he realized, perhaps quite as well as some of his critics, that progress in botanical science had left him somewhat in the rear and quite wisely he sought the aid of a younger man. Thus he and D. H. Scott, largely through the work of the latter, brought out three fine contributions as a supplement to the original *Organization* series, under the title *Further Observations. . . .* In a preliminary note to the first paper (Williamson and Scott, 1894) Williamson says:

> My morphological enquiries seem to have reached a stage that makes a more minutely careful examination of these questions of development and growth desirable; but before specially undertaking this, I saw clearly the extreme importance of doing so in combination with some younger colleague whose familiarity with the details of the physiology of living plants was greater than my own. Under these circumstances I have secured the co-operation of Dr. D. H. Scott, and the present paper embodies the results of our united investigations. The work has been carried out in the Jodrell Laboratory of the Royal Gardens, Kew. [P. 863]

The first two parts of the *Further Observations* are concerned with calamite and sphenophyll stems, roots, and cone structures; the third deals with *Lyginopteris* and *Heterangium*. They give us a fine connecting link between two great periods in the investigations of petrified Carboniferous plants. The following is a notation on this "transition period" in Scott's words:

> I first saw Williamson on February 16, 1883, when I attended his Friday evening lecture at the Royal Institution. . . . I did not, however make his acquaintance till six years later, when we met at the British Association Meeting at Newcastle-upon-Tyne, in 1889. This led to a visit to his home in company with Prof. Bower; it was on March 8, 1890, that I first had a sight of his collection. I find the entry in my diary: "Spent 7 hours over fossils, especially *Lyginodendron* and *Lepidodendron*, preparations magnificent." I at once became an ardent convert to the cult of fossil plants to which I had hitherto been indifferent, though I must in fairness admit that

> Count Solms-Laubach's *Einleitung* [1887] had done something to prepare the way. I well remember the state of enthusiasm in which I returned home from Manchester. A subsequent visit confirmed me in the faith, but it was some little time before I put my conviction into practice. In 1892 Williamson, then in his 76th year, resigned the Manchester Professorship and came to live near London. In the same year I migrated to Kew, and it was agreed that we should work in concert, an arrangement which received every encouragement from the then Director Thiselton-Dyer. Williamson first came to the Jodrell Laboratory on Friday, December 2, 1892. Then, and on many later visits, he carried a satchel over his shoulder, crammed with the treasures of his collections. For some months he came quite regularly once a week, afterwards less often. On these visits we discussed the work I had done on the sections during the interval, and sometimes our discussions were decidedly lively. In the end, however, we always managed to come to a satisfactory agreement. [Scott, in Oliver, 1913, pp. 258–259]

Thus ended the great initial push that Williamson gave to the study of coal-ball plants and began the masterful studies of Scott and others, which, in recent decades, have been especially fruitful in the United States.

I have encountered a few suggestions in my search through the literature and letters that Williamson may not always have given due credit to collectors who supplied him with specimens. One such comment appears in a letter from William Cash to D. H. Scott of August 6, 1910:

> When I took up fossil botany I made the acquaintance of the late Dr. Williamson & whatever specimens in my cabinet he wished for I *gave* to him—he would call upon me looking over the specimens and would say—"Cash, I *must* have these" & so of course he got them—scores & scores of them—for I felt that the best interests in *Science* were served by placing all I could in his hands & it seemed to me a presumptuous thing to think of my personally writing on the subject at that time.
>
> I treated the Scientists whom Dr. Williamson introduced to me in the same spirit (such as—Count von Solms Laubach—Dr. Hovelacque, Mr. Kidston & others) & indeed the Dr called me to task for being too liberal to them. [Scott letters]

Williamson was a great natural scientist without doubt, but his tremendous publication output, along with his other duties, also suggests that he must have depended rather heavily on others for aid in various ways.

Scott's contributions to botanical science were varied and numerous; if one had to pick "the greatest of them all" in our science he would certainly be a leading candidate. His technical writings as well as his many semipopular review lectures are as readable as any I know of; I have often urged my graduate students to read them for the information they contain and especially for his literary style.

Dukinfield Henry Scott (1854–1934) was born in London on November 28, 1854. The economic circumstances of his family were comfortable, and his early education was received from tutors at home. He showed a strong interest in natural history in his early teens and before he was sixteen had read some of the German works of Hofmeister on the higher cryptogams and developed an interest in plant structure from Griffith and Henfrey's *Micrographic Dictionary*. Reflecting on this early period he has recorded: "I owe much to those who made the great works of the German botanists accessible to English readers. . . . At the same time I was using the microscope so far as I could without any training. Freshwater Algae were my chief joy—I still think them the best possible introduction to Scientific botany" (Rendle, 1934, p. 84).

There followed a period of nine years when this botanical leaning was dormant. He read classics at Christ Church, Oxford, from 1872 to 1876; Seward says of his period: "There is good authority for suspecting that he devoted such time to the study of *Literae Humaniores* as he could snatch from angling for pike" (1934, p. 206). On leaving Oxford he obtained work at Euston Station in London, where for three years he pursued his intention to become a railway engineer. This interest apparently continued throughout his life, for when he retired to Oakley he took daily walks to a bridge to watch the London-to-Salisbury express trains and checked their time with considerable interest (Walton, 1959). Maurice Wonnacott has told me an amusing anecdote which suggests that it is best for all concerned that Scott decided to give up engineering. Scott gave his slide collection to the British Museum, and his notes accompanying it contained reference numbers pertaining to the location of particular specimens on the slides. These number-entries were frequently altered, and when Scott died W. N. Edwards obtained Scott's microscope to try to establish some order out of the numbers. He found, however, that its mechanical stage had no fixed starting point; the numbers were quite useless and the scope was returned to Scott's estate.

Between 1880 and 1882 Scott spent several extended periods studying in Würzburg, Germany, under the general direction of Julius von Sachs. I have selected a few comments from his "German Reminiscences" (1925) that tell a little of the university life there at the time.

> There were no students, in the ordinary sense, working in the laboratory, but only one or two, like myself, who hoped to become botanists.

> Students (chiefly medical) attended the lectures; on rare occasions Sachs gave a demonstration in the laboratory; that was all. [P. 10]
>
> The hours were long: we were supposed . . . to attend the 8:15 lecture, and work went on (of course with a long intermission for Mittagessen) till supper time at 8 o'clock. I remember, one day, leaving the laboratory about 6:30, I met Sachs in the garden and he asked me: "Machen Sie Feiertag, Herr Scott?" ("Are you taking a holiday"). Often we also worked on Sundays. [Pp. 11–12]

He also attended lectures by Karl von Goebel:

> In the following February, he reached the subject of Plant-palaeontology; except for a few references in Sachs's text-book, this was practically the first I had ever heard of fossil plants. There were only three lectures on the subject, evidently much influenced by the powerful authority of Renault. I was interested, though it was nearly ten years later before the study of the fossils really attracted me. [P. 13]
>
> [Prior to his oral examination for the Ph.D. degree] I had to call on all the Professors of the Faculty, and solemnly invite them to my examination. This formidable ordeal had to be gone through in dress clothes and a tall hat. . . . The same costume was required for the examination itself, which took place on July 20th, 1881, about the hottest day I have ever experienced. Happily, however, dress clothes are cool. [P. 14]
>
> The great advantage of working in Germany in those days was that one found oneself in the main stream of botanical progress. . . . The revolution in English Botany, carried out at about the time to which these reminiscences belong, was the direct result of German influence. . . . The chief characteristic of German university life, as I saw it, was the dominance of research over mere learning. [Pp. 15–16]

Based on these German studies Scott devoted the next ten years chiefly to anatomical investigations of living plants which made him not only a leading authority in the field but gave him a superb background for his later similar studies of the fossils. This research included studies of the lactiferous vessels of *Hevea*, the internal phloem and polystely in the dicotyledons, and secondary tissue in certain monocotyledons; in 1894 he brought out his book *An Introduction to Structural Botany*, which went through numerous editions and reprints.

When he returned to England in 1882 Scott took a position as assistant to Professor Daniel Oliver at University College, London, the assistantship having previously been held by F. O. Bower. A member of the class at that time was Francis Oliver, who became a lifelong friend of Scott's and whose work will be introduced on a later page. In 1885 Scott was appointed Assistant Professor under Thomas Huxley at the Normal School of Science and the Royal School of Mines. His teaching seems to have been well received; Mary Adamson, in a note in the *Times* for February 1, 1934, paid him this tribute: "All women should honour the memory of Dr. D. H. Scott, for he was the first lecturer on Botany at University College who allowed women to enter his class."

Scott gave up formal teaching in 1892 to accept a post as Honorary Keeper of the Jodrell Laboratory at Kew, where he served until 1906; something of this period is recorded by Walton (1959) and by Metcalfe in his *History of the Jodrell Laboratory* (1976). Scott was instrumental in introducing small classes in botany at the Jodrell, although it was against the policy of the time. One of the first classes included among its four students Miss Henderina Victoria Klaassen, who later became Scott's wife. She produced several botanical articles (under the name of Rina Scott), and she was one of the earliest, if not the first, to use cinematography in plant physiology (Metcalfe, 1976).

In 1906 Scott retired to his home, East Oakley House, near Basingstoke, although he continued with his research and writing unabated for many more years.

In 1900 he brought out his *Studies in Fossil Botany*, a book dealing chiefly with petrified Paleozoic plants that stemmed from a course that he taught at University College. The second edition came out in two volumes in 1908 and 1909, and the third in 1920 and 1923. Very few texts have been written with such astute selectivity, readability, and clarity of presentation. A comparison of the first and third editions shows how rapidly evolutionary paleobotany was advancing during this period. These advances have continued, but I had the third edition available on my desk through my entire career and I think that it will continue to serve students of fossil botany for some time to come. His *Extinct Plants and Problems of Evolution*, based on lectures given at the University College of Wales and which appeared in 1924, includes much of the same material, though the presentation is somewhat different, and also makes excellent and informative reading.

Among Scott's more important technical papers is a series that he started in 1897 under the title "On the Structure and Affinities of Fossil Plants from the Palaeozoic Rocks." The first of these was on the incredibly complex Lower Carboniferous cone *Cheirostrobus*. Other papers in the series included his fine description of *Medullosa anglica* in 1899, *Lepidocarpon* in 1901 and the cone *Sphenophyllum fertile* (1906). The last item is representative of the difficulties that one may get into in preparing ground thin sections of very small organs, and Scott was in error in his interpretation; the peel method has been a great aid with such structures and some years later Suzanne Leclercq published a revised version of the organization of the cone.

Dukinfield Henry Scott. From the *Annals of Botany*, 1935.

Scott served at various times as president of the Linnean Society and the Royal Microscopical Society and presided on two occasions over the Botanical Section of the British Association for the Advancement of Science. His presidential addresses are models for such occasions; they are readable and informative and cover a wide range of paleobotanical topics. There is not space to discuss them in detail here; references, which were compiled by W. N. Edwards and F. M. Wonnacott, are included in Oliver's (1935) biographical sketch of Scott.

In view of the continuity of the work that flowed from their pens I could not avoid dealing with Williamson and Scott as a unit, but it is now necessary to go back a little and bring in two other men: Frederick Ernest Weiss (1865–1953), a connecting link from Williamson to others that followed him in Manchester; and William Carruthers (1830–1922), a man of great importance in developing botany, and to some extent fossil botany, at the British Museum.

Weiss was born in Huddersfield on November 2, 1865; his father died when he was only three and his mother took her family to Germany, where education was less expensive and where Frederick attended the Heidelberg Gymnasium, and later to Switzerland. Upon returning to England Weiss entered University College in 1884, intending to go into zoology. He served as demonstrator for Professor Daniel Oliver in his fourth year and became acquainted with Scott and F. O. Bower. At Oliver's suggestion he shifted to botany, and then spent some time at the University of Strasbourg, where he studied under Solms-Laubach, among others. All of this, as might be expected, tended to direct his interests to botany and to some degree into fossil botany. In 1892, when Williamson resigned at the age of seventy-six, Weiss was appointed Professor of Botany at Owens College, Manchester.

Weiss was well thought of, and his selection as Williamson's successor proved advantageous for the continued development of botany at Manchester. Weiss held the chair for thirty-eight years and when he died in January of 1953, at the age of eighty-seven, Thomas wrote: "His passing severs one of the few remaining links between the biology of to-day and of that great period of high endeavour, great enthusiasm and high hopes, which have been so amply fulfilled. Among the young men who drew their inspiration from Huxley, Vines, and their entourage, few could have done more than Weiss to spread a knowledge of, and an interest in plants as objects of truly scientific studies" (1953, p. 606).

Weiss encouraged his Manchester colleagues W. H. Lang and John Walton. Most of Weiss's research was with living plants but he produced a few paleobotanical papers—on *Stigmaria* (which everybody had a hand in!), on the aerating tissues in the stems of the lepidodendroid trees, and on some well-preserved root apices of *Lyginopteris*.

William Carruthers was one of the first paleobotanists who was trained from the start as a botanist and geologist and worked professionally as such throughout his life. He made several important contributions to fossil botany and was one of the great leaders in developing that department at the British Museum.

Carruthers was born in Moffat, Dumfriesshire, on May 29, 1830. Although it was his original intention to enter the ministry he was influenced by Dr. John Fleming to take up scientific studies, which he did at Edinburgh. He also came in contact there with John Hutton Balfour (1808–1884), whose teaching must have stirred up his interest in fossil plants. In this context it

William Carruthers. From the *Geological Magazine*, 1912.

seems permissible to make a "supplementary diversion" in the sequence of our story; Balfour was for some years Professor of Botany at the University of Edinburgh and regius keeper of the Royal Botanic Garden. His biographer (Bettany, 1908) also notes that he was the first in Edinburgh to introduce classes for practical instruction in the use of the microscope. I know of only one short paleobotanical research paper of Balfour's which deals with the structure of coal, but fossil botany entered very prominently into his teaching. In the 1854 edition of his *Class Book of Botany* he devotes thirty pages to fossil plants; to the best of my knowledge this is the first significant introduction of paleobotany in a botanical textbook. And some years later (1872) he produced his *Introduction to the Study of Palaeontological Botany*, the first paleobotanical textbook in English. In the preface to this book he says: "If I can be useful in encouraging students to take up the study of Palaeontological Botany, and to prosecute it with vigour, I shall feel that this introductory treatise has not been issued in vain. As one of the few surviving relations of Dr. James Hutton, I am glad to be able to show an interest in a science which may aid in elucidating the 'Theory of the Earth'" (1872, p. ix).

William Carruthers was for a time lecturer on botany at the New Veterinary College in Edinburgh, but in 1859 he accepted an offer of Assistant in the Department of Botany in the British Museum of Natural History, then located in Bloomsbury; in 1871 he became Keeper of the Department. In 1880 the Botany Department moved to its much more ample quarters in South Kensington, and all of Carruthers biographers give him credit for the development of the great herbarium and botanical library. Asa Gray was a frequent visitor, consulting the early American collections preserved there, and he invited Carruthers to join him at Cambridge, Massachusetts, with the view of ultimately becoming his successor, but Carruthers decided to remain at the "B. M." He did, however, visit on this side of the Atlantic, attending the British Association meeting in Montreal in 1884 and traveling extensively in America to establish contacts in museums and make collections for his own Department.

In his biography of Carruthers, Britten (1922) describes him as an efficient and helpful administrator. Over the years botanists in general have had a tendency to become involved in professional squabbles, and Carruthers came to the British Museum when such a dispute was in progress between the Museum and the Royal Botanical Garden at Kew. The dispute seems to have centered around an attempt to transfer the Banksian Herbarium to Kew. Britten writes: "I entered the Department in September, 1871, when the storm was still raging, and, coming as I did direct from Kew, had the advantage of hearing both sides of the controversy. The sufficiency of Carruthers's rebuttal of the attack is shown by the fact that the Department remained untouched, nor was a later effort in the same direction more successful" (1922, p. 251). Britten also comments on Carruthers's willingness to aid visitors:

> I remember, for example, that we supplied specimens and drawings to the artist who was designing the laurel wreath which Tracy Turnerelli proposed to present to Lord Beaconsfield, and a fig-leaf for a sculptor who required that garment for a statue on which he was engaged; still more do

> I remember a large lady, with a small companion, who was a frequent visitor, to whom Carruthers lent at her request a volume—his own copy—of the *Genera Plantarum*, which she returned in the course of two or three days with a remark that she had found several mistakes in it. [1922, p. 250]

Carruthers had under his care in the Museum a large collection of fossil plants which in part had been gathered together by Robert Brown, the first Keeper of the Botany Department. Carruthers became interested and productive in several areas of paleobotany. One of his short but important contributions established the genus *Beania*, a cycadophyte seed organ which has been referred to by several people subsequently, including Harris (1941), who gives a fine summary account of the plant. It is a fairly common fossil in the intertidal beds at Gristhorpe on the Yorkshire coast where I have had the pleasure of collecting it in company with Maurice Wonnacott. Presumably the "Mr. Bean" for whom *Beania* was named, and who is mentioned in connection with several fossil plants in Lindley and Hutton's *Fossil Flora*, is William Bean II, as described in McMillan and Greenwood's biography (1972) of the Bean family. William Bean was a vigorous collector of plant and animal fossils and associated with the Scarborough group of naturalists that included William Smith, John Phillips, John Leckenby, and the Williamsons. Some of Bean's collections were sold to the British Museum in 1850.

One of Carruthers's most comprehensive works was his 1870 account of cycadophyte trunks in which he established the genera *Yatesia*, *Fittonia*, *Williamsonia*, and *Bennettites*. It deals with several others, includes a general history of cycadophyte trunks, and also contains the first account of petrified bennettitalean cones.

Paleobotanists are people, and have most of the usual good and bad points; it is not necessary to recount too many of their disputes, but a classic one between William Carruthers and J. W. Dawson over the nature of *Prototaxites* is especially interesting. Dawson established the genus in 1859, and since then it has been found rather widely distributed, stratigraphically and geographically, through the Devonian. It is known from petrified trunks up to nearly three feet in diameter which consist of a rather homogeneous matrix of two different sizes of tubes; the closest comparison seems to be with the structure found in the stipes of some of the large brown algae. Among the best recent accounts are those of Paul Corsin (1945) and Chester Arnold (1952). In his original description of 1859 Dawson noted that the structure of the large tubes presented a comparison with *Taxus*—until rather recently included in the Coniferales. In an article written in 1872 Carruthers pointed out that the structure was definitely not that of a coniferous-type wood but rather more like that of the holdfasts of the large kelps, and he substituted the seemingly more appropriate generic name *Nematophycus*, which is not a valid procedure according to our present rules. Actually Dawson had visited with Carruthers in England in 1870 and they had examined his material together, but Dawson was not convinced. Some rather strongly worded correspondence on the matter passed between them. Later Dawson relented to some degree and even went so far as to say that he had never implied that the fossil was of a coniferous nature and allied to *Taxus*. Arnold (1952) has given a detailed account of the dispute in his paper describing the superbly well preserved material of *Prototaxites southworthii*. Most of the specimens of *Prototaxites* that I have seen are rather poorly preserved, and I have some sympathy with Dawson's erroneous interpretation. There is still a good deal that remains to be known about this curious plant, such as its gross morphology, mode of reproduction, and habitat. And some of my colleages who work with living algae take a dubious view of an algal stem three feet in diameter!

Carruthers was a man of many interests and accomplishments. He served for many years as consulting botanist to the Royal Agricultural Society and was instrumental in developing seed-testing stations which greatly aided farmers in obtaining seed that was known to be of good quality. He was active in the Presbyterian Church throughout his life, being interested in Puritan history and biography, and he edited *The Children's Messenger* for forty-two years.

There were many others in Britain during the period encompassed by this chapter who contributed in various ways to the progress of fossil botany. I cannot include them all but I do want to devote a few lines to a trio of men who were responsible in their individual ways for great advances in the discovery and dissemination of geological and paleontological knowledge. They were contemporaries and rather close friends, they were outstanding scientists and teachers, and each was a unique character I would like very much to have known personally: John Morris (1810–1886), William Buckland (1784–1856), and Gideon Algernon Mantell (1790–1852).

In his sketch of the history of paleobotany written in 1921, Arber refers to a gap or lull of some five decades in the study of Carboniferous compressions, and he notes that the chief authority on such fossils at this time was John Morris, although his own publications were not conspicuous. I think that his accomplishments were somewhat greater than he has been credited with.

As a teacher of geology Morris was very influential. He seems to have been largely self-educated; he was born near London, attended several private schools, and was engaged for some years in Kensington as a pharmaceutical chemist. His interests and efforts shifted more and more toward geological pursuits, and by 1836 he was collecting materials for his *Catalogue of British Fossils*, the first edition of which appeared in 1843. It includes a generic-alphabetic list of British fossil plants, with locality data and corresponding literature citations. Publication was actually initiated in 1839 in a series of short sections in the *Magazine of Natural History*. Morris accompanied Sir Roderick Murchison on two European tours in 1853 and 1854 and he was urged by Murchison to be a candidate for the Chair of Geology at University College, to which position he was duly appointed, and served from 1854 to 1877.

Morris was a teacher first and foremost, and one who was well-liked and effective. He is said to have delivered some eleven hundred lectures during his career, to his university classes and to various outside organizations. One of his biographers says: "We have heard one who has frequently acknowledged publicly his indebtedness to Professor Morris, say, that it was often difficult for him to distinguish what was his own work, and what he owed to Prof. Morris, seeing that the Professor so freely communicated his knowledge in conversation, that it became incorporated with the author's own store" (anon., 1878, p. 483).

He collaborated with Thomas Oldham in bringing out the first significant systematic work on the fossil flora of India, *The Fossil Flora of the Rajmahal Hills*, in 1863; he described and figured Permian plants in the first volume of Murchison's Russian travels, and contributed the earliest paper on the fossil plants of Australia, which appeared in Strzelecki's account of New South Wales and Van Dieman's Land (1845). His shorter accounts tend to be more in the way of review or summary papers than original studies. His article "On Coal Plants" (1862) is a microhistory of paleobotany, and in 1841 he brought out a short summary of cycadophyte foliage types.

William Buckland was born at Axminster in Devonshire. He obtained a scholarship at Corpus Christi College, Oxford, in 1801, received the B.A. degree in 1805, and was admitted a Fellow of his college in 1808 and ordained a priest in the same year. His association with Oxford was long and vigorous; he was appointed to a chair of mineralogy in 1813 and a few years later to a readership in geology. It may be appropriate to say that his energy and enthusiasm enabled him to devote himself full-time and with great effectiveness to both theology and geology and it seems to me that he welded the two together as well as, or better than, any one before him.

He traveled extensively through southwestern England during 1808–1812, initiating the large natural history collections that he accumulated at Oxford. He had a particular fondness for the area around Lyme Regis, a lovely town on the Dorset coast; a biographer describes "his breakfast-table at his lodgings there, loaded with beefsteaks and belemnites, tea and terebratula, muffins and madrepores, toast and trilobites, every table and chair as well as the floor occupied with fossils whole and fragmentary, large and small, with rocks, earths, clays, and heaps of books and papers, his breakfast hour being the only time that the collectors could be sure of finding him at home, to bring their contributions and receive their pay" (Gordon, 1894, p. 8).

I have a particular fondness for this part of England, as well as sympathy for Buckland relative to a mishap that he encountered there. In a letter of 1817 he writes of being incapacitated by an ignited spark of iron from his hammer which went into the cornea of his eye and required several operations. My own experience in this respect is offered as an admonition to younger paleobotanists that haste and misguided enthusiasm may be dangerous: Some years ago several of us were examining a large open-pit coal mine in Kansas for coal balls; we encountered a fine looking and very large aggregate, and being some distance from our auto, where most of our digging equipment was stored, I used one hammer as a wedge—a very foolish thing to do. When I struck the head of the wedge hammer with another one a fragment flew off with great violence and embedded itself in the calf of my leg. The little fountain of blood that spurted out was more surprising than serious, but since I was bending over at the moment it could well have struck my eye. We went to a hospital in a nearby town where a physician placed me on a table preparatory to probing for the piece of metal. This procedure did not appeal to me; I decided against it and have carried the piece of metal about with me ever since.

Buckland's paleobotanical publications are not numerous. He described some fossil woods that had been collected along the Irrawaddy River near Rangoon, and probably his most important contribution was an article on the silicified cycadophyte trunks from the Isle of Portland, in which he mentions drawing upon Robert Brown's assistance. These trunks were known to the local quarry men as "bird's nests" because of the cavity formed at the top by the crown of leaves. Buckland compared them with *Cycas* and *Zamia* (as those generic terms were used at the time) but he noted significant differences between them and the living cycads and established the Cycadeoideae, with two species of *Cycadeoidea*. This paper is an important landmark in the early studies of fossil cycadophytes.

At the time of his appointment at Oxford, in 1813, he delivered a lecture, one of his best-known works, entitled "Vindiciae Geologicae; or the Connexion of Geology with Religion Explained." Its objective was to demonstrate the lack of any real conflict between the science of geology (and its findings) and the accounts of the Creation and Deluge as recorded in the book of Genesis. It impresses me as one of the best treatises of its kind for the time. He does lean rather heavily on Cuvier's concepts and he does hold to a universal deluge at about the time cited in the Bible, but drawing on his extensive knowledge of geology and paleontology he deals very effectively with matters such as the lack of plant or animal remains in the oldest known rocks, the evidence that rock strata generally were formed slowly and over a long period of time, and the succession of plant and animal life with its abundance of forms that are no longer living. It may be described as an updated version of Woodward's ideas of the importance of large amounts of time, but he does not go so far as Hutton, who saw "no beginning and no end."

As a lecturer Buckland must have been a real spellbinder, dispensing sound facts in a highly entertaining fashion. A couple of short notes from Gordon (1894) may be sufficient: "He compared the world to an apple-dumpling, the fiery froth of which fills the interior, and we have just a crust to stand upon; the hot stuff in the centre often generates gas, and its necessary explosions are called on earth volcanoes" (p. 28). And: "It is impossible to convey to the mind of any one who had never heard Dr. Buckland speak, the inimitable effect of that union of the most playful fancy with the most profound reflections which so eminently characterised his scientific oratory" (p. 34).

Much less time was available for his geological pursuits when he became Dean of Westminster in 1845. He was of necessity concerned with alterations in the Abbey and School; he was a leader in the drive to improve the sanitary conditions of the London water supply, and he had an interest in agricultural improvements. And along with Adam Sedgwick and Charles Lyell he was influential in the establishment of the Geological Survey of Great Britain.

Coming now to the last of our trio, few men in the history of science have driven themselves so hard through sixty-two years of life as did Gideon Mantell in order to achieve social prominence and scientific reknown. His ambition appears to have been so insatiable that in spite of all of his achievements he failed, at least in his own mind, to reach the goals he aimed at.

Mantell was born near Lewes in Sussex, educated in schools that were favorable to his father's religious and political principles, and then apprenticed to a surgeon of Lewes, James Moore. After studying for a time in London he returned to Lewes and ultimately succeeded to Moore's practice, which he continued to develop with great success. But it was not adequate to satisfy his yearning for social and scientific recognition, and he later moved to Brighton and then to London. In the course of the two moves he devoted more and more time to geology and less to his practice, with the result that the latter suffered and he was harried with financial problems to the extent that finally "his house with his collection of fossils was turned into a public museum and his distracted wife and children were forced to seek shelter elsewhere" (Curwen, 1940, p. v). Mantell kept a journal of his various activities between 1818 and 1852, and an edited portion of it was published in 1940 (Curwen, 1940). In part it makes fascinating, if rather tragic, reading of a man consumed with ambition and trying to do the work of several men in both geology and medicine. But with it all he was not a selfish man, he was a keen surgeon with a sense of fairness and sympathy for those less well off than himself, and at one time he went to considerable trouble to save a woman from the death sentence who had been convicted of poisoning her husband with arsenic; he demonstrated that the evidence was not adequate. He visited dozens of patients some days, traveling on foot when his horse was ill.

Mantell's published works are numerous and many of them comprehensive and I shall only touch on those that deal with fossil plants.

As there was apparently a continuing demand for the works of Parkinson and Artis, Mantell arranged *A Pictorial Atlas of Fossil Remains*, a selection of the plates from the works of these two authors, with annotations, which he brought out in 1850. The "Supplementary Notes" briefly summarize the works of Artis, Brongniart, Sternberg, and Martius. Judging from the frequency with which the *Pictorial Atlas* could be found in the second-hand book shops a few decades ago it must have been published in a rather large edition.

In his *Geological Excursions round the Isle of Wight* (1847) he gives a vivid description of the petrified plant remains to be seen there: "On one of my visits to the Island, the surface of a large area of the dirt-bed was exposed, preparatory to its removal, and the appearance presented by the fossil trees was most striking. The floor of the quarry was literally strewn with fossil wood, and before me was a petrified forest, the trees and the plants, like the inhabitants of the city in Arabian story, being converted into stone, yet still remaining in the places they occupied when alive!" (p. 396).

He described fossil plants in 1822 in his *Geology of Sussex*, and in 1846 he brought out a short paper on coniferous strobili from the chalk beds he was so well acquainted with.

These three Englishmen were united, as they apparently were in life, on the flyleaf of Mantell's *Pictorial Atlas*: The book is inscribed to William Buckland, Dean of Westminster, "as an expression of the high respect and affectionate regard of one who has for more than thirty years enjoyed the honour and privilege of his correspondence and friendship," and Mantell acknowledges "the valuable assistance of my friend, John Morris" in correctly naming the fossils.

6

Paleobotany in Britain: The Age of Seward

FOR about four decades starting with the closing years of the nineteenth century Sir Albert Seward (1863–1941) was the dominant figure in paleobotany. I do not think that his research contributions were as great as those of several others before and during his time, but his knowledge of botany and geology was wide in scope and vast in quantity, and nobody else has so effectively gathered together the accumulated knowledge of fossil botany, making it both readily and pleasantly available. He produced several great paleobotanists who in turn carried on with the high standards that he established, and his contributions in many ways to British education were enormous. Seward's "Age" was a great one, and in recognition of the legacy he has left us I will first devote a few pages to some aspects of his own life and labors.

He was born at Lancaster in 1863, and in 1883 he entered St. John's College, Cambridge, one of his contemporaries being the distinguished botanist A. B. Rendle. Hamshaw Thomas says of this phase of Seward's life: "Professor T. McKenny Hughes, who did so much for the study of geology at Cambridge, appears to have been responsible for directing Seward's attention to the largely unexplored fields of paleobotany. This led in 1886 to a year's study of fossil plants in Manchester under Professor W. C. Williamson, the founder of modern palaeobotany in Britain" (1941, p. 867).

He spent a year traveling and studying collections on the continent and his very retentive mind accumulated a huge quantity of information which he used later in many of his writings. In 1890 he was appointed University Lecturer in Botany at Cambridge and he continued in the Botany School for forty-six years, becoming Professor of Botany in 1906. He was a tutor of Emmanuel College for some years and later served as Master of Downing

College from 1915 to 1936. I am rather proud of this, having been a student at Downing, myself; according to Hamshaw Thomas, "As Master of Downing he was immensely popular in his college, he knew every one, dons,

Albert Charles Seward. From *Obituary Notices of Fellows of the Royal Society*, 1941.

students and servants; he invited them to his Lodge, and took an interest in their work and in their recreations" (1941, p. 871).

Marjorie Chandler has told me a little of her contact with Seward when she was a student at Newnham College, Cambridge, during the World War I. Seward was apparently a stern teacher but fair and kind. She writes: "At the

Botany School Prof. A. C. Seward was our chief lecturer, clad in khaki. He was a strict disciplinarian and I well remember his hand shooting out and a dreadful voice thundering 'Get out' to a man who had dared to smoke in his lecture and another serious rebuke to a woman student next to me who was copying my notes instead of making her own. Seward was friendly to the cause of women's education and entertained us to tea parties at the Master's Lodge at Downing" (personal letter).

I can attest briefly to Seward's consideration of younger people who were in the embryonic stages of becoming paleobotanists. He had retired from Cambridge about a year before I went there (1937–38) to study with Hamshaw Thomas, and was living in London in a fine apartment near the British Museum. I spent the vacation periods (rather long ones) that year at the Museum going through the fossil-plant collections and I saw Seward quite often as he was then working on the Mull flora. He was most cordial and friendly and one day brought a Mull specimen to my desk and asked my opinion as to its identity. It seemed quite clearly to be a marchantiaceous-related liverwort, and Seward expressed sincere satisfaction in my identification. I am sure that he was not really in need of my assistance and did it as a friendly gesture; to be treated on an equal basis by the foremost authority in the world on fossil plants was a kindness that I have never forgotten.

Seward of course received many honors. He served for a time as Vice-Chancellor of Cambridge University, received the Wollaston and Darwin medals, and had a long-time interest in the Cambridge Philosophical Society, the Scott Polar Institute, and the Fenland Research Committee. Few people have accomplished so much in all of the academic areas of service. Thomas has summed this up succinctly:

> Through his long tenure of the post [Professor of Botany], Seward lived a very strenuous life devoted to scientific and administrative work carried out with the greatest efficiency. He had the power of turning rapidly from one piece of work to another, and of apportioning his time so that in days of heavy administrative responsibility and with regular classes to be taught, he was usually able to devote some time to research and spent much of the evening writing his books and papers. [1941, p. 868]

The scope of Seward's knowledge of fossil plants is best revealed in his four-volume work *Fossil Plants*. The first volume (1898) contains a brief historical sketch and a long and informative chapter on the fossilization process, the bulk of the volume being devoted to the articulates. Volume 2 (1910) deals with the lycopods and ferns. Volume 3 (1917) gives a fine account of the pteridosperms and cycadophytes. Volume 4 (1919) takes up the ginkgophytes and conifers. It is still an invaluable work containing an abundance of information that is difficult to find elsewhere.

Seward is undoubtedly best known to botanists, geologists, and laymen alike for his *Plant Life through the Ages*, a six-hundred-page volume that first appeared in 1933 and has been reprinted several times. It reveals his special leaning toward plant geography and paleoclimatology. The first fifty pages constitute a brief introduction to geology and the remainder of the book takes up fossil plants according to geological ages.

For my own choice, his delightful style reaches a peak of excellence in his book *A Summer in Greenland*. Seward left Copenhagen on June 18, 1921, in company with R. E. Holttum, then of St. John's College, Cambridge. In the course of a cruise of a little more than two months he saw much of the west coast of Greenland. His small book is a record of the people, customs, geology, and geography; actually rather little of it is devoted to paleobotany. A few passages from it may entice the reader to seek the book and read more:

> The fossil-bearing rocks it was our aim to investigate are exposed along the shore and in the ravines of Disko and other islands and especially on the Nûgssuaq Peninsula. Most of them were deposited during the Cretaceous period; others are Tertiary in age. Slabs of rock detached with the aid of a pick-axe from the side of a ravine where the hills are made of a succession of sheets of sediment—the sands and muds of some ancient lake or lagoon—are found to be covered with the clearly outlined impressions of large leaves like those of the Plane or Tulip tree, fronds of ferns hardly distinguishable from species (of the genus *Gleichenia*) living to-day in tropical and sub-tropical countries; there are also twigs of Conifers, some of which are almost identical with those of the Mammoth tree (*Sequoia (Wellingtonia) gigantea* now confined to a narrow strip of the California coast), and massive stems of forest trees. None of the leaves preserved in the Greenland rocks have a greater fascination for the student of the past history of living plants than those of the genus *Ginkgo*. [1922, pp. 27–28]

The reaction of the local population to fossil hunters does not seem to have been so very different from that in any other part of the world. Seward says:

> One evening a sturdy little Eskimo returning from hunting, the dead body of a seal made fast to the side of the kayak, paddled to our motor-boat. . . . Being attracted by the native, we engaged him to accompany us for a few days and took his kayak on board. He soon became an expert fossil collector and expressed his satisfaction by singing doleful tunes that he had learnt in church. The natives generally regarded us with curiosity and, we were told, spoke of us as vagabonds or tramps. [Pp. 80–81]

Through his long career, and due to his great reputation, Seward received collections from many parts of the Empire, and beyond. He dealt with almost all of the major groups of plants and geologic ages, although his studies of the Mesozoic are probably most significant. One of his earliest works (1892) was the Sedgwick-prize essay "Fossil Plants as Tests of Climate." Among his more comprehensive studies are the two-volume *Wealden Flora* (1894, 1895) and the *Jurassic Flora* (1900, 1904), both based on collections in the British Museum.

A biographer has summed up rather well a significant aspect of Seward's philosophy:

> His preference for the large led him to maintain in systematic work that when in doubt it was better to unite genera or species than to separate them, and it would indeed be hard to find a more wholehearted "lumper." Not only did he unite what others had separated but he showed that he had little patience with the subtle intricacies of rules of nomenclature. His preference for what seemed to him the natural and sensible way led him now and again into controversy but never into ill humour. [Harris, 1941, p. 162]

We will return to Seward now and again in considering the work of his many students and contemporaries.

Quite possibly some readers will feel that I should have entitled this chapter "The Age of Seward and Kidston." The title I use is intended to point out the breadth of Seward's knowledge and influence, but quite clearly Robert Kidston (1852–1924) was one of the greatest contributors to our knowledge of the plant life of the past; since he collaborated closely with several other leading botanists in his major works I will bring them into the story as we move through Kidston's career.

Robert Kidston was born at Bishopton, Renfrewshire, on June 29, 1852. After his education at Stirling High School (Scotland) he was employed by the Glasgow Savings Bank but gave this up at the age of twenty-six, thereafter devoting his time almost solely to the investigation of fossil plants. He had less formal education than some of his noted contemporaries but probably worked harder on his own than most of them. His biographer notes that "he worked, with the strictest regularity, from morning til nearly midnight, and his energies were made to yield the maximum of results on account of his remarkable passion for order and method" (Crookall, 1938, p. 8).

During his bank days he attended lectures in Glasgow, some of them by Williamson, which were influential in directing his interest into fossil botany.

In 1879 he attended botany classes given by Hutton Balfour at the University of Glasgow.

Kidston's research output was tremendous; he seems to have recognized his own limitations, however, for he was not the accomplished anatomist that Scott was and we find him collaborating with such noted botanists as David Gwynne-Vaughan and W. H. Lang. In his earlier days he devoted a good deal of time to the study of living plants, and, in company with Col. J. S. Stirling he made an extensive survey of the flora of Stirlingshire, from which numerous publications resulted. This interest never ceased, although he gradually shifted his efforts in the direction of paleobotany. Kidston was a great collector, although he was aided by many others, including W. Hemingway and James Lomax. In this connection Crookall says: "The Kidston collection of plant incrustations, one of the finest in the world and probably the best which has ever been accumulated by a single worker, comprises over 7,000 selected fossils and includes many types and figured specimens. Some were obtained by Kidston himself, but the majority were carefully chosen from those collected by the numerous workers in Great Britain over a half a century" (1938, p. 10).

The collection was bequeathed to the Geological Survey of Great Britain and is housed on the top floor of the Geological Museum on Exhibition Road next door to the Natural History Museum. It is a model for any museum to follow. The specimens are superior, carefully labeled, and readily available for study.

I never had a course in paleobotany as such but one could hardly fail to absorb a great deal of knowledge about Paleozoic plants from the Williamson and Scott slides, along with all of the other treasures in the British Museum. I dug out my own course in fossil botany there, under the efficient tutelage of Maurice Wonnacott, and when I wanted a change of environment, or another kind of information, I walked across the courtyard to the Geological Survey building and studied such parts of the Kidston collection as suited my interests at the moment. Kidston's collection of more than three thousand slides was given to the Botany Department at the University of Glasgow.

Kidston's published works number about two hundred titles, his chief focus being on floristic accounts with particular attention to stratigraphic considerations. These works include studies of numerous coalfields—Lanarkshire, Ayrshire, Northumberland, Lancashire, Bristol, and others.

According to Crookall (1938), in about 1880 Dr. B. N. Peach, acting paleontologist to the Geological Survey of Scotland, began to submit Paleozoic plants to Kidston for identification, and this resulted in siphoning large collections to him for study. As his reputation increased he was asked to engage in comparable continental studies, among the more important of which were his study of the Carboniferous plants of Hainaut Province in Belgium (1911) and a monograph on the calamites of Western Europe in collaboration with W. J. Jongmans (1915, 1917).

All of this culminated in his monumental *Fossil Plants of the Carboniferous Rocks of Great Britain* (1923–1925). According to Crookall: "As early as 1901, Sir J. J. H. Teall, then Director of the Geological Survey, had invited Kidston to prepare a Monograph on the British Carboniferous floras: this was to be the chief work of his life, placing on record his unrivalled floristic and stratigraphical knowledge" (1938, p. 8). Publication arrangements were not completed until 1920, and the first part appeared in 1923. According to an inscription facing the title page of Part 1: "It is intended to issue this Memoir in about ten parts, each containing from twenty to thirty plates."

This is one of the most valuable works in the entire history of paleobotany and deserves much more space than I am able to give it here. The six parts that Kidston produced bear the stamp of his extensive knowledge of Carboniferous plants, as well as revealing detailed and exact stratigraphical and geographical information, and they are richly and effectively illustrated with photos and drawings. Citing a few exemplary topics may be more effective than trying to review the entire six parts: Part 3 deals with some of the intriguing Lower Carboniferous genera such as *Rhodea* and *Rhacopteris*, about whose anatomy and reproduction we would like to know more, and I am sure we will in the near future; Part 4 includes a wealth of information and illustrations on such fern and pteridosperm genera as *Crossotheca, Corynepteris, Oligocarpia*, and *Renaultia*; Part 5 continues with comparable information on such genera as *Asterotheca, Telangium*, and *Zeilleria*. The work is a storehouse of both geological and botanical information, and I found it most useful in teaching as well as a research tool.

Kidston did not live to complete the task, and Robert Crookall has more recently brought out six more parts (1955–1970). In the introduction to the first Crookall says:

> Robert Kidston's death occurred soon after he had completed the first section of his great monograph. Fortunately, his extensive collection of fossil plants, together with a large amount of manuscript material, illustrations, and photographic negatives were bequeathed to the Geological Survey. The manuscript included a number of descriptive notes of specimens preserved in his own collections and in the chief public and private collections in the country. This material was the basis for the continuation of the work, but it has been extended by the addition of specimens collected by Geological Survey officers in various coalfields.

Robert Kidston (left) and David T. Gwynne-Vaughan.
From a photo received from Professor William Chaloner.

Crookall's Part 1 includes *Alethopteris, Caulopteris, Lonchopteris*, and *Megaphyton*; Part 2 continues with *Neuropteris* and *Linopteris*; Parts 3 and 4 deal with the lycopods; Part 6 takes up a number of reproductive organs, as *Whittleseya, Boulaya, Aulacotheca, Calathiops, Megatheca, Trigonocarpus*, and others. Crookall continued the work in Kidston's excellent fashion and his six parts include many illustrations of type specimens as well as some of the classic restorations. The amount of work involved in such a production is enormous, and it should be mentioned that these six parts have benefited from the editorial assistance of S. W. Hester, W. G. Chaloner, and others.

In 1929 Crookall brought out a useful little book entitled *Coal Measure*

Plants which I think should be more readily available; it describes compression fossils and was intended as a nontechnical guide to the British Coal Measure fossils. Robert Crookall was engaged by the Geological Survey in 1926 and retired in 1952.

Since Kidston's strongest area of knowledge was not structural botany he collaborated with David Gwynne-Vaughan in several of his investigations of petrified plants. Their study of the fossil royal ferns (Osmundaceae) is one of the classics on petrified stems. In five parts (1907–1914) they trace the family from Tertiary species back to the Permian. The oldest members of the group were protostelic, that is, with a solid central core of primary wood; in the later members this differentiated into two distinct zones, the inner one eventually becoming a parenchymatous pith while the outer ring, at first continuous, becomes dissected as in the living species of *Osmunda*. Some of these early royal ferns were appreciably larger than present-day specimens. I had an opportunity in 1950 to add to the story in a minor way with a study of a Brazilian specimen (*Osmundites braziliensis*) in which the woody cylinder is twenty-seven millimeters in diameter; it was a small tree fern and probably of late Mesozoic age, although that is uncertain. The Osmundaceae are a venerable family that I believe may now be enjoying a high point in its long history. The cinnamon fern (*Osmunda cinnamomea*) and the interrupted fern (*O. claytoniana*) thrive in great abundance in New England in the half-shade of roadsides—one man-made ecological niche that seems beneficial. The royal fern itself (*O. regalis*) may be found in the same habitat, mixed in with the other two, but it generally prefers wetter places.

Two comprehensive recent studies that deal with the living and fossil members of the Osmundaceae should be mentioned. In 1962 Walter Hewitson issued a comparative morphological study of the living members of the family and in 1971 Charles Miller compiled an evolutionary survey of the family that includes the fossils.

David Thomas Gwynne-Vaughan was born on March 12, 1871, and thus was nearly twenty years younger than Kidston. He came from an old Welsh family and was descended from Sir Roger Vaughan who was killed at Agincourt. He received his botanical education at Cambridge and at the Jodrell Laboratory at Kew when Scott was Honorary Keeper. He carried out numerous studies of living plants—on the anatomy of the Nymphaeaceae, *Primula*, *Pandanus*, and certain fern stems—and he made botanical expeditions to the Amazon and to a remote Malayan area. He held teaching posts at Birkbeck College in London, Queen's University in Belfast, and Queen Margaret's College in Glasgow, where he began the joint studies with Kidston. They probably would have collaborated on additional projects, but Gwynne-Vaughan died on September 15, 1915.

Scott gives Kidston and Gwynne-Vaughan credit for producing the first significant account of the curious Cretaceous tree fern *Tempskya.* A considerable number of botanists have had something to say about this genus and its history. *Tempskya* is known from petrified trunks up to about a half meter in diameter, which probably attained a height of some six meters. The trunk consists of a compact matrix of siphonostelic stems and their small wiry roots. A trunk six inches in diameter may be composed of as many as 180 stems; these branch quite frequently, giving off two rows of small petioles toward the periphery of the trunk. The leaves were small and numerous so that the appearance of the plant in life must have been quite different from that of most tree ferns, which have a small number of very large fronds. The trunks have been found associated with *Anemia*-like foliage and in Seward's description of *T. knowltoni* from Montana he reported sporangial fragments scattered among the stems that suggest a relationship with the Schizaeaceae; however, in specimens from Idaho Ellen Kern Lissant and I found sporangia that are more like those of the Polypodiaceae. Thus the affinities of the genus are in doubt, but it was a common and widespread plant of the Cretaceous landscapes in North America and Europe.

Tempskya was first described under the name of *Endogenites erosa* by Charles Stokes and Phillip Webb in 1824. C. B. Cotta in his *Die Dendrolithen* of 1832 mentions it under the name *Porosus marginatus.* I have the second edition of this work available (1850) and two of Cotta's figures on plate 8 show in a crude but recognizable fashion the stem-root organization of the trunk; I believe that he was the first to suggest that it might be the axis of a fern. Franz Unger, August Corda, and August von Schenk, among others, also dealt with the plant.

Tempskya was first described from American deposits by Edward Berry in 1911, and Read and Brown brought out a comprehensive study in 1937 based on specimens from the northwestern states. I would like to mention my own entrance on the *Tempskya* scene in order to point out again the part that amateur collectors play in paleontology. Through a chain of helpful acquaintances I met Henry Thomas of Wayan, Idaho, in the early 1940's. He was a bachelor rancher, born and raised in a one-room sod house in Nebraska, who later moved to the Gray's Lake region of southern Idaho, where I met him. He had accumulated enough of the world's goods to enable him to live as he pleased, and much of his time was spent on periodic trips to San Francisco to bring back carloads of books to his cabin, and roaming the nearby hills collecting *Tempskya* specimens. He had many tons of these laid out on wood platforms in his back yard. I enjoyed two extended visits with him. We would ride his horses over the hills to dig here and there for the fossils, and then come out the next day with a wagon to collect our "loot." Some of the specimens were exposed on the ground of the sagebrush hills, but he would also frequently stop his horse, dismount, and point to a small gully and say "We will dig here." I do not recall that we ever stopped at such a place without finding several good specimens; how he picked these places I do not know. I thus gathered together a fine study collection, most of which eventually went to the Natural History Museum in Washington, and I had access to the hundreds of specimens that Henry Thomas had previously obtained. It was an opportunity to make a statistical study of a petrified plant that one does not often encounter. The preservation of the Idaho specimens was in many cases superb, with structures such as root hairs being well preserved, and the study that I made of this collection with one of my students, Ellen Kern Lissant, was most intriguing.

Of all Kidston's contributions to fossil botany he is probably best known for the study of the Rhynie chert plants that was carried out with William Henry Lang. Rhynie is a pleasant little village about thirty miles northeast of Aberdeen, Scotland. The chert bed lies a few feet underground in a field on the outskirts of the village. The original discovery was made by Dr. W. Mackie, who found specimens in a nearby wall and scattered through the field. Trenches were later dug and the chert bed located. The deposit consists of some eight feet of alternating bands of petrified peat and sand. Mackie issued a preliminary note in 1912 and his detailed geological account appeared in 1914, this including a brief description of the plants and a few illustrations.

Kidston and Lang give the following summary of the origin of the deposit:

> The whole history of the formation of the Rhynie Chert Zone, at least of that portion from which our specimens were taken, can be clearly read. One can in imagination see a land surface, subject at intervals to inundation, covered with a dense growth of *Rhynia Gwynne-Vaughani*. By the decay of the underground parts of *Rhynia* and the falling down of withered stems (for this plant had no leaves) a bed of peat was gradually formed varying from an inch to a foot in thickness. The peat was then flooded and a layer of sand deposited on its surface. Again the *Rhynia* covered the surface, and this process of the formation of beds of peat, with the deposition of thin layers of sand, went on till a total thickness of 8 feet had accumulated. . . . It may be mentioned that the presence of geysers or hot springs has been suggested by Dr. Mackie as an explanation of the occurrence of so many cherty developments in the rock series of Rhynie. [1917, p. 764]

At the time this was written only *R. Gwynne-Vaughani* and *Asteroxylon Mackiei* were known, but in the other papers of their series Kidston and Lang

(1917, 1920, 1921) described *R. major* and *Hornea lignieri*, as well as thallophytic remains.

These plants have been described and illustrated in many textbooks and there is no need to do so here; however, there is still more to be known about the Rhynie flora. In 1964 A. G. Lyon wrote a short paper on *Asteroxylon* which indicates that the plant bore its sporangia on short stalks distributed among the leaves. The exact nature of *Rhynia* is problematical; in recent years several investigators have made careful studies of the fossils that Kidston and Lang identified as *R. major*, and *R. Gwynne-Vaughani*. Distinctive features in the axes of *R. Gwynne-Vaughani* seem to be quite clearly archegonia and antheridia, which suggest that this "species" is in fact a gametophyte.

Several years ago Dr. Lyon showed me specimens of a distinctly different plant, simpler even than *Rhynia*, which he kindly allowed me to mention in a review I prepared in 1974. I hazard the opinion that the Rhynie deposit, like many others that have been referred to as "worked out," will, with continued collecting and study, yield more information about early land vascular vegetation. During the International Botanical Congress in Edinburgh in 1964 a field trip was arranged which included Rhynie. Several trenches had been dug into the chert bed which gave the participants a splendid view of this famous locality—a real tribute to the generosity of some of our British colleagues.

But Kidston and Lang's study of the Rhynie deposit accomplished much more than just giving us several new plants. Their work vindicated Dawson's account of his Gaspé plants of some seventy years earlier; such strangely simple plants did, indeed, exist, although the morphology of the day, based on what was known of living plants, did not include a vocabulary or concepts that were adequate to fully appreciate them. Even in 1917 it must have been apparent to any competent biologist that one must look to the fossil record for facts and theories to explain the origin and development of land vascular (woody) vegetation. Many new, strange, and fascinating plants have been described from Devonian rocks since that time and many more will be brought to light in the near future.

I had the good fortune to spend a few days in Professor Lang's laboratory in Manchester when I was a graduate student. He was most cordial and helpful to me and invited me to his home one evening where, among other things, he brought out some slides of the Rhynie plants. He showed me a slide with some especially well preserved specimens of *Asteroxylon*. The leaves contained cells in which there appeared to be a central nucleus and bodies arranged peripherally around the inside of the lumen. The effect was very much like that of a living *Elodea* leaf that we used to demonstrate the nucleus and chloroplasts to beginning botany students. I could see no reason to doubt the comparable identifications of these bodies in the *Asteroxylon* leaf, and Professor Lang told me that he had refrained from publishing a description of this remarkable example of preservation for fear that some of his colleagues would take it to be a fake!

During World War I Kidston made a paleobotanical contribution of a rather different nature. Since large quantities of *Sphagnum* were needed for surgical dressings, he became engaged in exploring the bogs; he worked as hard at this task as he did with his other fossil plants, as may be seen from a letter that he wrote to Scott in June, 1918:

> I am very busy—Last Wednesday I began the examination of all the Moors in the Forth Valley N of Stirling for Sphagnum. I have a motor at my disposal & leave about 9 in the morning and return about 7 or as near then as possible—I have got two good Moors for Sphagnum but of one the water & the transport may be difficult but this I am going to look into—on the other I hope to have a camp placed soon—But my commission is to examine the moors "North of the Forth"—a pretty big order . . . they do not place a camp unless 10,000 sacks of Sphagnum are believed obtainable. [Scott letters]

Kidston's collaborators spoke very highly of him. Lang (1925) says: "In my own case, and I am sure in Gwynne-Vaughan's also, the most important and valued influence in our mature scientific lives was the privilege of working with Kidston" (p. v).

The lives and work of several of our former colleages are closely entwined in the latter years of Kidston's time and for a couple of decades thereafter, and Lang played a central role. W. H. Lang was born on May 12, 1874, near Groombridge, Sussex. He entered the University of Glasgow when he was fifteen, took a bachelor's degree in science in 1894, and graduated in medicine and surgery the following year; he was registered as a medical practitioner but never actually engaged in practice. At Glasgow he became acquainted with Gwynne-Vaughan and also came under the influence of F. O. Bower, the leading authority on ferns at the time, and in the autumn of 1895 he went to the Jodrell Laboratory at Kew, this being the period when Scott was there as Honorary Keeper. His great knowledge of plant morphology was enhanced by a visit to Malaya and Ceylon in the autumn of 1899.

Lang served on the staff of the Botany Department at Glasgow for fifteen years, resigning in 1909 to take a professorship of cryptogamic botany at Manchester. After Kidston died, in 1924, Lang initiated a series of studies of plants from the Old Red Sandstone of Scotland. Although dealing with rather fragmentary material he added appreciably to our understanding of several Devonian plants, and especially *Zosterophyllum*. Lang was a considerate person; Walton recalls that when Lang went collecting in the Orkneys on Sundays, in order not to offend the local folk, he would conceal his geological hammer by carrying it with the head in his hand and the shaft up his sleeve.

In 1937 in his paper on the Downtonian (late Silurian–early Devonian) plants of England and Wales he established the genus *Cooksonia*, one of the smallest and simplest of early land vascular plants. This was a fitting tribute to Isabel Cookson, with whom Lang had collaborated in 1935 on a study of the *Baragwanathia* flora of Victoria, Australia. This created a considerable stir at the time as it was believed to be of late Silurian age and thus the oldest well-preserved assemblage of land vascular plants. More recent studies indicate that the age is probably early Devonian, but this in no way detracts from the importance of the flora. The chief element is *Baragwanathia longifolia*, a plant with axes up to six centimeters in diameter, bearing long slender leaves with sporangia scattered among them and attached either to their base or to the axis itself. In life it would seem that it must have looked like a large version of the living *Lycopodium lucidulum*. Other plants in the assemblage include *Yarravia*, with its terminal synangium of several linear-oval sporangia, and *Hedeia*, with its corymbose system of terminally borne sporangia. The *Baragwanathia* flora was described in two 1935 papers, one by Lang and Cookson and the other by Miss Cookson alone.

Isabel Cookson (1893–1973) is now best known for her work in palynology. She was born in Melbourne on December 25, 1893, and educated at the University of Melbourne. She made numerous trips to Europe including one to Cambridge in 1925 to consult with Professor Seward. She was serving as a mycologist in cotton research at Manchester when she met Lang in 1933–34. After the work with Lang on the *Baragwanathia* flora she became primarily concerned with microfossils and produced many studies on the pollen of the brown-coal deposits of Victoria. She also collaborated with Professors A. Eisenack and G. DeFlandre on hystrichosphaerid and dinoflagellate studies, which fall outside our main theme here. Miss Cookson continued to be very active in micropaleontological research until shortly before her death on July 1, 1973.

At the time I visited Professor Lang's laboratory in 1938 William N. Croft was a student there working on Lower Devonian plants from Wales. This resulted in a fine publication by Croft and Lang in 1942 that dealt with several of the early land vascular plants, including *Gosslingia*, *Cooksonia*, *Zosterophyllum*, and *Sporogonites*. Croft later brought out a paper on the charo-

Left to right: François Stockmans, William N. Croft, and Wilfred N. Edwards, at Stockholm, 1950.

phytes (1952), and an interesting study with E. A. George of some blue-green algae found in the Rhynie chert.

Croft joined the Geology Department of the British Museum in 1939 and remained on the staff there, with time out during the war with the Royal Engineers in India, Italy, and the Middle East. In the year he joined the Museum he was sent on an expedition to Spitsbergen and later spent a year in the Antarctic. His life was a busy one but tragically short, for he died at the age of thirty-eight in 1953. I liked Croft very much and his death was a distinct shock. I think he would not object if I recounted a rather humorous incident that took place during the International Botanical Congress in Stockholm in 1950. Croft and I were on an afternoon excursion among the islands near Stockholm; it was strictly a pleasure trip. The boat we were on stopped for a time and we wandered along the shore to a secluded rocky spot and sat down. It was a beautiful place and the day was very warm; I looked at the clear, cool water and suggested that we take advantage of the seclusion to have a quick swim. The water *was* cool and Croft reacted rather violently in the negative; he would have nothing to do with it! I did not know at the time of his polar experiences, where I expect he had had enough of cold water.

His biographer has written the following: "Croft's early death is a tragic loss, for he showed a rare blend of both critical and constructive ability, combined with great personal charm. . . . He was an accomplished technician, and his room was full of strange gadjets of his own devising, including a sewing-machine converted into a microscope and a bacon-slicer adapted to rock-cutting. The Croft parallel grinding apparatus is in use in many palaeontological laboratories" (Edwards, 1954).

It is appropriate here to add a tribute to a highly promising young woman, Seward's student Ruth Holden. I consulted her papers many years ago, since she started off as I did studying fossil coniferous woods; her initial contact in paleobotany was with E. C. Jeffrey at Harvard, while my own was a second-generation one with Jeffrey's student, R. E. Torrey.

Ruth Holden was born in Attleboro, Massachusetts, in 1890 and was a student at Radcliffe College, Cambridge, where she graduated in 1911. She made a brief visit to Cambridge, England, in 1912 and returned the next year as a student at Newnham College. She was an ardent field worker and collected fossil wood along the Jurassic coast, as well as on Prince Edward Island and in New Brunswick in Canada. She brought out two papers on fossil pines and other conifers from the Cretaceous clays of New Jersey, and also in 1913 an account of coniferous woods from Yorkshire; she also did some work with Indian conifers and cycadophytes. They were all studies of fine quality and well illustrated.

At the outbreak of the war in 1914 she was at first not permitted to work in a British military hospital because of her nationality but later was able to join the Millicent Fawcett Medical Unit that was equipped in Britain for service in Russia. In Seward's biographical sketch he records: "With the whole-hearted enthusiasm with which she carried out any work she undertook, she set to work to learn Russian and so successful was she that ultimately she was appointed interpreter and courier to the Unit" (1917, p. 155).

Much of her time was spent in Kazan, where she helped to prepare and equip a hospital for Polish refugee children, but she still found some time to study the fossil-plant collections at the University there. She also traveled between Petrograd (Leningrad) and Galicia distributing supplies to hospitals. In January of 1917 she became ill with typhoid fever at Kazan; this later became complicated with meningitis and she died in Moscow on April 21. Of her short but eventful and promising life Seward has written: "though Miss Holden was a student of exceptional originality and promise she was much more than that—a chivalrous and noble woman whom it was a privilege to count a friend" (1917, p. 156).

What I am calling "the age of Seward" in Britain is a period of numerous workers who were responsible for some of the great advances in paleobotany. One of them, Edward Alexander Newell Arber, was somewhat before my time but I have been interested in his work and have followed in my own research one aspect of his Devonian studies. Newell Arber was born on August 5, 1870, in London and died at the age of forty-eight in 1918. One of my first and most memorable experiences at Cambridge, when I went there to study, was being invited to tea by his widow, Agnes Arber, who was still living in Cambridge with their daughter, Muriel, also a good friend who is now carrying on the family tradition in a different branch of geology. Agnes Arber was one of the kindest people I have ever met and one of Britain's great botanists. Although she made a few minor contributions to fossil botany she is best known for her many morphological studies, including *Herbals, Their Origin and Evolution*; *Water Plants*; and *The Natural Philosophy of Plant Form*. On the several occasions that I visited her at home in Cambridge she served a special cake with tea that I particularly enjoyed; she remembered this and some years later after the war when I returned to Cambridge she had the closest item of pastry that was then available!

Newell Arber's health was not good in his early years and when he was fifteen he went to Davos, Switzerland, for about a year where he developed an interest, little short of a passion, in alpine vegetation. He entered Trinity College, Cambridge, in 1895, and after completing his formal education was appointed Demonstrator in Palaeobotany in the Sedgwick Museum, a position that he retained for the rest of his life. He devoted a good deal of his time to the identification and curating of fossil plants in both the Sedgwick Museum in Cambridge and the British Museum of Natural History. This was correlated with paleobotanical-stratigraphical investigations of the Kent Coal Field, the Yorkshire Coal Field, and the Forest of Dean. In 1905 he produced an interesting paper describing what he thought to be the sporangia of *Glossopteris Browniana*. They were associated with scale leaves of specimens from New South Wales, and his illustrations show the characteristic structures that are now known to be the sporangia of *Glossopteris*. Arber also prepared the British Museum's *Catalogue of the Glossopteris Flora*.

Newell Arber's passion for field studies seems to have been the great driving influence in his life. In her biography of him Agnes Arber quotes from one his letters: "A perfect day when one is in the field is one of the greatest things on earth. . . . My mania is quite a modest one. It is a desire to visit every spot in this country where fossil plants have ever been found. To gain that full power of knowledge which can only be got by having been to the place, seen it, photographed it and collected from it. When you have done this you have a 'grip' which is masterly" (1918, p. 428).

E. A. Newell Arber, at Saas Grund, Switzerland, 1908.
Courtesy of Muriel Arber.

Arber's field studies led him back again and again to two areas, the Swiss Alps and the coast of Devon. The former resulted in his *Plant Life of Alpine Switzerland*, a very readable account of the general ecology of the Swiss alpine flora. His studies in Devon produced *The Coast Scenery of North Devon* in 1911; this is a well-illustrated book which seems to me to be a good example of a regional geology written to inform and to incite further interest.

Newell Arber, like some others in his time and since, was rather carried away with the cycadophyte discoveries that were made in the early years of the century, especially the work of G. R. Wieland on the cycadeoids. This, as I understand it, was based on the presumed similarity between the cycadeoid "flower" (as shown in some of Wieland's well-known illustrations of 1906) and the magnolia-type flower. The striking divergence in the structure of the pollen-bearing organs has always seemed to me much too great to admit of a very close comparison, but this did not inhibit some of the more enthusiastic seekers of angiosperm origins. I do not mean to imply that the cycadophytes, in the broadest sense, are completely out of the picture. It is an intriquing group, and new information bearing on the origin and evolutionary patterns in the cycads has recently been produced by Sergius Mamay (1976); his memoir is based on many years of field work and laboratory study, and I think will be a classic for this area of plant evolution.

In collaboration with John Parkin, Arber wrote "On the Origin of Angiosperms" in 1907, an account that attracted considerable interest. They regarded the angiosperms as monophyletic, and captivated, I believe, to a considerable degree by Wieland's work, they postulated a hypothetical hemiangiosperm built along the plan of a *Cycadeoidea* cone but with much simplified "anthers."

Arber and Parkin tried to be cautious but they were clearly excited about the possibility of having solved the riddle. On October 21, 1906, Arber wrote to Scott: "You will be interested to hear that, just at the present moment when our wits are no doubt rusty, Parkin and I are suffering from the delusion that we have solved the problem of the origin of the Angiosperms! This idea is no doubt totally erroneous but at the present it has a cheerful influence" (Scott letters). They regarded the Bennettiteae as being very close to angiosperms, but the case was not sufficiently convincing, and the problem seems to be still with us.

As one who has a particular concern with the evolution of early land vegetation I regard Newell Arber's little book *Devonian Floras* (1921a) as one of the genuine classics in fossil botany. Agnes Arber brought her husband's manuscript to life with the aid of D. H. Scott and she has this to say of his last contribution: "I do not think that anything in his scientific life gave him a keener intellectual pleasure than the development of the idea—the *Leitmotiv* of the present essay—that the transition from the Algae to the Vascular Cryptogams no longer remains a matter of pure conjecture, but that, in the fossil plants of the Devonian rocks, we witness, actually occurring beneath our eyes, the passage from the Thallophyta to the Cormophyta" (Arber, 1921a, p. v).

Newell Arber would certainly be fascinated if he could know of the great developments in our knowledge of Devonian floras that have been made in recent years. In his book he divides the Devonian floras essentially into the "Psilophyte" flora of the early Devonian and the "*Archaeopteris*" flora of the later Devonian. I have dealt with the psilophytes elsewhere and I would like to insert here a short résumé of the *Archaeopteris* study which has shed a burst of light on our knowledge of the transition from pteridophytes to seed plants in the latter part of the Devonian.

The "fernlike fronds" of *Archaeopteris* have been known for well over a century and have been found in many localities in the eastern United States, Quebec, Ellesmere and Bear Islands in the Arctic, Ireland, and in the U.S.S.R., and they have been frequently found associated with the petrified trunks known as *Callixylon*. In 1960 Charles Beck showed that this typically pteridophytic foliage with its heterosporous mode of reproduction was borne on the *Callixylon* stems, which are typically coniferous in the anatomy of the secondary wood. As a consequence he established the new group Progymnospermopsida. Several other genera are now included within it. This has not answered all of the questions that one might ask about the pteridophyte-gymnosperm transition but it has blazed a rough trail through a dark forest where no recognizable path existed before. It is one of fossil botany's important contributions.

Archaeopteris was long known on the basis of presumed fernlike fronds, bipinnate, and attaining a length of about two-thirds of a meter. The "pinnules" in different species vary from being entire to being deeply dissected; the frond as a whole may be entirely sterile, partially fertile or wholly so. The name was created by Dawson in 1871 in place of *Palaeopteris* Schimper (1869), since that name was preoccupied; classic illustrations were given by Schimper in his *Traité de paléontologie végétale.*

Quite a few people over the years have been concerned with *Archaeopteris* and have offered their opinions as to its position in the plant kingdom. In 1883 Williamson wrote in one of the articles in the *Organization* series: "That the Devonian *Palaeopteris Hibernica* is a Hymenophyllous form appears to be almost certain." In a letter to D. H. Scott written in 1904, A. G. Nathorst said: "*what is Archaeopteris*? Until the microscopical structure of the supposed sporangia had been ascertained, it is perhaps not quite sure that they were marattiaceous ferns!" (Scott letters). Newell Arber wrote to Scott from

Stockholm on April 6, 1906: "I know nothing about *Archaeopteris* at all fresh. My opinion is worth little so there is no cause for alarm. What I saw on your estate in Ireland* led me to believe that so far as actual facts are concerned *Archaeopteris* is a fern pure and simple. Everything points to it. All the same, intuition forbids me to believe it. . . . Oh! for a discovery of structure. . . . The man who gets out the real nature of *Archaeopteris* deserves the V.C." (Scott letters).

Not until 1939, however, was reasonable proof of the heterosporous nature of *Archaeopteris* established. In that year Chester Arnold demonstrated two distinctly different spore sizes in sporangia associated with fertile fronds of *A. latifolia* from Pennsylvania. Later John Pettitt demonstrated that specimens probably referable to *A. jacksoni* (Dawson) from Quebec are heterosporous, and Phillips, Andrews, and Gensel described two heterosporous species from West Virginia.

In 1960 the single most important discovery was made by Charles Beck when he showed that portions of the fronds were attached to partially petrified axes showing the characteristic wood anatomy of *Callixylon*. Most recently, Carluccio, Hueber, and Banks, working with petrified specimens have shown that the *Archaeopteris* "frond" is in fact a modified branch system. The main and primary axes display the anatomy of stems, which implies that the ultimate units that have been called "pinnules" are more appropriately termed "leaves."

A few personal comments may be permissible before leaving this intriguing genus of fossil plants. Some years ago James M. Schopf, when prospecting along the Upper Devonian outcrops near Valley Head, West Virginia, found a promising deposit of plants. He kindly turned these over to Tom L. Phillips and myself, and on several excursions to the area we found two species of *Archaeopteris*, superbly preserved. The enclosing sediment, a relatively soft, fine-grained sandstone, enabled us to degage or excavate the "fronds" and demonstrate their three-dimensional form; we found that the ultimate units, formerly called pinnules were arranged in a dense spiral around the axis. One of the specimens bore leaves that are nearly entire, while in the other the leaves are quite deeply dissected. In both species the sporangia and their contained spores are well preserved, but because the microspores and megaspores of the two species proved to be almost indistinguishable they did not yield the information we hoped for on evolution within the genus.

* In two of his letters to Scott, Newell Arber refers to Scott's acquisition of the Kiltorcan estate in Ireland, but I have been unable to confirm this.

Newell Arber and Hugh Hamshaw Thomas (1885–1962) were close friends. For personal reasons I take particular pleasure in writing about Thomas and his work. My undergraduate work was done at the Massachusetts Institute of Technology with a major program in food technology, but I managed to take a paleontology course there with the noted invertebrate paleontologist Hervey W. Shimer. This was the beginning of my serious interest in fossils. I then spent a year at the University of Massachusetts with Ray E. Torrey, a fine plant anatomist who had done some work on fossil coniferous wood with E. C. Jeffrey at Harvard. Being in need of some kind of employment I next accepted an assistantship at Washington University in St. Louis. After two years there my mentor, Robert E. Woodson, Jr., shipped me off to Cambridge to study under Hamshaw Thomas.

Hamshaw Thomas was born in 1885 in Wrexham, Wales, and entered Downing College, Cambridge, where he remained for much of his career. He was a great sports enthusiast and always very much concerned with Downing's success in the university events. This did not seem to me quite in keeping with the rest of his character for he was a quiet man and often seemed to be lost in his own thoughts. Tom Harris knew Thomas well at Cambridge and has written: "All his working life was devoted to tutorial teaching and I believe his rather slow speech, quiet, thoughtful manner and habit of asking questions instead of telling answers must have made his 'supervisions' extraordinarily good. At any rate a large number of young men who had them have since done very well, and still remember him with affection, admiration and gratitude" (1963, p. 291).

I attended the lectures in Thomas's morphology course; his delivery was not brilliant, but I almost always left the room with some new thought or question. I think I was one of the few research students he accepted. Harris has said that Thomas did not wish for "disciples," and I am most grateful for my association with him although I might say that the aid he gave me came in the form of short chats in the laboratory, where he cast out ideas and suggestions that were valuable. As to material advantages I was rather on my own in the Botany School. I arrived in Cambridge in late August of 1937, when Dr. and Mrs. Thomas were getting ready for a vacation on the continent. I was therefore sent to the Natural History Museum in London to work for a month after we discussed a study of the secondary wood of the pteridosperms as a research project. The initial problem was to obtain suitable material to work with. He said he would be going to Prague and would try to obtain some pieces of the Chemnitz medullosas, which sounded like a fine start. When he came back to Cambridge he said, in his casual and slightly dreamy fashion, that he had been offered some of the desired material but had forgotten about my project! It was rather a let-down but he immediately wrote back to Prague and I did obtain some useful specimens.

The tradition at Cambridge with most of the leaders in paleobotany, such

Hugh Hamshaw Thomas. From *Biographical Memoirs, Fellows of the Royal Society*, 1952.

as Walton, Seward, Arber, Thomas, and Harris, had been predominantly with studies of plant compressions rather than petrifactions. This was quite evident when I found that the rock-cutting machine was an old phonograph that had been modified to turn a soft iron disc. The operating technique consisted in first roughly notching the edge of the disc by striking it with a jack-knife; the disc was then set in motion and a piece of chert smeared with a mixture of oil and diamond dust was held against it. Some of the diamond dust became embedded in the notches and one could then cut a small specimen, whereupon the charging process must be repeated. I used the peel method where the petrifactions allowed, with William Darrah's formula, but the expensive "parlodion" was not available in Cambridge and I substituted movie film that had had the emulsion washed off; some such films worked well and others did not.

But my struggles with primitive equipment and supplies and the difficulties in obtaining adequate study specimens were greatly outweighted by the general atmosphere in Cambridge; its great traditions, unique and lovely buildings and gardens, fine libraries, and distinguished men and women all contributed to my education in ways that I do not think were possible anywhere else at that time.

Thomas first worked with Arber in the Sedgwick Museum, but later moved across the courtyard to the Botany School. Two other associations also influenced the course of Thomas's career: He spent some time with Kidston collecting in the Scottish coal fields and absorbed something of Kidston's belief that much collecting must be done to assure that all available evidence has been obtained. To some degree this was probably detrimental, as Thomas was cautious to a fault and required no urging in this direction. And in 1911 he spent the summer vacation with Alfred G. Nathorst in Stockholm. This was a visit of considerable historic importance; Nathorst had started to work effectively in about 1900 with compression fossils in which the cuticle was preserved. Nathorst was well along in years at the time; Thomas, as one of his few disciples, carried on with the development of the technique. In Stockholm Thomas also met Thore Halle, who had done some collecting in the Yorkshire Jurassic beds.

When Thomas returned to England he initiated a collecting program in Yorkshire, apparently at the suggestion of Seward and encouraged by Halle. Thomas told me that others at the time felt he would be wasting his time along the Yorkshire coast; in his biography of Thomas, Harris says: "The very localities were forgotten, in fact he was told they were worked out. He told me he found the famous Gristhorpe Bed again when he was forlornly collecting a seaweed as a museum specimen. The seaweed was tough, the rock broke and he saw a fossil in the Gristhorpe Bed" (1963, p. 290.

This raises the interesting subject of collecting *per se*, which has changed historically along with the science of paleobotany. Harris observes that during Thomas's career "the disastrous notion prevailed in England that the proper subject of a fossil botanist's study was to be found in a museum; or it was specimens sent in by a geologist who wanted to be told their age" (1963, p. 289). The presence of fossils in a particular locality has always raised questions about whether there is any importance to be attached to the association of various organs that might belong to a single species. Most people who have collected at Gristhorpe have noticed an order in the way the various plant remains are found in different parts of the bed. Thomas commented on this in a letter to G. R. Wieland of January 13, 1920: "I gather from some of your work that you are inclined to regard evidence derived from the occurrence & association of fossils in the field as worthy of consideration. My somewhat intensive study of the Yorkshire plant beds has led me to regard both positive and negative field evidence as of the greatest value, though formerly I should have greatly discounted it,—like so many other Palaeobotanists who mainly rely on the collections of other people" (letter in possession of H. N. Andrews).

Paleobotanists such as Thomas and Halle brought in a new brand of collecting, in which the emphasis was on reproductive structures rather than pretty, well-preserved leaves and other specimens that look nice in museum cases. They developed new techniques for macerating, embedding, and sectioning the compressions or "mummifications" which yielded cellular detail in a way that had not been possible previously.

One of Thomas's early and very important papers, the result of his summer with Nathorst, was a study of cycadophyte cuticles, written with Nellie Bancroft, which demonstrated the striking distinction between the leaf cuticle structure in the cycads and that in the Bennettitales. But of all his research contributions Thomas is best known for the one on the Caytoniales, in which he demonstrated the relationship of the seed-bearing *Caytonia* "fruits," the *Sagenopteris* foliage, and the *Antholithus* pollen organs. In the introduction to this paper Thomas says: "The present communication records the discovery of fruits, seeds and groups of stamens, which have a strong claim to be regarded as parts of angiospermous plants, in the Middle Estuarine beds of the Jurassic series of rocks of the Yorkshire coast" (1925, p. 300). I do not believe that Thomas thought he had solved the question of the origin of the angiosperms; he knew that it was a complex affair and that the Caytoniales presented a contribution but not the entire solution. Nor was the importance of the Caytoniales lessened by Harris's later discovery of pollen in the micropyles of the *Caytonia* seeds. *Caytonia* must be regarded as a gymnosperm but it was very close to the angiosperm stage.

The Caytoniales study was followed by his discovery and description of the pteridospermous Corystospermaceae from the Triassic of Natal. This is one of the more significant contributions to our knowledge of the stream of evolution that, along one or more lines, flowed from the pteridosperms ultimately to the angiosperms.

He contributed other original reports on a variety of Paleozoic and Mesozoic fossil plants. And in his later years he wrote at some length on the "new morphology," which I think is best summed up as a shift from the "old morphology," based at least in part on the thoughts of Goethe, to a morphology based on the newly discovered simple land vascular plants such as *Rhynia*. In the later years he also concerned himself with the history of science and helped to found the British Society for the History of Science.

Thomas's services were not confined to paleobotany. At the start of World War I he joined the Officers Training Corps and was assigned to the Royal Artillery in France. In 1915 he was transferred to the Flying Corps, where he worked as the officer in charge of aerial photography, his maps being some of the first ever made in that way. His map making in the Near East is credited with having contributed very directly to the success of British military operations there. He also recognized the many peacetime uses for aerial photography in the study of vegetation patterns and geographical and geological features, as well as other applications that are in wide use today.

Among this remarkable group of young men who were either trained directly or strongly influenced by Seward in the early decades of the 1900's John Walton (1895–1971) must rank very high. Walton was born in Chelsea, London, and entered St. John's College at Cambridge in 1914. His study there was interrupted by the war, and being a Quaker he served with the Friends Ambulance Service in France and Belgium from 1915 to 1918, when he returned to Cambridge. He graduated in Botany in 1921 and then took a position as Senior Lecturer at Manchester. In 1930 he was elected to the Regius Professorship of Botany at Glasgow, where he remained for the rest of his academic career. He inherited the great tradition that was built by Bower, as well as the difficult financial problems of the early 1930's, and served the University ably in several capacities.

Walton's contributions to paleobotany fall into three categories: the development of new techniques, research with Carboniferous plants, and a fine textbook. He deserves much or most of the credit for the peel method of sectioning well-preserved petrifactions. I will be concerned only with the historical aspect here, as the technique itself is described admirably in Lacey's (1963) article "Palaeobotanical Techniques." Walton issued the first published report on the peel technique in 1928 and states that the method was developed in collaboration with R. G. Koopmans of Utrecht. He notes that they first used several cellulose compounds as a peel medium, the best being "Duroflex." However, Walton once told me that he first experimented with the method by ringing the smooth surface of a petrifaction with plasticine and, after etching the surface, pouring a warm gelatine solution over it and allowing it to dry—a very slow process. A further great advance in this method was made by later workers when sheet film was introduced in place of the original solution method. The peel method saves much time and renders possible the preparation of serial, or nearly serial, sections of large or small petrifactions. Walton also developed the balsam-transfer method, in which a compression fossil is affixed in balsam to a glass slide, the rock matrix then being dissolved away with suitable acids, leaving the "back side" of the fossil visible. I believe the first description of this technique appeared in 1923, but there is also a well-illustrated account in the proceedings of the Heerlen Congrès de Stratigraphie Carbonifère in 1928. A similar technique, the "Ashby cellulose film-transfer method," was introduced shortly thereafter; it has been described by Lang and Lacey.

Walton made lasting contributions to Carboniferous paleobotany, including several papers on fossil liverworts, and I regard his study of the hollow arborescent trunks from the Island of Arran as an interesting and astute investigation. He demonstrated very nicely the ontogenetic sequence of the numerous stelar fragments found in the partially preserved trunks. And the two editions of his text *An Introduction to the Study of Fossil Plants* have served many students; it deals largely with Carboniferous plants but is one of the most readable and best illustrated textbooks even today.

I have especially fond memories of John Walton which I expect are shared by many others of my generation. He was a man with a rather formal and slightly stern exterior, but it did not take long to get beneath this and find an extraordinarily kind and helpful person. He was the son of E. A. Walton, a leading artist of the time, and inherited both interest and ability from him. In addition to his regular duties as the Professor of Botany at Glasgow he served as Honorary Curator of the University Art Museum; this was, however, no honorary post in the usual sense since he functioned as gallery administrator of the day-to-day tasks and was responsible for the addition of several notable collections. On my first visit to Glasgow, when I was a graduate student, I met N. W. Radforth, who was then studying with Walton. Radforth has told me many interesting things about him during our long friendship. On Walton's classroom artistry:

> his blackboard sketches were masterpieces. His father had been a Royal Artist. Son John was equally gifted but apart from the simple yet expressive reconstructions his published work displayed, he rarely sketched. He could draw skillfully and rapidly equally well with both hands in action. At the lecture's end the panorama on the blackboard was enchanting. I remember many a post-lecture period of quiet "reverence" spent contem-

John Walton, at the Botanical Congress, Edinburgh, 1964.
Photo by Theodore Delevoryas.

> plating the mosaics he produced for us. . . . Each blackboard reflected a fresh vision of the convincingly unfolded past; then it would be unhappily retired into extinction. [Personal letter]

On his keenness as an observer in the field: "Once on a canoe trip in the Georgian Bay country of Ontario, I confided that I had never seen *Isoetes* outside a pickle jar. He glanced at the shallow river bottom beneath us and deftly extracted a beautiful specimen from the silt, much to my embarrassment—for the site was close to where I had pitched my tent many times" (personal letter).

I was working in Cambridge in February of 1971 and waiting for the cessation of a general transportation strike in order to keep a meeting date with the Waltons in Dundee. There were a few things that I wanted to discuss in connection with my history of paleobotanists but this final meeting was not to be. His wife, Dorothy, who was Sir Albert Seward's daughter, informed me that he had died on February 12.

At this point I wish to bring in two people whom I cannot claim as close friends but whom I knew and admired, and who were helpful to me—two important contributors to the annals of paleobotany who may best be identified as great "period pieces": Marie Carmichael Stopes and William Thomas Gordon.

At least three book-length biographies and numerous magazine articles have been written about Marie Stopes (1880–1958) and most of this literature deals with her books on sex and pioneering birth-control work in Britain. Some of it is hardly flattering—even quite unfair—as far as presenting a balanced picture of her achievements. I believe that I am competent to judge her botanical work, which most "popular" writers have generally avoided. It is precarious to try to sum up a person's character in a few words, but the following excerpt from the preface to *The Trial of Marie Stopes*, edited by Muriel Box (1967), may be reasonably close:

> Its* author was perhaps a little mad—or perhaps it was simply that she was wholly unlike any other human being ever recorded in history or fiction. Her vanity was so colossal, so uninhibited, and so unashamed, as to be positively endearing to those who knew her. Having done so much for other people's personal relationships, she mismanaged all her own. She was compassionate, headstrong, tactless, public-spirited, humourless, intellectually distinguished and wholly lacking in aesthetic taste. I count it a privilege to have known her. I remember her with affection and I shall never forget her. [P. 9]

Marie Stopes was born in Edinburgh in 1880, although her family moved to London shortly thereafter. Her father was intensely interested in archaeology and her mother in university education for women. She received her early education from them at home and did not attend a formal school until she was twelve. Her beginnings in school were painful; she had trouble with Latin and Greek, which her mother had tried to teach her without success. She managed to get through the University College (London) matriculation exams but her Latin mistress commented that she had no business in doing

* In reference to Marie Stopes's book *Married Love* (1918).

so! Marie retorted: "I did it by writing very clearly everything I knew and saying nothing about the things I did not know" (Briant, 1962, p. 31).

She entered University College, London, and, encouraged by Francis Oliver, she turned to botany as a major subject, with chemistry second.

Marie C. Stopes, at Leatherhead, 1950.
Photo by James M. Schopf.

Oliver's judgment was justified, for she won a gold medal and placed second in her zoology work. Upon graduating in 1903 she went, as so many others did from England, to Munich, where she studied for two years and obtained her Ph.D. degree. Her research in Munich was chiefly with cycad seed morphology although her first published paper dealt with a cordaite leaf.

When she returned from Germany she was appointed a lecturer at Manchester University—the first woman lecturer on the staff there. Her interest in fossil plants had become quite strong; she met Captain Scott at a dance one evening, told him of her interests, and tried to persuade him to take her and his wife on his next Antarctic expedition. He had to reply in the negative, but he visited her at the University later to acquaint himself with her botanical interests. Briant notes that "when he was found dead in the Antarctic, there were discovered near him some pieces of fossil plants" (p. 48).

In 1906 she presented a carefully worded plan to the Royal Society proposing to visit Japan in search of Cretaceous angiosperms, and she predicted where she would find fossil plants. She obtained the necessary funds and was in Japan from August 6, 1907, to January 24, 1909. Her various biographers have emphasized her attachment to a Japanese botanist, Professor Kuyiro Fujii, whom she had met in Munich, implying that she had followed him to Japan. However strong this interest may have been, she certainly did not neglect the primary object of her mission. She traveled widely through Japan, often alone, and was not merely the first foreigner that many Japanese had ever seen but the first foreign woman and scientist in the bargain. In his short biographical sketch of her, William Chaloner says: "This was no small adventure, and in the days when women in academic work were still a novelty, and in the field of geology almost an impropriety" (1959, p. 119).

Her record of this year and a half, *A Journal from Japan*, is a fascinating and readable book. The chief paleobotanical result of her trip was an article written with Fujii on petrified plant remains from the Cretaceous of Hokkaido. These occur in nodules and include a rather wide array: fungi, fern sporangia and stem fragments, coniferous and angiosperm woods, and a specimen with a trilocular ovary. The mode of preservation is interesting and suggestive of what else might be found with continued searching.

One of Marie Stopes's more important and comprehensive papers is the one on the origin and distribution of coal balls written with D. M. S. Watson. It was initiated in collaboration with James Lomax, but apparently the divergence of opinions between the authors and Lomax made it necessary for Stopes and Watson to present their views separately. They demonstrated that coal balls are representative samples of the vegetation, often beautifully preserved, that composes the surrounding coal. They showed that the size, shape, and occasional continuity between coal balls imply that they were formed in the place in which they are found. The account also includes chemical analyses of coal balls and a description of their distribution in coal seams as well as of the nature and kinds of materials that are preserved.

Marie Stopes spent some time in the United States and apparently did a fair amount of field work; she wrote me on May 31, 1950: "I did myself find Coal Balls in America, and before [Adolph C.] Noé, but like so many things I have, I never published about them."

A brief divergence here may be permissible to record a few words about David Meredith Seares Watson (1886–1973), known to his friends as "D.M.S." Watson entered the University of Manchester in 1904, intending to study chemistry and go into industry. His biographers say:

> He records that in his first years his subjects were mathematics, chemistry, physics and geology. At that time students were left much to their own

> devices (he saw a demonstrator in practical chemistry only twice in a whole term) and thus he had 'time to read extensively, attend classes in other subjects and study in the museum.' He had been given some pieces of coal balls by a taxidermist, sectioned them, found them to contain the seed *Lagenostoma physoides*, and became known to Professor F. E. Weiss and Dr. Marie C. Stopes of the Department of Botany, famous for its cryptogamic studies. On the advice of Professor Boyd Dawkins he switched from chemistry to geology as his main subject, and even found time to attend classes in botany and a few in zoology. [Parrington and Westall, 1974, p. 484]

I had but a fleeting glimpse of Watson once in the refectory of University College and he impressed me as a rugged individual. I think he was and is probably best known for his work with fossil vertebrates. His biographers say, in reference to his field work:

> He travelled in the Karroo for ten months, mainly by a two-wheeled covered trap drawn by one horse, making a most valuable collection that was to keep him busy for many years. . . . The next summer Watson revisited South Africa, making further valuable collections. It was there that he recalled finding, among the relicts left by Alexander Bain, one of the great early collectors from the Karroo, a snout which, after washing, he recognized as the missing piece from a specimen described long before. He had previously (1911) collected the missing part of the skull of Seeley's *Diademodon entemorphonus* collected about twenty-five years earlier. His visual memory was phenomenal. [1974, pp. 485–86]

Watson's paleobotanical contributions were rather few; in addition to the coal-ball study he produced a good paper on the small heterosporous cone *Bothrodendron mundum* in 1908.

Marie Stopes's paleobotanical studies were numerous and diverse. In 1912 she published a well-known paper on early Cretaceous dicot woods, and in 1918 an article on Cretaceous bennettitalean cones. Her work with Cretaceous plants perhaps reached a culmination in the two-volume *Cretaceous Flora* published by the British Museum. She was responsible in part for several studies on the structure and chemistry of coal, of which the most important was her 1918 *Monograph on the Constitution of Coal* which introduced such terms as "fusain" and "vitrain." This was written in collaboration with the eminent British chemist Richard V. Wheeler, and D. M. S. Watson says it "gave her an international reputation, for it was, in a modified form, accepted as the basis for the classification of coals at the Third International Congress on Coal in Heerlen in 1935" (1959, p. 153).

Marie Stopes's book *Ancient Plants* was one of the early efforts, and a very successful one, to bring fossil botany to the general reader. Shortly after the war, in 1920, she began to correspond with paleobotanists in various parts of the world to try to find out who was still alive and active. I think this was, in a modest way, the first effort to do anything about an international information organization in paleobotany.

She did little with fossil plants after about 1920, when her interest and energies turned to birth-control problems. She had written a book entitled *Married Love* in 1918 which went through many printings and was translated into numerous languages; this brought her fame of a very different kind from the results of her studies with fossil plants.

On March 17, 1921, she opened the first birth-control clinic in the British Empire, with the aid of her second husband, Humphrey Verdon Roe. This was still a time when one hardly dared to suggest such a move without real fear of the most violent criticism. Few people have suffered the abuse that came to her during the years that she initiated and developed this work. Kenneth Allsop, in reviewing Keith Briant's biography of Stopes, says: "a generation which enjoys for granted the benefits of her pioneering may find incomprehensible the hostility, fear and hatred she enflamed. . . . In those times you might socially have got away even with spitting in church or defiling the Union Jack—but not with uttering the name Marie Stopes in respectable company. She was reviled as immoral, monstrous, obscene, attacked from the pulpit and in poison-pen letters, and her life was threatened" (*Daily Mail*, May 24, 1962).

I had some correspondence with Marie Stopes in the 1950's about American coal-ball studies, which still seemed to interest her very much. And in the summer of 1950, prior to the Botanical Congress in Stockholm, she invited James Schopf and me to lunch at her home in Surrey and it was a most pleasant afternoon. As we were looking through some dust-covered fossil-plant slides in her carriage house, which had been damaged in the war, I noticed an odd fourth volume of Seward's *Fossil Plants*, which I needed to complete my set. I made a comment to this effect, and she immediately dusted it off, autographed it, and handed it to me. She expressed a hope to get back to the Natural History Museum to do more work in paleobotany, but this was not to be.

She was the first woman to contribute significantly to the progress of fossil botany and one who also benefited humanity in general. She was a great woman, and to me it was a pleasure to have known her.

In recent decades paleobotanists have contributed immensely to our knowledge of the evolution of vascular plants, from the simplest of early Devonian times to the first seed plants which have been found at the top of the Devonian and in more abundance in the Lower Carboniferous. We still need much more information, especially from the early Carboniferous horizons, but the work of W. T. Gordon and Albert Long have laid important foundations in this area.

William Thomas Gordon was born in Glasgow on January 27, 1884, although his parents soon moved to Edinburgh where he attended the University. He received the B.Sc. in 1906, studying geology under James Geikie; he later spent two years at Cambridge then returned to Edinburgh, where he served as Lecturer for three years; in 1914 was appointed Lecturer in Geology at King's College in London, where he remained. Gordon is remembered for his work with petrified fossil plants from the Lower Carboniferous of southern Scotland and for his great knowledge of mineralogy. He was very kind to me when I was a graduate student and I well remember an afternoon that I spent in his laboratory at King's College. I was somewhat consoled to find that his lapidary equipment was almost as primitive as what I was using at Cambridge, and my own rudimentary apparatus in St. Louis. He had modified a bicycle to serve as the source of power for his cutting and grinding wheels, and he demonstrated it for me. It was a rare sight to see this distinguished professor perched there on the bicycle, pedaling away, not getting very far geographically but accomplishing a good deal for the cause of fossil botany.

Gordon's contributions were chiefly concerned with plants found petrified in volcanic ash deposits. In the introduction to what I believe was his first paper (1908), which dealt with a new *Lepidophloios*, he says:

> As Carnegie Research Scholar in Geology under Professor James Geikie, . . . at Edinburgh University, I have entered upon a systematic examination of fossil plants from Pettycur, Fife. Though many plants have been described from this locality, nothing systematic, as far as I know, has ever before been attempted. The main objects of this research are to endeavour to connect the various strobili obtained at the locality in question with the stems on which they were borne; to describe any new species met with, and to give some account of the mode of occurrence of the material in which the plants are enclosed. [P. 443]

Among his other studies of the Pettycur fossils is one on the coenopterid fern *Metaclepsydropsis* and a pteridosperm stem, *Rhetinangium arberi*, which he named in honor of Newell Arber. In 1935 he produced a comprehensive paper on the genus *Pitys*, and in his last two papers he described a primitive pteridosperm stem, *Tetrastichia bupatides*, which may have borne comparably primitive seeds, described as *Salpingostoma dasu*. I will return to these studies on a later page in connection with the more recent works of Albert Long. Gordon also wrote (1935) a very informative sketch of the history of certain aspects of paleobotany under the title "Plant Life and the Philosophy of Geology"; this was his presidential address of 1934 to the Geology Section of the British Association for the Advancement of Science.

I am greatly indebted to Mr. E. O. Rowland of the Geology Department of King's College for some interesting reminiscences of Professor Gordon during the time he served as his chief laboratory steward. The comments that follow are taken from a spoken record that Mr. Rowland taped at my request; a typed copy of the complete recording is preserved in the U.S. Geological Survey's Paleobotanical Library in Washington.

Gordon was a bachelor and one with some distinctive attributes. Mr. Rowland recalls:

> I first met Professor Gordon in 1946 when I applied for a job as his steward in King's College. He was Professor of Geology. I met this rotund, red-faced gentleman in his office, which left a very great impression on my mind because he was sitting in his shirt-sleeves and the office was piled high with unopened envelopes, and all kinds of what would almost be described as debris in the room, including of all things the luggage rack from a railway train, and across the center of the room was a piece of string, and hanging on this was a shirt which he had obviously just washed. . . . Now Professor Gordon was obviously a character, but he was one of the nicest characters that it has been my pleasure in life to meet.

Professor Gordon was a noted authority on and a great collector of gemstones and was frequently called upon to lecture on them. Mr. Rowland tells several anecdotes relative to this phase of Gordon's career:

> Finishing a gemstone lecture, for example, where he would have anything up to . . . fourteen or fifteen thousand pounds [sterling] of diamonds in a small lunch case, he would hand them to me on the bus and say "You take them home and bring them in in the morning." And then I said, "You know, isn't it a bit of a risk?" and he said: "Well I look like a farmer and you don't look like a very prosperous person so who the ——— is going to pinch your lunch case!"

Not having any children of his own Gordon especially loved to entertain the younger ones, and he took a special interest in the welfare of his students. He had a film projector which he took to birthday parties at the homes of his

W. T. Gordon of King's College, London.
Photo received from E. O. Rowland.

colleagues and showed the children films of Charlie Chaplin and Donald Duck. As for the university students, Rowland recalls that "one particular student had for some days been suffering from a cold. Professor Gordon called him into his office, produced a large bottle of white emulsion and proceeded to give this student a very large table-spoonful. He then instructed him that he was to report three times a day; and religiously, three times a day, for a couple of weeks, either Professor Gordon or I dosed this student up."

One final excerpt from Mr. Rowland's tape brings Marie Stopes back into the story:

> He [Professor Gordon] was always raising funds for the University and for various charities and he put on a science week. The science week was open to members of the public and he managed to persuade some very eminent speakers to give a series of lectures, including Dr. Marie Stopes. Now Dr. Marie Stopes, in England, as you know, was well known for her books on sex. And so we had a very large lecture theatre packed with the general public who were treated to an hour's talk on . . . coal!

Either Professor Gordon inveigled Marie Stopes into this little bit of duplicity under false pretenses—or she did have a sense of humor.

Although it brings us somewhat beyond "the age of Seward" it will lend significant continuity to my story to continue here with a look at the great series of discoveries that has been made in the Scottish Lower Carboniferous since Gordon's time by Albert Long. Long spent two years (1937–1939) in Lang's laboratory in Manchester, and then for financial reasons he found it necessary to seek some kind of employment. It was not easy to do this as a paleobotanist in the 1940's, and he had been rejected for military service because of a foot injury he sustained some years before in a shooting accident. He says of this period: "I applied for various posts which I felt would have given me the chance to realize my hopes but all to no avail. In my disillusionment I turned to my first love—entomology, and built up gradually an apiary of about 30 stocks of bees while collecting lepidoptera, trichoptera, and other insect orders" (1976, p. 179).

He did find employment, however, in two different boys' schools and later a position as Assistant Science Master at the Berwickshire High School in Duns, where he started in January of 1945.

His initial paleobotanical work was with Upper Carboniferous coal ball plants, and for several years after 1939 he explored the old Lancashire localities, many of them mines that had long ceased to operate; I rather suppose that he has a greater knowledge of such places in Britain than any one else alive today. Long's most important contribution in this period was his 1944 paper on the prothallus of the seed *Lagenostoma ovoides*—a remarkable example of fine preservation which added a significant link to the *Lyginopteris* story.

I started to correspond with Albert Long in 1959 when I first heard of his work with Lower Carboniferous plants in Berwickshire and, like a number of other colleagues, had the pleasure of meeting him at the time of the Botanical Congress in Edinburgh in 1964. He was then living in Gavington near Duns in Berwickshire. In his reminiscences he says: "My first attempt at collecting fossil plants in Berwickshire was made in 1945. I cycled to the coast near Cockburnspath and dug compressions from shale near Cove Harbour" (1976, p. 180). However, he credits Peter Barnard with getting him started on his great work with Lower Carboniferous plants:

> It was soon after the Suez crisis of 1956 and while preparing to resume my moth collecting after the first petrol shortage that I received a letter dated 17th February 1957 which changed the tenor of my life in a most unexpected way. This letter was from P. D. W. Barnard of Birkbeck College, London, and at that time a complete stranger to me. His letter acted like a catalyst and spurred me to undertake a search of the Langton Burn near Gavinton. . . . As a result of this letter I first searched the Langton Burn upstream to Langton Glen but without success. Then I decided to search downstream from the road bridge near the Red Brae below Langton Church. When I reached Hanna's Bridge I remembered and rediscovered the blocks first seen in 1951 when fishing. As soon as I got the first specimens of *Calymmatotheca kidstoni* Calder the "penny dropped" and immediately I realised that here was something of great interest and significance. [P. 181]

When several of us visited Albert Long in the summer of 1964 his "laboratory" was a small shed at the back of his house. His larger specimens had to be taken into Edinburgh to be cut, but it was here that much of his early work with Lower Carboniferous seeds was accomplished. The great series of studies of these Lower Carboniferous plants for which he is now so justly famous began in 1960 with his description of *Genomosperma kidstoni* (Calder) with its integument of eight free lobes, open and with no micropyle as such. There followed descriptions of other seeds, some of which were found partially enclosed in rudimentary cupules:

> Of the eleven species of Lower Carboniferous seeds now described from Berwickshire none shows a true narrow micropyle. The nearest approach is *Stamnostoma bifrons* which still retains a salpinx. In all the others the pollen was probably received directly by the salpinx (with the possible

Albert G. Long (left) with William S. Lacey and Arthur A. Cridland at Gavinton, 1960. Long's "laboratory" where his early studies of the Lower Carboniferous fossils were conducted is directly behind the trio.

> exception of *Genomosperma latens*). From this it can be inferred that the development of a salpinx and its variation in size and form was an evolutionary phase preceding the establishment of the integumental micropyle. [1961, p. 416]

In several of his more recent studies Long has started to put together some of the scattered parts of these plants into whole forest trees. These studies to a considerable degree center around the great *Pitys* trees, which have had a long and chequered history. Witham described one of the localities where he found the petrified trunks and in his historic work of 1833 he described two species (his original spelling was *Pitus*): *Pitys antiqua*, the Lennel Braes tree from the banks of the River Tweed in Berwickshire, and *Pitys primaeva*, the Craigleith tree found in a quarry near Edinburgh. Lindley and Hutton as well as Hugh Miller were concerned with the *Pitys* trees, and Seward devoted several pages to them in volume 3 of his *Fossil Plants*.

Thanks to several of Long's contributions in 1962 and 1963, *Pitys* began to emerge as a genus of large forest trees bearing a primtive type of "fernlike" foliage of the *Diplopteridium*, *Sphenopteris*, or *Sphenopteridium* type. The fertile fronds or branches were those of *Tristichia ovensi*, which have a triradiate protostele and some secondary wood; these bore microsporangia of the *Telangium* type and have been found associated with the seeds and cupules of *Stamnostoma huttonense*. Thus it would seem at this point that they must be classified in the Pteridospermae and very possibly are a line that evolved from the late Devonian progymnosperms.

In 1966 Albert Long gave up his teaching post and took a position as Deputy Curator at the Hancock Museum in Newcastle-upon-Tyne, with much better facilities to continue with his Lower Carboniferous studies. My wife and I had the pleasure of spending a weekend with Albert and his hospitable wife in late January of 1976. He took me out one day in the direction of Carlisle, following Hadrian's wall much of the way, to a cold and windy moorland north of Gilsland (Cumberland), where we hiked for a couple of miles to a partially frozen brook where he had located several stumps and trunk fragments of *Pitys*; the largest specimen, probably the basal part of one of the trees, measured about seven feet in diameter; they were indeed large trees.

I think that Long is one of the keenest field observers that paleobotany has ever had and he is cautious but astute in putting isolated parts together. In his earlier studies of the Berwickshire fossils he kept strictly to the facts. As he wrote in a letter of February 15, 1964: "Personally I feel theorizing is better kept in correspondence and discussion lest it becomes accepted as fact." Perhaps this stemmed in part from an admonition from Professor Lang, who once told him: "People are not interested so much in what you believe as in what you have evidence for." It is good that Long has departed somewhat from that policy, for his general theoretical considerations concerning the evolution of ovules and carpels (1966) is a valuable contribution.

Long also has a strong interest in the history of the area in which he has worked and has written very useful summaries of the paleobotany of Berwickshire over the last two centuries. He is a man who is completely dedicated to his scientific work and one with a keen sense of what is significant;

yet he is generous with his time and I have enjoyed his friendship and his informal and spirited correspondence for nearly two decades.

It is necessary and appropriate to refer rather frequently to the great collections and services that have been made available to so many of us at the British Museum of Natural History. For close to a half century the fossil-plant collections there were administered by Wilfred Norman Edwards (1890–1956) and Maurice Wonnacott. Edwards was very helpful to me on many occasions, and I am indebted to Maurice Wonnacott for a good deal of my early education in paleobotany and for a great many favors during the forty years that we have known each other.

Edwards was the first paleobotanist as such to be appointed to the Museum staff; he joined the Department of Geology in 1913, shortly after he had graduated with honors from Christ's College, Cambridge. However, the war interrupted his start and from 1914 to 1918 he served with the Royal Army Medical Corps in the Balkans as an assistant to Army surgeons. He returned to the Museum in 1919, and at about the same time Wonnacott, as a very young man, also joined the staff of the newly established Paleobotanical Section, under Edwards's direction.

Edwards became Deputy Keeper of the Department in 1931 and Keeper in 1938. He had a tremendous knowledge of fossil plants and of paleobotanical literature, and I believe it is appropriate to record that his greatest contribution, a genuinely enormous one, was in administering the collections and making them available to the many visitors who came from all parts of the world. He did, however, contribute directly to several areas of paleobotany, with articles on Tertiary plants from Southeast Burma, the cuticle of *Psilophyton*, a *Laccopteris* from North Africa, and a study of *Glossopteris* from South Victoria Land, as well as more comprehensive studies of cuticular characters in recent and fossil angiosperms, and the geographical distribution of fossil floras. Probably his best-known publication is a guide that he prepared in 1931 for a temporary exhibit illustrating the early history of paleontology. It is a most useful and informative booklet and still something of a best-seller in the Museum Bookshop.

Edwards was an enthusiastic and unconventional traveler. Among his various journeys was one that he made from Cape Town to Cairo, walking a good deal of the way and including an ascent of Kilimanjaro. He carried with him a good supply of string, making "string-figures" with his fingers which apparently served as a passport to ensure his safe-conduct in out-of-the-way places.

The closest thing to a formal course in paleobotany that I have received has been guidance over many years from Maurice Wonnacott of the British

Maurice Wonnacott, on the Yorkshire coast, 1950.

Museum of Natural History. Probably no one has worked for so long a time (nearly fifty years) in what might be called the main stream of paleobotanical traffic leading to the Museum. I first met Wonnacott in 1937, when I was a student at Cambridge and spending my vacation periods at the Museum. He is a great field man and a splendid companion. He took me on a trip to Yorkshire—we repeated the experience some years later—where we lodged at Gristhorpe Hall, the old manor house of Gristhorpe Village, a few miles south of Scarborough. We would walk a couple of miles each morning to the beach below the Red Cliff rocks and, when the tide was out, dig in the rich plant deposits under the beach rubble and seaweeds, and each afternoon struggle back up the cliff path with our packs filled with fine specimens of the mid-Jurassic flora. Just as Hamshaw Thomas had "rediscovered" the "worked out" fossil beds of the Yorkshire coast in the early years of the century, so Wonnacott opened them up again in the mid-thirties and assisted T. M. Harris in his investigations, which have yielded several scores of publications, and I think it is safe to say that the end is not in sight.

Wonnacott's collecting activities have been numerous; in his younger days he made what was probably the last collection (owing to the erection of sea walls, terracing, and other "improvements") of well-preserved leaves from the Eocene beds at Bournemouth, and he has made several extended trips to Spain gathering collections of plants, fishes, crustaceans, and insects from the Kimmeridgian beds in the foothills of the Pyrenees.

Wonnacott was very helpful to me when I was preparing the *Index of Generic Names of Fossil Plants* (Andrews, 1970), and he has played an important role in paleobotany in an editorial capacity, seeing through the press such renowned works as Reid and Chandler's *London Clay Flora* and the several volumes of Harris's *The Yorkshire Jurassic Flora*. His own publications include three volumes of the *Fossilium catalogus* series, as well as assisting John Walton with several of the Annual Reports on British Palaeobotany.

I would like to insert a few lines here about a man, Henry Smith Holden (1887–1963), whose paleobotanical publications were few but significant and of high quality; he was a fine teacher and his other contributions to humanity were unique. I met him several times in the back study room of the British Museum where we both worked; he was much my senior in both years and knowledge; he was friendly and helpful, and I was always glad to see him come in.

One of the most intriguing and now best known of Carboniferous plants is the coenopterid genus *Botryopteris*, and Holden produced a fine paper on the morphology of *B. antiqua* Kidston in 1962. In 1930 and 1931 he had written two papers on *Ankyropteris corrugata*, and of the first of these D. H. Scott wrote to him: "I think it an admirable piece of work. You have dealt with all the questions which had occurred to me and have added new results of importance. I am impressed by the beauty of the photographs. As illustrations of the perfection of preservation they are marvellous. . . . It is a great pleasure to see it so excellently done" (Wonnacott, 1964, pp. 230–231). He produced several other papers on the coenopterids as well as two studies of *Scolecopteris* with D. H. Scott and collaborated with A. Bentley in a textbook on pharmacy. The latter leads us to another phase of his life work. He was not primarily a paleobotanist and is best remembered, I should think, for his teaching in botany and for his police work.

Holden was born in Castleton, Lancashire, in 1887 and educated at Manchester, where he graduated with honors in botany. He was then appointed assistant lecturer at Nottingham and eventually became Professor and head of the Biology Department. He served as a hospital bacteriologist during World War I and gradually became more and more involved with the identification of biological materials in police investigations. In 1936 he was appointed Director of the East Midland Forensic Science Laboratory in Nottingham and was later invited by the Home Office to take charge of the more important laboratory that was planned for the Metropolitan Police at Hendon and later transferred to New Scotland Yard.

T. M. Harris was an undergraduate at Nottingham at the time Holden was a lecturer there; he gives Holden much credit for his early training and says that he "saved him" from a medical career which he had contemplated up to the age of twenty.

In all of his investigations for the police Holden was as scrupulously careful in the acquisition and use of information as he was with his paleobotanical research, and he always insisted that both sides of any court case should have any information that he had been able to obtain in his own investigations. Some of the cases that he was concerned with are recounted in an article in the *Medico-Legal Journal* for 1952.

Henry Holden was a good plant anatomist and paleobotanist, and, as Tom Harris has said of him; "He was a man of fiercely independent views and hated affectation and especially scientific conceit. He expressed himself freely and plainly without any fear of a row. . . . He had a real kindness for any in trouble (including those accused of crime) and nearly killed himself trying to eliminate all risk of miscarriage of justice through a slick inference from his evidence" (1963, p. 175).

In my efforts to report meaningful sequences in this story of the fossil botanists and their work I have found numerous paths along which our science has developed. Such paths may be quite distinct for a time but they inevitably become intertwined. Some contributors can be dealt with readily as individuals, while others, whose studies were more diverse, must be brought in more frequently. A few—a very few—have produced great works and small ones and have been influential as teachers and researchers over a long period; their technical contributions and their philosophy are felt in many places by many people and of this select group I think Tom Harris is the most respected and loved among us today. He has introduced many of the young into the world of plant life and he has guided and influenced many of us who are close to being his contemporaries.

One of my most memorable experiences was a visit that Mrs. Andrews and

I made in March of 1971 to Chusan, the Harris home, located a few miles outside of Reading. When he retired from his professorship Tom Harris and his wife acquired a one-acre tract adjoining a large country estate; the main building had been made into several apartments, and the Harrises had removed the stable and built in its place a house in keeping with the main building and the surrounding countryside. The land itself is surrounded by

Tom M. Harris, at Rhynie, Scotland, 1964.

fields and woods that add much to the setting. But in one acre I have never seen so many different plants arranged so delightfully and informally; it is dominated by a great *Sequoia*, and although it was a cool and foggy March day many of the plants were coming into flower: mostly ornamentals—herbs, shrubs, and trees—but with edible plants tucked in here and there, and the vegetable area proper occupied much of the lower slope where a few palms argued for the mild south English climate! We walked along the paths that Tom Harris had roughed out, listening to him tell the most intimate gossip about all the plants along the way.

Tom Harris was born in 1903 in Leicester, received his early education at Leicester and at Bootham School in York, and went on to University College, Nottingham. He intended to be a medical student but an interest in plants that had been growing for some time was greatly accelerated by Henry S. Holden, as I have noted above. He next won a scholarship to Christ's College, Cambridge, and there met Seward, thus entering into the "age" we are recounting. He left Cambridge in 1935 to become the Professor of Botany at Reading, remaining there until his retirement in 1968. He was elected to the Royal Society in 1948, served as its Vice-President in 1961, and was President of the Linnean Society from 1961 to 1964. He also spent a short period at the Sahni Institute in Lucknow and a year at the University of Accra, in Ghana, where he organized a teaching program suited to the local needs.

At the time of our visit Tom Harris met us at the railroad station in his somewhat battered Morris and we drove to the University faculty club for lunch. This was in the same room where he had been interviewed years before for the professorship. There were two other candidates, both of whom were biochemists, and Harris gave much credit to his Greenland experiences (see below) for obtaining the post; they were apparently more interesting and comprehensible to the interviewers than the credentials of the other two! He also told us of the "mistake" that started him on his great study of the East Greenland flora, and this brings us back to Cambridge.

Seward, because of his interest in the Tertiary-Cretaceous plants of West Greenland, had arranged to have some collections from that area sent to him from Copenhagen. When the box arrived it was immediately evident that the specimens were not from the Disko Island region, as expected. They were part of a collection made on Hartz's expedition to East Greenland in 1900, probably having been gathered by some sailors who were paid a little extra for their efforts. Harris worked up the collection and the results were published in 1926. Shortly thereafter Lauge Koch, from the Danish Geological Survey, visited Seward and in turn became acquainted with Harris. As a result Koch invited Harris to accompany him and Dr. A. Rosenkrantz on his next trip north. They left England in August, 1926, for a stay of one year along the northeast Greenland coast. They landed at Scoresby Sound without trouble although this is a notoriously difficult coast, foggy and with pack ice tending to pile up along the shore. Harris and Rosenkrantz worked in the field until October, traveling along the coast in a motor boat as long as the sea was open, most of the fossil outcrops being only a few hundred yards inland. From November through February they were obliged to work inside, returning to field work in March. Harris informed me of a problem that I believe others have encountered in the Arctic, that is, the impossibility of splitting frozen rock. On south-facing slopes thawing would take place to a depth of about a meter, and half that much on a north slope. So they would work the available rock and then return in a month or so when further thawing and made more rock available.

The result of this trip was *The Fossil Flora of Scoresby Sound, East Greenland*, which came out in five parts from 1931 to 1937. Two features contribute especially to the greatness of this study: Harris's skillful use of the maceration method and his numerous drawings—which are a distinctive and informative feature of almost all of his books and papers.

The Scoresby Sound flora comprises several fern families including the Osmundaceae, Matoniaceae, and Marattiaceae; curious leaves of *Neocalamites carcinoides*; a possible *Glossopteris*; many cycads and Bennettitales, with an abundance of *Nilssonia incisoserrata* and the *Beania* seeds; *Taeniopteris*, and numerous ginkgophytes and conifers, as well as some problematical genera; there is also a very informative section on cycadophyte stomatal structure. From the standpoint of detailed field exploration, the magnitude and diversity of the flora, and the care that went into the study and description this is probably the greatest single work that we have on the fossil floras of the Arctic.

Harris more or less picked up where Thomas had left off, devoting many years to the exploration of the Jurassic rocks of Yorkshire. And from the beds that had long since been "worked out" he produced a series of short papers starting in 1942 that appeared in the *Annals and Magazine of Natural History*. Among his more comprehensive studies is a fine account of the *Beania-Androstrobus-Nilssonia* assemblage, bringing together these organs and demonstrating their probable unity, thus revealing a mid-Jurassic plant with many points of resemblance to modern cycads such as *Zamia*.

Starting in 1961 he began to bring together a vast amount of knowledge in his monumental *Yorkshire Jurassic Flora*, a five-volume work of which four volumes have reached me at this time. The first volume deals with the bryophytes and pteridophytes and is a fine source of general information on the ferns, especially the Marattiaceae, Osmundaceae, Matoniaceae, Schizaeaceae, and Dicksoniaceae. Volume 2, dealing with the Caytoniales, Cycadales, and pteridosperms, appeared in 1964; the summary account of the Caytoniales is especially useful. Volume 3, on the Bennettitales, appeared in 1969; this is a large and beautifully illustrated account of the foliage and to some extent the reproductive organs of that group. Volume 4 deals with the Ginkgoales (by Harris and W. Millington) and the Czekanowskiales (by Harris and J. Miller). It is my understanding that the fifth and last volume will present the Coniferales and thus complete this great work which will certainly serve as the standard reference on the plants of the Jurassic for a long time. It is appropriate to add that the work of many people finds its way into the sources of this great account, and we are especially indebted to Maurice Wonnacott for editing and guiding it through the press.

In his introduction to the first volume Harris, unlike most of his predecessors, points to the future rather than the "worked out" past. This is an exemplary admonition, well worth recording: "vigorous collecting should begin again generally, and more powerful methods might be used. There must be rich beds in the upper parts of the great cliff section south of Whitby; . . . I am sure that spectacular advance will be made by collectors with fresh ideas, and I am convinced that it is the collector rather than the locality which is exhausted" (1961, p. 3).

The most piquant tribute to the great Yorkshire deposits that I have encountered in the literature was written by Peter Murray as long aso as 1829:

> The interesting deposit at Gristhorpe Bay may be considered as a vast herbarium, of which the leaves opening to the readiest observation, offer every facility and pleasure in the examination; and not, as is the case with the generality of coal plants, surrounded with dirt, and darkness, and perils, imbedded in the roofs and sides of mines; and they resemble so many fine drawings in Indian ink, or the shadows of delicate foliage by moonlight cast upon a smooth and white ground or wall. [P. 312]

7

Francis Oliver and the Pteridosperms

"It was rather late, and there was no steady light for the microscope, only an electric bulb on a flex from the ceiling swinging to and fro, and a microscope that limped and jumped about. But I was morally certain I'd got the thing, only I couldn't return next day, as I was due at the British Association at Belfast. . . . " Thus did Francis Oliver, in a letter to Seward, describe his discovery of the identity of the glands on the seeds *Lagenostoma* and the stem *Lyginopteris*, a discovery which crystallized the concept of the pteridosperms or seed ferns. Oliver had been going through the slides of the Williamson collection, which are housed in a first-floor room of the British Museum, looking out into the rather gloomy back-yard area. A good many paleobotanists have visited and studied in that room with the same poor lighting and probably the same crippled microscope that Oliver used. The equipment may have been something short of ideal, but the slide collections, especially those of Williamson and Scott, held countless treasures of the Coal Age forests, and the pleasure of working there was always enhanced by the warm cordiality of the keepers, W. N. Edwards and Maurice Wonnacott. It seems fitting that so great a discovery should have been made in a laboratory so humble and comfortable.

The identification of the gymnosperm group Pteridospermae is probably the greatest contribution that fossil botany has made to our knowledge of evolution in the plant kingdom. It is a long story to which many people have contributed; Francis Oliver was astute enough to recognize a particularly vital link in a long chain of observations and discoveries. It is one of the finest examples of how an important morphological concept evolved over a period of many years, and I will depart somewhat from my usual format to relate some highlights of the story.

As we have seen, a considerable part of the paleobotany of the early nineteenth century was concerned with Upper Carboniferous floras. The "fern" foliage was a conspicuous element and was largely responsible for the tropical climate that many people postulated. It was not long, however, before the more astute workers began to wonder about the "fern" leaves, there being two points that especially perplexed them: much of the fernlike foliage lacked spore-bearing structures typical of the ferns, and in many places seed compressions were found in abundance associated with the leaves. There were other explanations. Oliver (1906) cites a rather curious one that was presented at the meeting of the British Association at Birmingham in 1849: "R. Austin, Esq., exhibited a specimen of a Fern from the English Coal-Measures bearing abundance of fructifications. He did this as the rarity of the occurrence had led him to suggest, at a previous sitting of the section, that in these latitudes the ferns of the coalbeds did not fructify on account of the low temperatures in which they existed" (p. 235). In 1883, Dionys Stur, in noting the absence of fructifications on certain types of fernlike foliage, expressed his opinion that they should not be classified with the ferns, and consequently he excluded *Neuropteris*, *Alethopteris*, *Odontopteris*, and others from consideration in his memoir.

Suspicions about the nature of the fernlike foliage were accompanied by a considerable amount of evidence from petrified plants which suggested the presence of a unique and previously unknown major group of plants. Scott, in 1909, stated that Williamson appeared to have been the first to recognize the presence in the Carboniferous flora of plants combining fern and cycad characteristics. He referred particularly to Part 13 of Williamson's *Organization* papers in which *Heterangium* and *Kaloxylon* (the root of *Lyginopteris*) are described. Of the affinities of these plants Williamson says: "Possibly they are the generalised ancestors of both Ferns and Cycads, which transmitted their external contours to the former and their exogenous modes of growth [in reference to the presence of a cambium] to the latter types. In considering this possibility, we must not forget that in *Stangeria* we have a still living plant in which the stem of a Cycad bears fronds, the leaflets of which retain the dichotomous nervation of a true Fern" (1887, p. 299). Bernard Renault carried this aspect of the matter a step further in demonstrating that the fronds of both *Alethopteris* and *Neuropteris* belonged to the petioles that were called *Myeloxylon*, and the latter were shown by Weber and Sterzel to be the petioles of the medullosan stems.

Some of this information was formalized by Henry Potonié, who established the class Cycadofilices in his textbook in 1898; the group included fossils such as *Lyginopteris*, *Heterangium*, and *Medullosa*. René Zeiller in his text *Éléments de paléobotanique* of 1900 says that the Cycadofilices are intermediate between ferns and cycads and in some ways more like the latter;

Francis W. Oliver. Portrait by F. A. de Biden Footner, from *Biographical Memoirs, Fellows of the Royal Society*, 1952.

and he notes that they were without any close relationships among living plants. He included in the new group *Cladoxylon*, *Heterangium*, *Lyginopteris*, *Calamopitys*, and *Medullosa*.

In 1903–1904 a "precipitation" of contributions brought the pteridosperms—seed-bearing plants with fernlike foliage borne on stems of gymnospermous anatomy—clearly into focus. It seems almost as though a contest had been declared to decide who would be the first to settle the "affair of the fernlike foliage," and I will comment on what seem to be the more important results.

The French paleobotanist Cyrille Grand'Eury, a great collector and observer, had been concerned with the "fern" foliage for a long time. In reference to his work Scott says:

> The argument from association, though of small weight by itself, has considerable value, when critically employed, as supporting more direct lines of evidence. We own to M. Grand'Eury, the veteran palobotanist of St. Étienne, an extensive investigation of the distribution of Neuropteridean fronds and associated seeds in the upper and middle coal-measures of France (Grand'Eury, '04). He is not only led to the conclusion that the Neuropterideae generally bore seeds, and were, to use his words, primitive Cycadinae with the fronds of ferns, but is able, as he believes, to assign definite types of seeds to particular species of the genera *Neuropteris*, *Alethopteris*, *Odontopteris*, *Linopteris* and others, on the evidence of constant and exclusive association. [1905b, p. 15]

To add a little to Scott's summary, Grand'Eury's 1904 report tells of finding *Pachytesta gigantea* associated with *Alethopteris grandini* at St. Étienne, of the close association of a *Pachytesta* and *Alethopteris* in the valley of the Gier, and of *A. serlii* associated with a small *Pachytesta* seed at Bully-Grenay. One other report in these early years of the pteridosperm concept that may be mentioned is Grand'Eury's 1905 discovery of numerous small seeds attached to *Pecopteris pluckeneti* foliage.

Actually, as far back as 1890 Grand'Eury was apparently convinced of the seed-bearing nature of much of the "fern" foliage. The following comment is taken from his *Géologie et paléontologie du Bassin Houiller de Gard* of that date: "Along with Mr. de Solms, concerned as he was with finding the vegetative organs to which are attached the seeds which are so numerous and varied in the Upper Carboniferous terrain, we think that certain cycad types have borne seeds with pteridophyte foliage, such as *Stangeria*, which was at first compared to *Lomaria*" (my translation).

In a paper read before the Royal Society on December 3, 1903, Robert Kidston described several specimens that were obtained from the "10 feet

Ironstone Measures" of Cosely, near Dudley, England, and sent to him by Mr. W. H. Hughes; these were seed compressions referable to *Rhabdocarpus*, attached to stem fragments which also bore the foliage *Neuropteris heterophylla*. Kidston says: "These seeds, if separated from the foliage, would find a place in the genus *Rhabdocarpus*, and differ little from *Rhabdocarpus tunicatus*, as figured by Renault, or *Rhabdocarpus subtunicatus*, as figured by Grand'Eury, except in being narrower in proportion to the length" (1904, p. 4).

In this country, on December 10, 1904, David White reported specimens obtained from a railway cut in West Virginia which consisted of seeds borne terminally on fragments of fernlike foliage. The foliage was referable to the genus *Aneimites* and the seeds were of a type previously known under the generic name *Wardia* (in honor of Lester Ward). White gave these specimens the new binomial *Aneimites fertilis*. He says:

> On examining the specimens in which the peculiar little rhombic winged fruits (*Wardia*) usually found at other localities in association with the Pottsville *Aneimites* were not only still attached to their stalks, but in actual union with fragments of pinnae unmistakably bearing the small reduced pinnules of the species described above. . . . The fruits of *Aneimites* can hardly be regarded as other than true seeds, and the group of hitherto supposed ferns to which they belong is therefore to be referred to the Pteridospermae of Oliver and Scott, the "Cycadofilices of Potonie." [1904, pp. 326, 329]

The collector and slidemaker W. Hemingway had a few observations and thoughts to contribute to the story. On April 20, 1903, he wrote to Scott: "I am pleased to receive your letter. Yes, the *point* in my letter is the association of *Lagenostoma ovoides* with the Halifax & Bullhouse form of *Lyginodendron*. I have found them together in a number of instances at these places" (Scott letters). And on July 4, 1904:

> You will remember some remarks I made to you some time ago as to the probable fruit of N. heterophylla, in which I referred a seed somewhat resembling Trigonocarpus Parkinsoni to this plant. This seed has a different basal scar to T. Parkinsoni & splits into 3 segments—these separated segments have been recorded by Kidston as Polypterocarpus N. sp. Now I believe I shall soon be able to produce evidence to show that these & the seed actually found attached to N. heterophylla are one and the same, & if so it will make me to have been the first to indicate the true seed of this plant. [Scott letters]

Francis Oliver, perhaps the real hero of the pteridosperm story, was born on May 10, 1864, the son of Daniel Oliver, Keeper of the Herbarium at Kew; thus botany was a family tradition. Francis was something of a rebel as far as the "establishment" was concerned. A biographer says of his younger days at Bootham School in York: "Oliver recalls how one year he made a collection of 'foreign bodies,' consisting mainly of insects, from the dishes served in the school refectory, which was shown at one of the yearly exhibitions, and was an early indication of Oliver's tendency to have a 'tongue in his cheek' in relation to constituted authority" (Salisbury, 1952, p. 231).

After Bootham he entered University College, London, where he was influenced by the teaching of his father and by D. H. Scott. He later went to Trinity College, Cambridge, and also spent two long vacations studying under Edward Strasburger in Bonn, where he became acquainted with A. F. W. Schimper and August von Schenk, both plant geographers who made significant contributions to paleobotany.

Oliver's studies in paleobotany were largely concerned with fossil seeds; his papers are not especially numerous but are of good quality, and several were done in collaboration with D. H. Scott. The latter two decades or so of his career were devoted to ecological studies of living vegetation, conducted in large part on the Brittany coast and in Norfolk, where he developed a field station for the Botany Department of University College. Salisbury gives him credit for initiating dynamic physiological studies. I recall that in my first year of graduate study at St. Louis, Edgar Anderson assigned me, as a seminar topic, to report on the work Oliver had done with *Spartina townsendii*, a natural hybrid that Oliver recognized for its possible use in land reclamation.

The interest that led to Oliver's pteridosperm discovery had been aroused by Scott's 1896 course of lectures at University College, and he took time to visit and obtain seeds from some of the famous French localities and to study the Williamson slide collection. The Williamson collection was at that time in the custody of the Geology Department of the British Museum; although it was available to everybody, Oliver says it was "morally Scott's" as he had lent a sum of money to the Museum for its purchase to prevent its being moved away from London.

When Oliver observed the distinctive glands of identical structure on the *Lagenostoma* seed cupules and *Lyginodendron* stems and petioles, he took the matter to Scott. At that time Oliver did not feel that he would be able to convince the world, single-handed, as to the validity of his discovery. Scott rejected the offer at first, feeling that Oliver should receive all of the credit, but later agreed to a joint report.

In their initial communication to the Royal Society in May, 1903, Oliver and Scott stated that

> a re-examination of the Palaeozoic seeds, placed by Williamson in his genus *Lagenostoma*, had revealed unexpected points of agreement between the structure of the envelopes of certain of these seeds on the one hand and that of the vegetative organs of *Lyginodendron* on the other, and the conclusion was drawn that the seed *Lagenostoma Lomaxi* could have belonged to no other plant than *Lyginodendron* [which in turn bore the fern-like *Sphenopteris* foliage]. [Oliver and Scott, 1904, p. 194]

The full report appeared in the *Philosophical Transactions* in 1904, having been read on January 21 of that year. They did not actually find the seeds attached, but the evidence seemed conclusive. In their own words: "The results of the present investigation have placed the whole question in a new light. If our evidence be accepted—and short of the proof of the continuity it could scarcely be stronger—it follows that *Lyginodendron*, so far as the female fructification is concerned, had definitely crossed the boundary between Cryptogams and Spermatophytes; as regards its seeds it was as true a Gymnosperm as any known Palaeozoic plant" (Oliver and Scott, 1904, p. 237). Two pages later they established the pteridosperm group: "It appears to us that the presence in the Palaeozoic flora of these primitive, Fern-like Spermatophytes, so important as a phase in the history of evolution, may best be recognised by the foundation of a distinct class which may suitably be named *Pteridospermeae*" (1904, p. 239). It is appropriate to mention that they acknowledged the assistance of Marie Stopes in searching through the Williamson and other collections.

One of the more significant and interesting figures in the Scott and Oliver report of 1904 is figure 34 on plate 10, which shows in logitudinal section a young seed within its cupule, which in turn bears the characteristic glands. In a letter to me of January 2, 1946, when he was living in Egypt, Oliver was kind enough to offer the following comments concerning this and other aspects of his discovery that I believe are worth recording:

> This important preparation I found under the following circumstances. Lomax (pere) cut and took trial sections from an enormous number of coal balls. Specimens he rejected as of no commercial value he just threw into a crate standing handy in his workshop. Knowing of this I had Lomax send me from time to time boxes filled with these "Rubbish" sections—thousands of them. About one in a thousand contained a good thing. . . . the ovule, fig. 34 in Oliver and Scott was found in just such a rubbish

section. It was the best preparation in a bunch of 1000 and I was just about to run off to the botany lab tea. Having looked it over generally under a low magnification and finding nothing of interest I was about to throw it away, when my eye noticed something at the edge which had been missed under the microscope. Reexamining I found this beautiful object—a real treasure. . . .

What really must have led me into fossil botany was its sporting aspect. I was early aware that certain horizons contained quantities of fossil seeds which as a rule bore no evidence of their origin. Also that many "Filicineae" were defective in the matter of their reproductive organs. Was it possible these so-called "Cycado-filices" achieved the seed-habit?

Sensational discoveries may start a stampede, and I think that there was a tendency shortly after 1904 then to regard any "fernlike" fossil as probably referable to the pteridosperms. It is well to be cautious at such a time. Seward had this admonition to offer in his introduction to the ferns in the second volume of his *Fossil Plants*: "Like the earlier writers who described fossils as *lusus naturae* fashioned by devilish agency to deceive too credulous man, the discovery of seed-bearing plants with the foliage of ferns threatened to disturb the mental balance of palaeobotanists" (p. 282).

To the best of my knowledge the finest and most complete article that has been written concerning the entire history of the "pteridosperm concept" is René Zeiller's "Une Nouvelle Classe de gymnosperms: les ptéridospermées" (1905). It begins with the early reports of presumed "ferns" by men such as Lhwyd and Scheuchzer and traces the progress of our knowledge as it unfolded up to the critical years of 1904–1905. I suggest it for those who want a more detailed account than I am able to give here.

I would like to close this chapter with two comments: one by Scott concerning the status of these seed-bearing "ferns" and one by Kidston concerning their origins.

> The name Cycadofilices designated a group, only known at the time by its vegetative characters, which hovered in the gap between Filicineae and Cycadophyta without showing any decided leanings to either side. The class name Pteridospermeae represents a more advanced stage in our knowledge, and indicates plants which we know to have been already definitely Spermophytic, though retaining many marks of a Filicinean origin. [Scott, 1908, p. 403]

> When the Pteridosperms first assumed a definite position in botanical science, I believe it was generally accepted that they had been derived from Ferns. More recent investigations have removed any ground for

holding this view. They may have sprung from a common ancestor in remote ages, but even this is a mere speculative assumption. [Kidston, 1923, p. 21]

8

Birbal Sahni and the Development of Paleobotany in India

OF all the men and women who were taught or influenced by Sir Albert Seward, none has left a greater impact on his own country or enjoyed greater respect and admiration from his colleagues than Birbal Sahni. Those who have known him will never forget that they knew a great man, one whose intelligence and humanity enabled him to cut across all of the usual social and intellectual obstacles. He himself contributed immensely to several facets of fossil botany, and he inspired and enabled many of his countrymen to carry on along the paths that he established in India.

Several events, both personal and paleobotanical, are responsible for this chapter's emphasis on India and on one man. It was my good fortune to have been able to spend a year with my family at Poona University as a Fulbright Lecturer in 1960–61. Although Professor Sahni had died nearly ten years before, it was a rather good time for a visiting paleobotanist to be in India. My primary task was to teach a course in paleobotany at Poona and help to develop that subject, which had been instituted by my good friend Professor T. S. Mahabale. I had a large class (as fossil-botany classes go) and several of the students later came to this country. It gave my children a chance to see at first hand a part of the world that is very different from their own, and I believe that my wife contributed much with her culinary talents as she has done in several other countries that we have lived in. Good food helps considerably in establishing good international relations.

In the winter I spent some weeks traveling in northern and western India, lecturing and making the acquaintance of some of the leaders in Indian botany. A few days were spent in Delhi and I was greatly saddened to learn of Professor P. Maheshwari's death not long after that; he was a fine scientist and I especially looked forward to the publication of his projected book on the gymnosperms. A few days were spent at Dehra Dun, where I lectured at the Oil and Gas Institute. My two sons, who were traveling on their own in western India, met me there and we had a mid-January weekend together at the hill station of Mussoori, above Dehra Dun. It was very cold, with only a microscopic fireplace in the one hotel that was open. Dehra Dun and Mussoori were favorite vacation spots in the days of the British Raj, but Mussoori seemed to be rapidly falling into disrepair. I next took the train to Aligarh and spent some time with Professor Chowdhury and his group and I remember the beautifully landscaped buildings of the University there. My next objective was Lucknow, where I had been invited to give the Seward Memorial Lecture. It was my general observation that Indian trains operate on exceptionally precise schedules but one can have a great variety of experiences on them. Several of the Aligarh staff members took me to the station to board the night train to Lucknow and they gave me a fine send-off. However, when the train arrived, a Delhi-Calcutta express, we found that a woman with several children was firmly established in my compartment, and preparing the evening meal in the bargain. She was in no way intimidated by the exhortations of the botanists and several train officials, and I was eventually assigned accommodations in another car. The final result was that the man who was sent to meet me in Lucknow did not find me since my name was posted on the door of the compartment that I did not occupy. My stay in Lucknow was most pleasant; I met many people there, including K. R. Surange, and renewed my acquaintance with Mrs. Sahni. Richard Kräusel was there at the time and we both were presented with diplomas of honorary membership in the Palaeobotanical Society. I next went to Allahabad where I spent some time with Professor D. D. Pant and several of his students, and I shall return to this on a later page.

I do not recommend India as a place for tourists who have only a week or two at their disposal. Visiting only the large cities such as Bombay, Delhi, and Calcutta is apt to result in a biased viewpoint of what is a large and colorful but complex country. With many months available—preferably a few years if one is aspiring to be an "expert" on the country—it can be an enlightening experience. One must spend some time in the villages, and there is of course a fabulous wealth of historical sites—stone carvings and structures of centuries long gone—that can only be appreciated by seeing them. For my own choice the most memorable experiences were those spent in the hill stations: towns such as Simla in the north and Darjeeling in the southeast, with their breathtaking views of the great snowy masses of the Himalayan mountains stretching for a thousand miles along the northeast border of India. We once hiked for two days out along the Indian-Nepalese border to a cluster of buildings called Sandakphu; at nearly twelve thousand feet Sandakphu is quite cold in November, but we were rewarded on a very frosty morning with a fine view of Mt. Everest some ninety miles away.

The first publications I am aware of that deal with Indian fossil plants appeared in the early decades of the nineteenth century. Brongniart includes *Glossopteris* in his *Prodrome* of 1828, and J. Forbes Royal described some fossils in his "Illustrations of the Botany and Other Branches of Natural History of the Himalayan Mountains and of the Flora of Cashmere." Much later Newell Arber reexamined the Royal plants in an article that appeared in the *Geological Magazine* in 1901, and Charles Bunbury wrote on fossils from Nagpur. Between 1876 and 1886 Ottokar Feistmantel produced a monumental series of studies on the Gondwana plants, but very little was done from that time until a new surge of life was brought to Indian paleobotany by Birbal Sahni, and it is his era that I am primarily concerned with here.

Birbal Sahni was born on November 14, 1891, at Bhera, a small town in the western Punjab. His father was a professor of chemistry at Lahore. Birbal obtained a B.Sc. degree at the University of the Punjab in 1911 and then went to England, where he entered Emmanuel College, Cambridge. He spent much of the next eight years there working as a student under Seward, and before he left they were collaborating in paleobotanical research projects. I will return to the "Seward era" of Sahni's life and its effect on the great work he engaged in when he returned to India, but I would like to insert a homely, but I think significant, experience that I hope will introduce him as the exceptional and delightful human being he was.

Professor and Mrs. Sahni visited St. Louis in 1947, when I was at Washington University, and spent several days with my students going over our collections, which were not very extensive at that time. He made a profound impression on my students, partly for the breadth of his knowledge and for the sincere interest that he displayed in what they were doing.

One evening Mrs. Andrews and I invited the Professor and Mrs. Sahni to dinner at our home. It was, as I recall, a rather cold December night, and it seemed to me that the Professor was slow about removing his overcoat. It is pertinent to note that the Sahnis had no children but they were especially thoughtful of others. As I was introducing the Professor to my own three he carefully opened his coat and there, nestled snugly under one arm, was a most lifelike little monkey. My children were enthralled, and I must say that my wife and I were astonished that he was able to carry the little creature about as he did. It soon proved to be a sort of puppet that he very cleverly manipulated with his hand. In his essay "Palaeobotany in Great Britain," John Walton refers to the monkey:

Birbal Sahni, in a train engine on the way to fossil plant localities in the Rajmahal Hills. Courtesy of the Birbal Sahni Institute of Palaeobotany.

The late Professor Birbal Sahni was one of Seward's most grateful and enthusiastic adherents. When he had to return to India after the years he spent in Cambridge, he was so apprehensive of the strain of parting that he bought a toy monkey, which fitted the hand so that its arms and head could be made to move, and took it to the station with him. He produced it just before the train left and its antics made the parting less emotional and more tolerable. Many years later, on his last visit to this country a year before his death, he had the monkey with him. It had become rather tattered and my wife performed some plastic surgery on it with a needle and thread. [1959, pp. 241–42]

Sahni remained in England from 1911 to 1919. Although the war years disrupted the work of nearly everyone in Cambridge, as elsewhere, Sahni's stay coincided in part with a vigorous period of paleobotanical progress. Hamshaw Thomas, Newell Arber, and Seward were at Cambridge; F. W. Oliver, D. H. Scott, and W. T. Gordon were in London. Thore Halle and Rudolf Florin in Sweden, Henry Potonié in Germany, and Paul Bertrand in France were all turning out their important works. Sahni became acquainted with most of these people—a galaxy of some of the greatest in the history of paleobotany.

Sahni's initial research was concerned with certain distinctive pteridophytes and gymnosperms. His first paper, which came out in 1915, was entitled "Foreign Pollen in the Ovules of Ginkgo and Its Significance in the Study of Fossil Plants." He then engaged in a study of the anatomy of the somewhat problematical and presumably primitive living vascular plant *Tmesipteris Veillardi*, which had been collected by R. H. Compton in New Caledonia, and Sahni demonstrated similarities in its anatomy to the Devonian plant *Asteroxylon*. In reference to this study Hamshaw Thomas says in his obituary: "Sahni concluded that the Psilotales approached even nearer to the Psilophytales than had hitherto been suspected. This conclusion has been generally adopted by subsequent writers" (p. 266). There is a continuing interest in these two groups, the fossil and the living, and it is my impression that full accord has not yet been reached concerning their interrelationships.

Sahni also studied the conifer *Acmopyle Pancheri* and the yew *Taxus baccata* (1920), in which his interest was centered on the structure of the seeds and seed-bearing parts. His detailed account of *Acmopyle* reveals Sahni as a well-informed and astute morphologist at an early age. He recognized the unique nature of the yew and one or two other related plants that had long been included in the Coniferales, the conifer order: "I have ventured to exclude from the phylum Coniferales the genera *Taxus*, *Cephalotaxus*, and *Torreya*. Both the general morphology and the internal structure of the fructifications lead me to the conclusion that the family Taxineae is too distinct to occupy a place among the Conifers, and that it should rank as an independent phylum, Taxales" (1920, pp. 287–88).

Botanists have long been interested in the classification of the yews because of their distinctive seeds, which are partially enclosed in a succulent scarlet cup, so very different from the hard brown cones that bear the seeds in other common conifers such as the pines, spruces, and hemlocks. But the foliage and wood structure of the yews is quite like that of the "true" conifers.

Sahni had the courage to formally divorce the yews from the conifers proper. I think it is interesting to note that two of the leading present-day morphologists (also two of my most long-standing botanical friends) accept Sahni's decision: Harold Bold in the third edition of his *Morphology of Plants* places the taxads in a distinct class, the Taxopsida; Kenneth R. Sporne regards this as going a bit too far, and in his book *The Morphology of Gymnosperms* he places them in their own order, the Taxales. The fossil record of the taxads is not very well documented, but Rudolf Florin concluded that they have long existed as a distinct group of gymnosperms.

In view of his early work with some of the more primitive living pteridophytes it is not surprising that Sahni looked to fossil plants in a somewhat similar category. Starting in about 1918 he initiated a series of investigations of some of the coenopterid ferns. Numerous other paleobotanists have been attracted to this problematical "group" of plants; we know a good deal about them but their classification is still far from being clearly established. I think it was courage and a special bent toward botanical detective work that contributed to Sahni's success in this direction. One example from his studies of continental collections will suffice: About 1880 a Swedish engineer picked up a silicified plant specimen near Pawlodar in western Siberia; the specimen ultimately found its way to Germany and was cut into several pieces. One of these reached K. G. Stenzel and was described under the name *Asterochlaena*; another piece was named *Rachiopteris ludwigii* Leuckart and Schenk by August von Schenk, the specific name being in recognition of a man named Ludwig of Darmstadt who obtained the fossil from its original collector. After Ludwig's death it passed into the hands of a Chemnitz druggist named Leuckart. In an article in the *Philosophical Transactions* in 1930, Sahni notes that, in addition to determining the identity of the two pieces cited, he found others from the original specimen in Dresden, Chemnitz, and Breslau. This division of a choice specimen into several pieces by professional mineral dealers or slidemakers was not uncommon at the time, but it is somewhat less common to have the parts recognized as such and brought back together again!

Shortly after Birbal Sahni returned to India in 1919 he initiated a series of ambitious and highly successful research programs with the numerous students and colleagues he attracted. These programs fall into four distinct categories: investigations of the unique petrifactions from the Tertiary deposits in the central part of the country; studies of Mesozoic plants from the Rajmahal Hills of western India; revisions of previous work and new studies on the late Paleozoic *Glossopteris* flora; and studies of Pleistocene floras and problems in the north. Microfossils were also included along with the macrofossils. Sahni wrote a detailed and very readable review of these programs in his

Recent Advances in Indian Palaeobotany (1938). I will touch on a few of the more interesting aspects of this surge of activity.

The Tertiary deposits present aspects of special interest. Sahni notes:

> An observant traveller landing at Bombay, who crosses the Western Ghats by the Great Indian Peninsular Railway, cannot fail to notice the low, flat-topped hills which dominate the landscape as far as Nagpur and beyond. These terraces represent the weathered surfaces of these lava flows or "traps" at different levels in a great basaltic series, aggregating several thousand feet in thickness. Here and there, between the traps, are preserved lenticles of freshwater sediments, occasionally mixed with beds of volcanic ash. These so-called Intertrappean deposits appear to have been mostly laid down in small temporary lakes which must have been formed at intervals by the damming up of streams by intermittent outburts of lava; and they contain a most valuable record of the floras and faunas of those times. [1938, pp. 58–59]

In mid-March of 1961 Professor T. S. Mahabale and several others took me on a field trip to see the Intertrappean beds of the Deccan. We stopped at Nagpur, in central India, for a day while I gave a lecture, and the next evening at about ten we took a narrow-gauge train which rattled along all night at a leisurely pace toward the north. About six the next morning we arrived at the villages of Markahandi and Udadan, which straddle the railway, just as the day's activities were getting underway. The people of the villages were getting up. Others were retiring: I was intrigued to see a short distance away a rather large tree with what appeared to be big black birds circling around it. They proved to be large bats, and in about a half hour they had settled down and looked like so many black footballs hanging from the branches. I had seen them previously at a village near Poona but not in such abundance.

After an hour or more of negotiations with a boy and his bullock cart, relative to payment and the route we proposed to take, we started off. Shortly thereafter bullock, cart, and driver, with our water and food supply, parted company with us in order to avoid some difficult terrain. We continued on foot and encountered the chert in place and in some abundance as fragments scattered through the fields. It is an interesting type of fossil material and it was my impression that with equipment for large-scale excavating much more fossil material might be found there. I am afraid that my enthusiasm was slightly dimmed by the heat, which was well over 100°, and our lack of water. We stopped at noon in the little village of Mohgaon Kalan at the house of an elderly lady who was a friend of Professor Mahabale, resting in the shade and enjoying a cool bath at the village well. The bullock cart with our lunch on board finally caught up with us.

The cherts were long thought to be of Cretaceous age, based at least in part on studies of the animal remains. The deposit had actually been known for nearly a century when Professor K. P. Rode of Banaras collected at Mohgaon Kalan in 1930 and turned his specimens over to Sahni, who in turn visited it in 1931. The cherts contain some interesting monocotyledons including palm wood, *Nipa* fruits, and two genera of conifer cones (*Indostrobus* and *Takliostrobus*). In 1943 Sahni described some silicified fruits which he named *Enigmocarpon* and assigned to the Lythraceae. In 1941 he described the reproductive structures of one of the water ferns, *Azolla*, which display quite remarkable preservation of minute details; his photos of them have been reproduced in several other publications.

Of the Jurassic fossil plants from the Rajmahal Hills in western India Sahni says: "This was the first Indian flora to be described in detail, and it is a classical flora because it has served as a basis for botanical comparison and geological correlation for many allied floras in other parts of the world" (1938, p. 42). Fossil plants had been known from the vicinity of Bihar in the Rajmahal Hills for a long time. They were described as far back as 1863 by Thomas Oldham and our friend John Morris. Other studies followed by Ottokar Feistmantel. In more recent years G. V. Hobson of the Geological Survey of India discovered petrified plant specimens in the Amrapara District of Bihar. They were brought to the attention of Sahni and his associates at Lucknow, who subsequently found additional fossils at Nipania. A considerable number of petrified plants have been described to date, of which the assemblage known as the Pentoxyleae seems most interesting.

The Pentoxyleae consist of associated leaves, stems, and reproductive organs that reveal a unique group of plants known now from the investigations of several of the Lucknow paleobotanists. A. R. Rao concentrated on a study of the anatomy of the leaves, which had long been known from impressions as a species of *Taeniopteris*. They were found to have vascular bundles of a cycadean type but the stomatal structure was more like that of the Bennettitales. B. P. Srivastava was able to correlate these leaves with a unique type of stem, which he named *Pentoxylon*, characterized by a ring of five closely aggregated steles with secondary wood of a coniferous type.

Srivastava was born in 1904 at Panna in central India, studied at Agra College, and entered the University of Lucknow in 1925 where he took a master's degree and served as a demonstrator. He then took a post as Professor of Biology at the College of Surat, and as time permitted he returned to Lucknow to carry on with the Rajmahal fossils. He described two species of seed-bearing organs of a distinctive type that was probably borne on the *Pentoxylon* stems. It is most unfortunate that his life was cut short; he died in 1938 at the age of thirty-four.

The microsporangiate organ of *Pentoxylon* was in turn identified somewhat later by Vishnu-Mittre; it consists of a ring of branched appendages bearing clusters of terminal sporangia. All in all this is a plant or plant group with an amazing assortment of highly distinctive features. In 1948, at a meeting of the Botanical Society of America in Chicago, Professor Sahni gave a summary account of the Pentoxyleae as it was known then and said in his introduction: "Some discoveries in science help, or appear to help, in the solution of old standing problems; others—and these are perhaps the more interesting—seem to create new difficulties in our path. My object here is to draw attention to a recently recognized group of plants which defies classification and presents a new problem in our understanding of the evolution of gymnosperms" (1948, p. 47).

The renowned *Glossopteris* flora presents a third major line of research that occupied the attention of Sahni and his associates, as well as many other paleobotanists in other lands. This is the flora of Gondwanaland, the presumed southern continent that broke up, as Alfred Wegner postulated early in the present century, with the approximate area that is now called India being pushed north to its present location. Sahni gives a good account of the status of the problem in his *Recent Advances*. He had become involved in this area of research when he was still in England and collaborated with Seward in a revision of the Indian Gondwana plants that was published by the Geological Survey of India in 1920. He was the first paleobotanist to devote serious attention to the matter since Feistmantel left off back in 1886.

The *Glossopteris* flora is not an especially large one in numbers of species and, as the name implies, is centered around the dominant plant or plant group, *Glossopteris*. One would be hard pressed to name a genus of fossil plants that has been the center of so much discussion and controversy and the subject of so many publications. My own first-hand experience with the flora is very limited. In company with Professor D. D. Pant and some of his students, I had an opportunity to collect at a coal mine north of Calcutta in 1961, and it is my recollection that the *Glossopteris* leaves were abundant and that there was little else. I also chatted with Edna Plumstead once when we were both visiting at the British Museum. I found her interpretations of some of the distinctive organs that she found attached to the leaves a bit puzzling but I think she deserves tremendous credit for initiating a new phase in *Glossopteris* studies. Whether Mrs. Plumstead was right or wrong in all of her interpretations is of little matter at this point; she showed a democratic spirit in her 1952 article in allowing several of her colleagues throughout the world to contribute their own opinions.

I do not propose to go into the *Glossopteris* story in detail; it is a task that would certainly require a rather large volume, but a brief review of reviews seems appropriate, with particular reference to the contributions of the Sahni school of workers. Professor Pant has given us a very useful and readable summary in his article "The Plant of Glossopteris" (1977). The complexity of the matter may be appreciated in some degree from Pant's statements that more than thirty form genera of fructifications are now attributed to *Glossopteris* and allied leaves, and that more than a dozen genera of fructifications have been described as being attached to *Glossopteris* leaves or the *Glossopteris*-like scales and bracts. Although much of the vast accumulation of information remains to be sorted out, we now have a fairly clear picture of what *Glossopteris* looked like in life.

Glossopteris was named by Adolphe Brongniart in 1828 and was also studied in the nineteenth century by several others, including Bunbury, Feistmantel, Zeiller, Oldham, and Etheridge; in the present century the number of authors who have dealt with it in one way or another is legion. Among the Indian workers, K. R. Surange and Shaila Chandra wrote an interesting and informative series of papers for *The Palaeobotanist* describing both male and female fructifications. D. D. Pant has contributed much to the development of the story, including a significant paper on *Glossopteris* fructifications from Tanganyika. J. Sen, P. K. Maithy, H. K. Maheshwari and others have also added important pieces of brick and mortar to the *Glossopteris* edifice.

When Birbal Sahni returned to India from England in 1919 he held the Chair of Botany at the University of Banaras for a year, spent another year at the University of the Punjab, and in 1921 was appointed Professor of Botany at Lucknow, where he remained for the rest of his life. He attracted students and colleagues in considerable numbers in the course of organizing and directing his monumental research programs. However, this was not enough; he had long envisaged a paleobotanical institute that would function as a central point for research on fossil plants in India. As is so often the case, funds were not readily available, and Professor and Mrs. Sahni resolved to use their own resources to initiate the project. The result was the establishment at Lucknow of the Birbal Sahni Institute of Palaeobotany, a unique organization where research on many facets of the plants of the past of India could be advanced. Modest government grants were eventually forthcoming and on April 3, 1949, the foundation-stone was laid by Prime Minister Jawaharlal Nehru. Birbal Sahni had achieved success in this, perhaps the core of his many objectives, but within a week, on April 10, 1949, he died of a heart attack. No country has ever had a more effective leader in the development of its science and culture in general. To quote the introduction to a booklet on the Institute that was issued on the occasion of its opening on January 2, 1953: "Not merely in its name the Birbal Sahni Institute of Palaeobotany commemorates its founder, but it is in itself a lasting monument to this man, who was one of the most prominent sons of modern India and a leading personality in the scientific life of his own country, and who, through his scientific achievements and his personal qualities, gained an established position in the first rank of international botany, palaeobotany and geology."

9

The Early Years in North America: Steinhauer to Bunbury

QUITE in contrast to the interest in fossil plants in Britain and on the continent in the latter part of the seventeenth century, it is not until after the mid-1800's that we can point to very much of significance in North America. Then the studies of such leaders as John W. Dawson, John Newberry, Leo Lesquereux, and Lester Ward made up to some degree for lost time. In a country still largely unexplored and unstudied as far as natural history was concerned, it is understandable that there was enough aboveground to occupy the attention of naturalists; the plants and animals entombed in the rocks below had to wait. There were, however, a few sparks of interest in the early years of the nineteenth century; since the sparks were fanned by the men noted above, it is worth having a look at them.

The earliest significant American publication on fossil plants that I have encountered is Henry Steinhauer's "On Fossil Reliquia of Unknown Vegetables in the Coal Strata" (1818). This was quite well received by the paleontological "establishment" in Europe, and it looked as though America had made a start. Steinhauer's article is often cited as the first of importance in this country; this is true in the sense that it was published here, but it is something of a paradox in that apparently all the fossil plants he dealt with were brought with him from England, chiefly from the Yorkshire and Somerset coal fields.

His knowledge of the plants was rudimentary but his article contains several good points. He is credited by several later writers (Arber, 1921b) as being the first to use binomials in describing fossil plants. He arranges them into four classes: Lithoxylon (fossil wood), Lithocarpi (fossil fruits), Lithophylli (fossil leaves), and a fourth for fossil flowers if such existed. His work consists of a description of ten species of Carboniferous plants under the generic name of *Phytolithus*. With one exception these are compressions or casts of Carboniferous stems (*Lepidodendron*, *Sigillaria*, *Calamites*) and stigmarian rootstocks.

Steinhauer was quite well acquainted with the important literature of the time; his study is detailed and carefully prepared with good illustrations, but he was inclined to accept a cactaceous affinity for at least some of his fossils, following a prevailing opinion of some of the European workers.

Henry Steinhauer was born at Haverford-West in South Wales on February 28, 1782. His strongly religious education was acquired in Yorkshire and later in Germany. In 1811 he moved to London and then to Bath, where he acquired a considerable circle of literary and scientific friends and frequently assisted the pastor of the Moravian congregation. He came to this country in 1816 and was placed in charge of the Moravian Seminary at Bethlehem, Pennsylvania. He apparently was dying of tuberculosis when he arrived but was able to contribute much to the life of the school in the two years that were left to him. He gave lessons in French, drawing, and botany; his lectures on natural history seem to have been especially well received. His innovations were not confined to the classroom; as in most schools the students looked forward to vacations and it is noted that "Brother Steinhauer also introduced the summer vacation in the month of July,—a season hailed with joy by the young ladies as a terminus to confinement to classroom and books, and a promise of many pleasures and delights among their friends at home" (Reichel, 1870, p. 191). He died in early 1818, the year his "Fossil Reliquia" was published, at the age of only thirty-six. He contributed to the start of paleobotanical studies in this country and it is unfortunate that he did not have time to explore the resources of his new home.

Studies of fossil plants in the United States were rather sparse for the next two decades, but they were not entirely wanting. In 1821 Ebenezer Granger described some plants found near Zanesville, Ohio; they are not named but he included two plates of good illustrations showing specimens of *Sigillaria*, *Lepidodendron*, and a few others. He refers to Steinhauer's work and has the following to say about his own fossils:

> I am satisfied they are mostly of a tropical growth. It would be gratifying to us to be informed of the species of plants to which they or any of them belong, or to which they bear the strongest analogy.
>
> I cannot forbear suggesting, that a botanical description of the vegetable remains found in different latitudes and longitudes, and which it is said always accompany the coal strata, may lead to very important results. They may at all events, afford some evidence, whether the poles of the earth have at some remote period been changed. [1821, pp. 6–7]

In 1821 Zachariah Cist wrote a paper in the form of an open letter to Alexandre Brongniart which included a few notations on his collection of fossil plants from the vicinity of Wilkes-Barre, Pennsylvania. Cist's interest was primarily in coal, but his contribution seems important in that it provided perhaps the first instance in which an American collection found its way to the leading authorities in Europe. His description is interesting if somewhat naive:

> The mass of the impressions are in the argillite immediately in contact with the coal, although they are common in the coarse sandy shist above it, and occasionally are found in the sand stone strata which alternate with the coal. There are above a dozen species of fern. A frequent impression, is that of a broad-leaved, apparently, aquatic plant, probably a sedge, with a transverse thread across the leaf at every three or four inches. The leaf is sometimes found of the breadth of six and even seven inches. Another very much resembles the leaf of the Indian corn (zea mays), or rather that which comes to us in boxes of tea. Occasionally very perfect specimens of flowers of a stellated form occur, and rushes and a variety of singularly formed plants and leaves, the originals of many of which are probably now lost. There are also numerous impressions resembling the bark of trees, or lichen attached to the bark. [1821, p. 6]

Apparently a considerable collection was sent to Paris, where it came to Adolphe Brongniart's attention. Ward (1889) notes that

> The plants thus sent are enumerated in [Adolphe] Brongniart's Prodrome, 1828, and described and illustrated in his *Histoire des Végétaux Fossiles*, 1828. The principal species are *Calamites Cistii*, *C. Suckowii*, *Neuropteris Cistii*, *Pecopteris polymorpha*, *P. gigantea*, *P. punctuata*, *Sigillaria Sillimani*, *S. obliqua*, *Sphenophyllum emarginatum*, *Lepidodendron mamillare*, *L. varians*, *L. aculeatum*, *L. Cistii*, *Stigmaria intermedia*, *S. tuberculosa*, *Annularia fertilis*, and *A. longifolia*, all from the Wilkes Barre coal beds. [Pp. 862–63]

In 1836, Dr. S. P. Hildreth (1783–1863) produced a long article on the general geography and geology of the country immediately west and east of the Ohio River between Pittsburgh and Portsmouth, Ohio. It includes considerable information on the location and extent of coal beds with several references to the associated fossil plants. It is especially noteworthy in that twenty-four of the thirty-six plates are woodcuts of Carboniferous plants from various localities in the area, although none of them is very precisely documented. In an appendix Samuel George Morton provided a very brief explanation of the plate figures. I believe that Hildreth's article is the earliest profusely illustrated study dealing with American fossil plants; numerous common Carboniferous genera are shown, such as *Alethopteris*, *Stigmaria*, *Asterophyllites*, *Lepidodendron*, *Calamites*, *Annularia*, *Pecopteris*, and *Neuropteris*.

Both Hildreth and Morton had some knowledge of the work of contemporary Europeans. Hildreth had access to the first six parts of Brongniart's *Histoire*, and in an introductory note Morton says: "With respect to the vegetable remains, I have compared them with the figures given by Sternberg, Brongniart, and Lindley, but in many instances without success. This department will require much more time, and a diligent comparison of specimens" (Hildreth, 1836, p. 149). Actually a good many of the specimens illustrated are not identified at all. It would seem neither Hildreth nor Morton had very much botanical knowledge; Hildreth compares one of his petrifactions with the root of a pokeweed (*Phytolacca decandra*) and he had a special fondness for drawing comparisons between the trunks (probably arborescent lycopods) and modern palms.

However, there can be no doubt as to Hildreth's strong *interest* in fossil plants as evinced by the fine set of illustrations. The interest, for better or worse, seems to have been partly theological and partly scientific; he says: "How beautiful and how valuable are the means which the all wise Creator has provided for the comfort and the happiness of man. Vast magazines of iron, salt and coal . . . were laid up in store for his use, before he was 'yet formed from the dust of the earth'" (1836, p. 108).

In a rather lengthy article of 1837 Hildreth presents, I think for the first time in the U.S., several descriptions of new species with binomials. Three of his fossils are sigillarian stems, although described under three different genera. He is consistent in his adherence to the presence of palms, for he figures two seeds which appear to be casts of medullosan seeds, with the comment that "this nut is probably the fruit of some antediluvian Palm" (1837, p. 29).

Samuel Prescott Hildreth was born in Methuen, Massachusetts, on September 30, 1783; he attended Phillips Andover Academy, studied medicine in his father's office and with a Dr. Thomas Kittredge, and topped this education off with a course of lectures at Harvard College. He began the practice of medicine in Hampstead, New Hampshire, but on learning of a good opening in Ohio he mounted his horse and rode west, reaching Marietta about a month later. He was a successful physician, treating his patients with the methods of the day such as bleeding, purging, and sweating, but he seems to have been on the lookout for new developments. His interests in natural history were broad; he was a collector of shells, insects, and plants as well as fossils. But his biographer (Matthews, 1932) says that his greatest service was probably as a historian. He sought out information from diaries and some of the early settlers themselves, and recorded it in several books and articles.

Samuel George Morton was born on January 26, 1799, in Philadelphia. His father died when he was very young and he lost his mother at the age of seventeen, at which time her physicians, being impressed with Samuel's abilities, aided him with the study of medicine. He attended lectures at the medical department of the University of Pennsylvania where he received his M.D. degree in 1820; this was broadened by a period of study at Edinburgh University. Like Hildreth, he contributed to several areas of natural history, including geology and vertebrate paleontology; among his publications is one dealing with the fossil collections brought back by the Lewis and Clark expedition. Another major interest consisted in the collection for comparative studies of a large suite of human skulls which Louis Agassiz declared was worth a journey to America to see. One may perhaps wonder how large an audience would travel the Atlantic with this objective!

The gap between the time of Steinhauer and the highly productive era ushered in by Leo Lesquereux and his contemporaries is also filled to some degree by the observations of the great nineteenth-century British geologist Charles Lyell and his friend Charles J. F. Bunbury. Lyell made two extensive tours through the eastern United States and Canada, the first in 1841 and the second in 1845–46. These experiences are recorded in considerable detail in his *Travels in North America* . . . (1845) and *A Second Visit* . . . which appeared in 1850. These volumes make fine reading and reveal Lyell as a great observer of people and places and a most hardy traveler.

Lyell (1797–1875) was born in Scotland of a family that was quite well off and encouraged his early interests in natural history. He graduated in classics at Oxford in 1818, then studied law, which he practiced for a short time. At Oxford he had attended Buckland's lectures and came to know Gideon Mantell quite well, but of special importance was his long and close friendship with Charles Darwin, who was twelve years his junior.

Lyell was appointed Professor of Geology at King's College, London, in 1831, where his efforts at modernization do not seem to have met with complete approval. A biographer says: "Lyell's first course of lectures was delivered in 1832, and his last in 1833. To begin with he was able to throw these lectures open to the public, including ladies; but before long the College authorities excluded the latter, because their presence 'diverted the attention of the young students.' Attendance dropped disastrously, and Lyell resigned. He felt that his proper place in life was that of gentleman-scientist-author, without strings" (Bailey, 1959, pp. 132–33).

Perhaps Lyell's greatest contribution was his insistence on recognizing the natural phenomena acting on the earth today as having been essentially the same in the past. And although Darwin depended heavily on Lyell's knowledge of geology and their friendship was a close one, Lyell did not come to accept Darwin's ideas of evolution until quite late in life.

This brief introduction to Lyell may be expanded somewhat with a few extracts from the journals of his American travels. On his first visit we find him at Blossberg, Pennsylvania, on September 7, 1841: "I had now entered Pennsylvania, and reached one of the extreme north-eastern outliers of the great Appalachian coal-field. . . . It was the first time I had seen the true "Coal" in America, and I was much struck with its surprising analogy in mineral and fossil characters to that of Europe" (1845, p. 61). He goes on to discuss the problem of the morphology and affinities of *Stigmaria*—which gave everyone a bit of trouble at this time.

A little later in the autumn he was in New England, and I find a nostalgic interest in his visit to the coal mines of southern Massachusetts and Rhode Island, where I first collected fossil plants as an undergraduate student. He mentions finding specimens of *Pecopteris plumosa*, *Neuropteris flexuosa*, *Sphenophyllum*, *Calamites*, and others. He also visited the famous Cretaceous cliffs of Gay Head on Martha's Vineyard: "Late in the evening I reached the lofty cliffs of Gayhead, more than 200 feet high, at the western end of the island, where the highly-inclined tertiary strata are gaily coloured, some consisting of bright red clays, others of white, yellow, and green sand, and some of black lignite. . . . I collected many fossils here, assisted by some resident Indians, who are very intelligent" (1845, vol. 1, p. 256). It may be noted that the Cretaceous age of the Gay Head cliffs was only recognized several decades later through the studies of David White.

At the cliffs of South Joggins in Nova Scotia he reports: "I saw the erect trees at more than ten distinct levels, one above the other; they extend over a space from two or three miles from north to south, and more than twice that distance from east to west" (1845, vol. 2, pp. 187–88).

Travel was a bit different at that time; on his second visit (1845–46), in the vicinity of Tuscaloosa, Alabama, he notes: "We traveled in a carriage with two horses, and could advance but a few miles a day, so execrable and often dangerous was the state of the roads. Occasionally we had to get out and call at a farm-house to ask the proprietor's leave to take down his snake fence, to avoid a deep mud-hole in the road. Our vehicle was then driven over a stubble field of Indian corn, at the end of which we made our exit, some fifty yards on, by pulling down another part of the fence" (1850, vol. 2, p. 70).

Much of the fossil-plant material that Lyell collected on his North American tours, as well as some sent by J. W. Dawson, went to Charles J. F.

Bunbury (1809–1886). This is partly because Bunbury was a competent naturalist and partly because his wife and Lyell's wife were sisters. Bunbury, like Lyell, did not seem to have any serious financial problems; he was born in Messina, Sicily, in 1809, his father being a high-ranking army officer in the Mediterranean command. After completing his education at Trinity College, Cambridge, he traveled rather extensively as a serious amateur naturalist. He visited Brazil and the River Plate, later took a voyage to South Africa and in 1853 he accompanied Lyell to Madeira and Teneriffe. Hooker gives Bunbury much credit for organizing the Carboniferous collections in the museum of the Geological Society of London.

Although Bunbury did not actually visit this country himself, he gave us some of the earlier and best papers on American fossil plants, based on specimens Lyell sent him. In 1846 he described some fertile fernlike foliage from the Carboniferous of Frostburg, Maryland; and in the same year he published a short paper on the fossil plants of the coal fields of Tuscaloosa, Alabama.

I think that Bunbury was a good observer and astute at dealing with the facts. Two instances in point are worth noting: First, like so many others he too became involved with *Stigmaria*! He says:

> The observations of Mr. Dawson on the fossil roots having the characters of Stigmaria are curious and important, as corroborating the statement communicated by Mr. Binney to the British Association in June 1845, respecting the tree with similarly marked roots which was discovered at St. Helens in Lancashire. . . . He [Dawson] appears however to have satisfied himself, that one of the Stigmaria-like specimens now before us did actually proceed in the manner of a root from the base of the stem [identified as a *Sigillaria* by Dawson]. . . . It has been urged, that the symmetrical arrangement of the scars is a fatal objection to the idea that these bodies were roots, which never emit their fibers with any degree of regularity; but, unless we suppose that Mr. Dawson has been deceived by appearances, this argument, drawn from the analogy of existing plants, must yield to the positive result of observation. It would not be the first instance in which the progress of discovery has revealed striking exceptions to what had been supposed general laws of structure. [1845a, p. 136]

Second, he saw fit to take issue with Brongniart's "island distribution" concept of Carboniferous floras:

> Those parts of Europe and North America in which the coal-fields were accumulated, may have existed at that time in the state of islands, like those of the present Pacific Ocean; but it would be rash to infer, as M. Adolphe Brongniart seems disposed to do, that no extensive continents

> at that time existed in any part of the globe. If, in all departments of geology, it is necessary to advance with caution, and to avoid dogmatism and rash generalizations, it is more especially necessary in the department of Fossil Botany, where so much of the evidence we possess is fragmentary and imperfect. [1846b, p. 90]

Bunbury's contributions were not confined to the North American collections; drawing on his own travels, he also wrote some significant papers on fossil plants from other parts of the world. In 1861 he described a collection from the late Paleozoic of Nagpur, India, which included specimens of *Glossopteris*, and in 1851 he described a few plants from the Jurassic of Yorkshire, apparently most of them from the Bean collections.

10

North America in Mid-Century: Lesquereux, Newberry, Dawson

ALTHOUGH the men and the work described in the last chapter gave some semblance of a start to North American paleobotany, a very different era was ushered in a few years later by Leo Lesquereux and John S. Newberry in this country, and by J. W. Dawson in Canada. If not the actual founders of the science on this side of the Atlantic, they initiated and developed studies of great importance which began to attract worldwide attention from botanists and geologists. To a considerable extent the great wealth of knowledge of the floras of the North American past, which continues to grow each year, began with these three men.

Leo Lesquereux (1806–1889) is best known to paleobotanists for his *Description of the Coal Flora of the Carboniferous Formation in Pennsylvania and throughout the United States*. The atlas of eighty-five beautifully executed plates appeared in 1879 and the three text volumes in 1880 and 1884. It presented the American Carboniferous plants in a fashion that compared favorably with the better works being produced in Europe. William Darrah (1969) notes that some 60 percent of the specimens described are preserved in the collections of Harvard University. Darrah also gives much useful information on the people who worked with Lesquereux and their collections dealing with American Pennsylvanian paleobotany, that I cannot include here.

There is some significance in Arnold's comment (1969) that the impressive magnitude of Lesquereux's *Coal Flora* tended to slow down American Carboniferous plant studies in that it gave the impression that there was little left to do. But if such was the case we have now recovered from the effect! Lesquereux was one of the great founders of fossil botany in North America and his life story is one of the most remarkable and tragic in the history of science.

Leo Lesquereux. Courtesy of the National Academy of Sciences.

Lesquereux was born on November 18, 1806, in the village of Fleurier in the Swiss canton of Neuchâtel. His father was a manufacturer of watch springs with a small factory employing four or five people. Leo evinced a love of nature from a very early age which took him on adventurous excursions into the nearby mountains. An accident on one of his climbs had a considerable bearing on his entire later life.

> To scale the most difficult summits and to gather the rare flowers that grew there, were among his early ambitions and pleasures. He must have been a daring climber. On one of his excursions, when about ten years of age, he met with an accident of so dangerous a character that his escape from death seems almost incredible. He had climbed the mountain that towers above Fleurier, but by a misstep he fell over the edge of a cliff, down the steep mountain side. He struck first upon a projecting ledge and was rendered insensible by the fall; from this point he rolled limp and unresisting, his descent being occasionally checked by branches of trees or shrubs, to the borders of the meadowland far below. When picked up there, he was found fearfully bruised and lacerated, but no bones were broken. For two weeks he lay unconscious, but at the end of six weeks he was on his feet again. [Orton, 1890, p. 285]

A lasting effect from this fall was an impairment in his hearing which gradually became worse. Some years later, he went to Paris to be operated upon; the operation was badly done and the result was total deafness when he was about forty years old. I have a letter that Lesquereux wrote on March 12, 1884, to Professor H. G. Seeley of the Geological Society in London, relative to a small grant of money that had been made to him. It relates in his own words the great handicaps that he worked under most of his life:

> Though I have devoted more than 30 years to the study of the vegetable Paleontology of North America I well know that the result of my researches do not come up to the present attainments of Geological Science. The deficiency has been caused by my ignorance, certainly but also by the great difficulties which I encountered in pursuing my work. You will have the kindness to excuse me from entering into some personal details.
>
> As I am totally deaf, and therefore unable to occupy any scientific position with permanent emoluments, and as I am without means or income of any kind, I have been forced to give a great deal of time to different kinds of work, outside of scientific pursuits in order to get support for myself and family. Beginning my researches on the vegetable paleontology of this country [in] 1850, I had to proceed gradually by association to different geological survey[s] as an assistant, sometimes without other pay [than] that of my travelling expenses. And I had then to procure all what

> was necessary to get some use of my researches: books, specimens, etc. For out of the great centers of population of the East, the public libraries of America have not any scientific books. I had therefore to build up my own library, my herbarium, etc.

In his memoir on Lesquereux, Lesley (1895) says, "I have been present when Lesquereux talked with three persons alternately in French, German, and English by watching their lips. The interview would begin by each one saying what language he intended to use" (p. 210). It may be noted that he never heard the English language spoken, having learned it after he became deaf, and it is not surprising that Lesley should add "his pronunciation of it was curiously artificial and original."

Lesquereux went to school at Neuchâtel when he was thirteen. He became acquainted with his fellow students Arnold Guyot and August Agassiz, the brother of Louis Agassiz who was to play an important role in Lesquereux's life. He later went to Eisenach, earning his way tutoring French, and one of his students was Sophie, the daughter of General von Wolfskell, whom he later married. Upon returning to Switzerland he obtained a position as teacher and later principal at the high school at Locle, but within three years his hearing had become so poor that he was obliged to give up teaching. He then found it necessary to work at rather menial tasks in his father's watch factory. His spare time was spent in the mountains, especially studying mosses. The national government at this time had an interest in obtaining more information about peat bogs with the object of using peat as a fuel. Lesquereux plunged into this work with great enthusiasm, succeeded in winning a twenty-ducat gold medal, and was given a commission by the king of Prussia—aided in part no doubt by the fact that his wife, of noble birth, had been a bridesmaid at the king's wedding.

Several significant events transpired as a result of his peat studies. He became acquainted with Louis Agassiz, who then held the Chair of Natural History in the Academy of Neuchâtel. He was able to travel rather widely in Germany, Sweden, Denmark, Holland, and France and to become acquainted with a wide range of scientists. It was apparently during these journeys that he made his first contact with other sorts of fossils:

> The impression produced on Lesquereux's mind by the greatest European collection of fossil plants at that time, in the Museum of Strassburg, was never effaced: He says: "I felt as if I had been transported into another world, and could scarcely leave the large room where the specimens were exposed. I said to Schimper how happy a man should be with such an admirable vegetation to study! I did not dare to ask him for even a small piece of one specimen, although I should have prized the smallest as a

treasure. Was it a remembrance of some former life, or a prevision of what was to come to me in the hereafter?" [Lesley, 1895, p. 206]

Lesquereux's bog studies, however, did not lead to a permanent position, and the political changes in Europe in the late 1840's did not help. He followed Agassiz to America. Of this journey he says: "I came to that promised land a poor emigrant family, having an abominable voyage of sixty days in the entrepont (steerage) of a sailship, together with 300 companions of misery, the most terrible experiences of my life" (quoted by Lesley, 1895, p. 209). Among Lesquereux's later writings is a book entitled *Lettres écrites d'Amérique destinées aux émigrants*, written to serve as an aid or warning to his fellow countrymen as to what to expect. One biographer refers to it as a valuable survey of America for the middle of the century and recounts the following amusing incident: In Boston, "as Lesquereux was enjoying his first walk in the American Athens, smoking a cigar, he was stopped by a policeman (whom he did not at first recognize as such) who confiscated his cigar and fined him two piasters (dollars). Indeed, in 1850, it was forbidden in the streets of the city as this might incommodate the Athenian ladies!" (Sarton, 1942, p. 100).

Lesquereux arrived in Boston in 1847, forty-one years old, accompanied by his wife and five children, deaf, and with no knowledge of English. He worked for a time for Agassiz classifying a collection of plants that Agassiz had brought back from his Lake Superior expedition. At the close of 1848 he was invited to Columbus, Ohio, by William Sullivant, the leading American bryologist, to assist him in his studies of mosses. He worked with Sullivant for two years. They published two editions of *Musci Exsiccati Americani* (1856, 1865), and Lesquereux also had much to do with the great work of Sullivant's life, the *Icones Muscorum*.

As Lesquereux's reputation increased he obtained commissions from several state geological surveys as well as the national one. He studied the coal floras of Ohio, Illinois, Kentucky, Arkansas, and especially Pennsylvania. In 1868 he initiated a study of more recent floras of the western states. An important paper on Cretaceous plants appeared in the *American Journal of Science*, and F. V. Hayden employed him to work up collections from the Territories, the results of which appeared in the annual reports of the U.S. Geological Survey for 1870–1874. This employment, which undoubtedly was what Lesquereux enjoyed most, was not adequate to maintain his family, and in Columbus he established a small watchmaking and jewelry business to make ends meet. Three of his sons were also engaged in the business.

Lesley says:

> At this time (1851) I became acquainted with Lesquereux, as he sat day after day on the anthracite coal-tips turning over each piece of waste slate in search of plant impressions. His patient zeal was a wonder to my impatient and restless nature. The broiling sunshine, the chilly wind, the soaking rain were alike disregarded by him. The evening brought him no repose, for his bag of specimens was exhibited, re-examined, discussed, and sometimes figured then. [1895, p. 210]

Sarton describes Lesquereux as "modest and kind, happy to do his work quietly, unobtrusively, in his own little corner and to leave to others more greedy of money, of power, or of fame, the credit or glamor, the excitement or the advertisement after which they lusted" (1942, p. 101). But honors if not wealth did come to Lesquereux. He was a member of nearly a hundred learned societies and was elected to the National Academy of Sciences the year after its constitution. We will meet him again in this account of the fossil hunters.

This is an opportune point at which to introduce one of the many "background" workers in paleobotany—certainly one of the most important fossil-plant collectors of all times, Ralph Dupuy Lacoe (1824–1901). I have drawn most of my information concerning him from David White's memoir of 1901.

White notes that "unlike most self-made men, Lacoe was a man of culture and refinement. He was conscientious, studious, and methodical in his scientific as well as his business affairs, while at the same time he was artistic in his tastes" (p. 514).

Lacoe was educated in a country school where he learned the carpenter's trade; he developed a modest estate by cutting railroad ties and gradually developed other business interests. In 1865, when he was forty-one, poor health led him to reduce the intensity of his business activities, and he began to devote much of his time to the out-of-doors and nature study. He started by collecting shells in Florida. He soon made the acquaintance of J. P. Lesley, the Pennsylvania state geologist, and Leo Lesquereux, with whom he developed a warm friendship that lasted until Lesquereux's death in 1889.

Lacoe's collecting was systematic and vast in scope. He worked the Devonian and Carboniferous formations in an area bounded by Georgia, Arkansas, Illinois, and Rhode Island, and later in the western states, using hired collectors along with his own efforts. He published but three short papers, one on fossil insects, another on plants, and the third on reptile tracks. White says:

> His purpose was to systematically gather and put before the most eminent specialists the raw material which should contribute to our knowledge of

> the nature and characters of the plant and insect life of the ancient epochs; which should show the horizontal and vertical distribution of the types and their significance regarding genetic sequences and stratigraphical characteristics, and which should throw light on the questions of continental relations and climatic conditions. [1901, p. 511]

White notes that the Lacoe collections by 1891 more than filled the entire upper floor of the Pittston (Pennsylvania) First National Bank building and it was decided to remove them to the Natural History Museum in Washington, D.C. By 1895 Lacoe's fossil-plant and fish collections had been moved; the insect, myriapod, and crustacean collections were sent to Washington in 1899. The collections included some one hundred thousand specimens of plants. Lacoe must have been the kind of collector that most paleontologists dream about—intelligent and generous with his vast collections and financially able to help with publication costs. It may not be irreverent to call him the joy of a researcher and (as to the quantity involved) the despair of a curator. Few collectors have contributed so much to fossil botany.

Nearly contemporary with Lesquereux, at least in their periods of paleobotanical productivity, was a man of quite different character and economic status, John Strong Newberry (1822–1892). Newberry and Lesquereux were colleagues of a sort, although the relationship does not seem to have been entirely cordial. His articles and books, like many of those from the earlier days, are not often referred to now, but on several counts he was an important pioneer in American botany, geology, and paleobotany.

Newberry was born in Windsor, Connecticut, on December 22, 1822, but his family soon moved to a large tract of land that had been acquired by his grandfather a few miles south of the present site of Cleveland, Ohio. The land included extensive coal deposits which were mined by his father, who also engaged in other industrial operations, securing for his family a very comfortable living. I can think of no other paleobotanist who started out in life with his own coal mine! Newberry was attracted to the fossil plants in the roof shales at the age of eleven, and he developed an early interest in natural history roaming through the forests and fields of his father's extensive tract of land. Added inspiration also probably resulted from a visit to the Newberrys by the noted geologist, James Hall, when John was nineteen years old. Since medical schools at the time presented the most opportune way of obtaining some semblance of a scientific education, he attended Western Reserve College at Hudson, Ohio, where he was graduated in 1846, and in 1848 he received the M.D. degree from the Cleveland Medical School. In the fall of 1849 he went to France to further his medical education. While there he

John S. Newberry. From the *Bulletin of the Torrey Botanical Club*, 1893.

frequented the School of Mines and Botanical Garden in Paris and listened to lectures by Adolphe Brongniart.

He was back in Cleveland in 1851 and maintained a medical practice, but it seems evident that he did not intend to devote his life to medicine. He had been conducting studies in several areas of science and had accumulated a considerable library, which was used for a time by Lesquereux; indeed, one of his biographers credits Newberry with having aided Lesquereux considerably in his early work with fossil plants (Fairchild, 1893).

Perhaps the wanderlust overcame Newberry; at any rate he decided to terminate his medical practice in 1855 and accept a position as botanist and geologist on a U.S. government expedition to northern California and Oregon. The party reached San Francisco on May 30 and began field work in connection with a route for the Pacific Railroad, reaching the Columbia River in early October. The group that Newberry was with left the field in November to return to Washington where he spent the following year working on his report, which is contained in the sixth volume of the Pacific Railroad Reports. In 1857 he was back in the west serving as physician and naturalist with Joseph C. Ives's Colorado exploring expedition. From the Gulf of California they journeyed up the Colorado River in a small steamboat, reaching the mouth of the Black Canyon on March 6. Newberry accumulated extensive natural-history collections on the expedition; he prepared its geological report (published in 1861) while the modern plants were worked up by Asa Gray, John Torrey, and George Englemann.

Once again he was back in the field in 1859 with the San Juan exploring expedition under Capt. J. N. Macomb. They started from Santa Fe, New Mexico, in July, going up the Rio Chama, across the continental divide, into southwestern Colorado and southeastern Utah to a point near the junction of the Grand and Green Rivers, and returned to Santa Fe in November. Newberry must have enjoyed these ventures, rugged as they were, but he wrote to F. V. Hayden, probably in a facetious vein, on February 10, 1858, when he was on the Colorado River:

> I should be very happy to be one of your pleasant circle at the Smithsonian this winter. . . . I am doomed to pass the entire winter and spring doing the hardest kind of field duty with few of its pleasures or rewards. Day after day we slowly crawl up the muddy Colorado—, confined to a little tucked up, over-loaded, over-crowded steamer with no retreat from cold, heat, wind or drifting sand, and nothing but the monotony of an absolute desert to feast our eyes upon, with nothing but bacon and beans and rice and bread *and sand*—or rather *Sand and Bacon*, etc. to eat, sleeping on shore with a sand drift, eyes, nose, mouth, ears, clothes and bed filled

with sand—with almost everyone discontented and cross. [Waller, 1943, p. 333]

Newberry's scientific studies were interrupted for several years by the Civil War, and although he was attached for a time to the War Department as Assistant Surgeon, he spent much of this period as a member of the U.S. Sanitary Commission. He served with enthusiasm and was something of a pioneer in the problems of effectively distributing hospital supplies to the sick and wounded, both friend and foe.

In September of 1866 Newberry was appointed to the Chair of Geology and Paleontology in the School of Mines at Columbia College in New York, a position he held for the remaining twenty-six years of his life. A biographer says of this period: "His extensive private collection in geology and paleontology was purchased by Columbia College, and was the beginning of the geological museum which under his affectionate care has become one of the best in America. It is especially rich in fossil fishes and fossil plants, the two groups of his particular interest, and in collections illustrating economic geology, necessitated by the character of the instruction" (Fairchild, 1893, p. 12).

Newberry's contacts with the other leading paleobotanists of the day do not seem to have been as thoroughly cordial as might be wished. Personalities, then as now, entered into scientific matters, and they certainly affect the course of events, although it is not always easy to decide how much. In a letter to Fairchild, Ward offers the following estimate of Newberry's work:

> Dr. Newberry was a great geologist, without which qualification no one can appreciate the full significance of fossil plants. He never spoke of them without evincing a lively consciousness that they were once real and living plants, and that they belonged to the great record which time has made of the events which have transpired in the history of the earth. . . . He was no species-monger, and not prone to found species on insufficient material. His descriptions were all governed by strong common sense, and, unlike many other paleobotanists he never forgot that he was dealing with real things.
>
> Dr. Newberry was not a good botanist; he had once been, but had neglected to keep pace with the science. Moreover, he seemed to have very little interest in the more important principles of botany. He was utterly indifferent to questions of classification, and to judge from his published papers one order of arrangement was as good for him as another. [Fairchild, 1893, p. 17]

His relations with Lesquereux do not seem to have been very close. They had several scientific disagreements and perhaps the social and economic gulfs between them were too much. In reference to one of their apparently few meetings, a biographer says:

> They may have been disappointed in each other from the start. Newberry was tall, sure of himself, with sharp, perhaps critical eyes. He may not have known that Sophie, Lesquereux's wife, was the daughter of Baron von Wolfskell. Glancing at the humble surroundings in the little house at Fourth and Mound Streets, in Columbus, Newberry may have been more abashed by their poverty than alert to their intellectual aristocracy or Sophie's ancient lineage. If Newberry spoke in English she would have to act as intermediary and translate to German or French, so that her deaf husband could lip read. If Newberry spoke French, he may, in spite of his two years abroad, have done so haltingly, thus adding embarrassment to his somewhat aloof manner of separating himself from his hosts. [Waller, 1943, p. 340]

A man's economic background inevitably influences his outlook on life and his associations with others. Lesquereux's total deafness, which rendered it impossible for him to hold positions that were open to Newberry, and the poverty that resulted largely from his affliction shut him off from the world that Newberry was able to live in and enjoy. Lesquereux wrote to J. P. Lesley about Newberry: "He is a born American, a rich man and is sustained by great political influence" (quoted in Waller, 1943, p. 341). These were, of course, the things Lesquereux could never attain. It is sad that his bitterness was so intense.

Newberry's published works were varied and voluminous. He wrote well over two hundred articles and books, as recorded by Fairchild (1893), who says that Newberry's beautifully illustrated (1857) report *The Botany of Northern California and Oregon* includes a chapter on the trees of the region that is a classic in American forestry. His writings include many articles on Ohio geology—after he became established in New York Newberry retained his Ohio residency and served for three years as Director of the Geological Survey of Ohio when it was formed—as well as aspects of economic geology ranging from Vermont marbles to the coals of Colorado. In paleontology he produced numerous studies of fossil fishes, including a rather well known monograph on those of the Triassic rocks of the Connecticut River valley. His works on fossil plants include several accounts of the Carboniferous floras of Ohio, the Cretaceous floras of the western states, the flora of the Amboy clay (New Jersey), and many popular works and reviews of other important studies of the time.

Before leaving Lesquereux and Newberry, I would like to insert a few lines on the life of Charles H. Sternberg (1850–1943), another great collector whose fossil plants, in large quantities, passed through the hands of the two paleobotanists. Of particular interest are the years 1867–1875 when Sternberg collected Cretaceous leaves in Ellsworth County, Kansas, many of which were described and illustrated in Newberry's *Later Floras of North America* and Lesquereux's *The Cretaceous Flora*.

Sternberg's autobiography *The Life of a Fossil Hunter* (1931) is a readable and classic example of the problems and hardships encountered in field work a century ago. He was born in Middleburg, New York, and spent his early days in the valley of the Susquehanna, where he developed an intense love of natural history that drove him on through much of his life. His family moved to Iowa in 1865, and for several years after 1867 he lived with an older brother who was a surgeon at Fort Harker, Kansas. Of this period he has written: "From 1867 to 1875, and many years since, I tramped over the hills of Ellsworth County and those adjoining, in search of new localities of fossil leaves. Carrying home on my back fifty pounds or more, if successful, light hearted and happy. Or as often happened an empty collecting bag, with weary feet. From collections I made from 1867 to 1869, I sent the choicest ones to the Smithsonian Institution" (1903, p. 312).

Sternberg's first meeting with Leo Lesquereux is interesting in the light of what we know about Lesquereux. As Sternberg describes it:

> I learned in 1872 that Prof. Leo Lesquereux had visited my locality I called Sassafras Hollow, south of Fort Harker, on Thompson Creek, and was a guest of Lieutenant Benteen, commander of the Post. Seizing my sketches and the first horse I could get I started posthaste for the fort. I was ushered into the presence of the commander, to find he was giving a reception to his honored guest. I was introduced to the professor by his son, who spoke in French. His father was very deaf. . . . When I produced my sketches he took me away from the other people, into a corner of the room, where face to face I told him of my discoveries. He understood every word I uttered and though he spoke in broken English and could not talk to others in the room except through an interpreter, I understood him well. His eyes sparkled when he examined my sketches. "This is a new species, and this and this," he said. "This is described from poorer material in The Cretaceous Flora and already illustrated. Oh, how I wish I could have had all this material for the Flora. I knew nothing of it or of you." [1903, p. 312]

Clearly, Lesquereux had some facility with spoken English when aided by quiet surroundings and subject material with which he was acquainted and in which he was intensely interested. Sternberg goes on to state that *The Cretaceous Flora* did include plants that he had previously given to a friend and which one way or another had reached Lesquereux.

I have noted elsewhere what seems to be an intuitive facility on the part of some field workers in paleontology. It is perhaps due to the fact that some people have an extremely acute sense of observation, although this explanation does not really satisfy me. Sternberg relates an incident which may be of significance in this connection:

> I have a vivid recollection of the discovery of another locality. One night I dreamed that I was on the river, where the Smoky Hill cuts into its northern bank, three miles southeast of Fort Harker. A perpendicular face in the colored clay impinges on the stream, and just below this cliff is the mouth of a shallow ravine that heads in the prairie half a mile above.
>
> In my dream, I walked up this ravine and was at once attracted by a large cone-shaped hill, separated from a knoll to the south by a lateral ravine. On either slope were many chunks of rock, which the frost had loosened from the ledges above. The spaces left vacant in these rocks by the decayed leaves had accumulated moisture, and this moisture, when it froze, had had enough expansive power to split the rock apart and display the impressions of leaves. . . . shut up in the heart of the rock for millions of years.
>
> I went to the place and found everything just as it had been in my dream. Two of the largest leaves known to the Dakota Group were taken from this place. . . . Probably my eyes saw the specimens while I was chasing an antelope or stray cow and too much occupied with the work in hand to take note of them consciously, until they were revealed to me by the dream, the only one in my experience that ever came true. [1931, pp. 18–20]

Much of Sternberg's activity in later years was devoted to collecting vertebrate fossils; he worked for men such as E. D. Cope and his fossils are preserved in many museum collections throughout the world.

At about the same time that Lesquereux and Newberry were laying the foundations of serious and continuing paleobotany in the United States one of Canada's greatest naturalists, John William Dawson (1820–1899), was equally busy with this and many other tasks in his own country. Dawson's life was a fascinating one and certainly exemplary for paleobotanists of any age; he was the first to give us some clear understanding of the earliest land vegetation—a vegetation so strange that his studies were largely ignored for more than a half century. The most interesting and productive decade of my own career has been devoted to a study of these early vascular plants in much the same areas of southeastern Canada and northern Maine where Dawson explored and collected. Several of us in recent years have carried on the work that he started, and the end is not in sight; it is a remarkable chapter in the study of the earth's earliest land vascular vegetation. I will therefore take the liberty to emphasize this aspect of Dawson's contributions and include a brief story of the recent developments that have been laid on the foundations that he prepared over a century ago.

Dawson was born in Pictou, Nova Scotia, on October 13, 1820. His father was a bookseller with a taste for study which undoubtedly had some influence on his son's development. John began making collections of fossil plants at about the age of twelve in the Nova Scotia coal measures. His first encounter is recorded in his memoir *Fifty Years of Work in Canada*:

> It happened, when I was a mere schoolboy, that an excavation in a bank not far from the schoolhouse exposed a bed of fine clay-shale, which some of the boys discovered to be available for the manufacture of home-made slate pencils. So we used to amuse ourselves occasionally by digging out flakes of the stone, and cutting them into pencils with our pocket knives. While engaged in this occupation, I was surprised to find that one of the flakes had on it what seemed to be a delicate tracing in black, of a leaf like that of a fern. I was at the time altogether ignorant of geology and of fossil plants, but was greatly struck by this unexpected discovery. I can remember, as well as if it were only yesterday, the effect on my mind of this new and mysterious fact, which was the beginning of many similar discoveries that have been among the chief pleasures of my life. . . . But the strangeness of the fact dwelt in my mind, and I was puzzled by the question whether they were real leaves or not, if real, how they came to be in the stone. [1901, pp. 34–35]

He studied at Pictou College, where, at the age of sixteen, he read his first paper to the local natural-history society. Its title, "On the Structure and History of the Earth," was perhaps prophetic of the rather wide scope of Dawson's research, as well as the teaching and administrative activities that characterized his entire life. From Pictou he went to Edinburgh, where he received a master of arts degree when he was twenty-two, having the good fortune to work under such well-known men as Forbes, Balfour, and Jamieson. He returned to Canada in 1842 and in the summer of that year he met Charles Lyell, who was most favorably impressed with the young Canadian naturalist. A second visit of about a year was made to Edinburgh in 1846. In 1849 he went to Halifax to give a course of lectures on natural history, and the next year he was appointed Superintendent of Education for Nova Scotia, which established him well on the path of geologist and educator that he followed for the rest of his life. The position obliged him to travel a good deal through the province and allowed him to accumulate much of the data for his monumental work *Acadian Geology*.

John William Dawson. Photo from *American Geologist*, courtesy of the Peabody Museum of Natural History, Yale University.

On a visit to Canada in 1852 Lyell spent considerable time geologizing with Dawson. He recorded that on September 12

> Dawson and I set to work and measured foot by foot many hundred yards of the cliffs, where forests of erect trees and calamites most abound. It was hard work, as the wind one day was stormy, and we had to look sharp lest the rocking of living trees just ready to fall from the top of the under-

> mined cliff should cause some of the old fossil ones to come down upon us by the run. But I never enjoyed the reading of a marvelous chapter of the big volume more. We missed a botanical aide-de-camp much when we came to the top and bottoms of calamites and all sorts of strange pranks which some of the compressed trees played. [Adams, 1899, p. 552]

How often have many of us looked for the elusive "tops" and "bottoms" of fossil plants! It is understandable that they should be missing in the case of large forest trees, but it has puzzled me many times that they should be so reluctant to expose themselves in the case of smaller plants.

Perhaps the most significant year in Dawson's life was 1855, when he was offered the position of Principal of McGill University in Montreal. Accepting it must have taken considerable courage and faith in himself. One of his biographers describes the job as follows:

> When Sir William assumed the principalship of McGill University, it was a day of small things. The financial condition of that institution at that time made it necessary for him to undertake the duties of several laborious professorships along with those of administration. The revenue then amounted to only a few hundred dollars. There were only eight instructing officers and with the exception of the faculty of medicine, the courses were most unsatisfactory. [Ami, 1900, p. 3]

The personal qualifications for the post were rigorous; the Principal must be young, capable, and with modern ideas; he must be of a religious frame of mind and tolerant, for the university, although Protestant, was undenominational in its teaching—and he certainly was faced with a task of long hours and low pay. That Dawson supplied all of these qualifications, and others in the bargain, is evident from the fine job that he did in the development of McGill University. Many honors fell to him in later years: election to the Geological Society of London and the Royal Society, the presidencies of both the British and American Associations for the Advancement of Science, and a knighthood in 1883.

Dawson's bibliography is a very long one and I believe that Ami's biographical sketch (1900) contains the most complete list. Of particular interest here are those writings that deal with general geology and the fossil plants of the Devonian and Carboniferous. His great work *Acadian Geology* appeared in several editions, and many men would have considered it sufficient for one lifetime. *The Geological History of Plants* is also worth mentioning. I sought for some time to find a copy of his autobiography *Fifty Years of Work in Canada* (1909), which does not seem to be very well known. It records some interesting information on the times that he lived through but is rather heavily weighted with his religious views—so much so, in my opinion, that its value as a historical record is diminished.

In following my policy of trying to recount in some detail a few important aspects of a man's work, rather than summarizing the whole, I wish to say a little of Dawson's work with the Devonian floras, where I believe he made one of his most significant contributions.

A few passages in his *Fifty Years* present the beginnings, so far as I am aware, of Devonian plant studies on the Gaspé. He says:

> Logan, [Sir William Logan, for many years head of the Canadian Geological Survey] I first met in 1841. He had come to Nova Scotia to familiarise himself with the carboniferous rocks, as developed there, in the interest of the Canadian Geological Survey. In the autumn of 1843 he again unexpectedly presented himself, in rough and weather-stained attire, and explained that he had spent the summer in Gaspé, where it had been reported that coal had been found, and had commenced there his great survey of Canada, in the hope—not destined to be realised—of the discovery of productive coal measures. . . . His visit was short, but we spent many hours over his notes and drawings of fossils, which showed that he had been studying rocks older than those of the carboniferous system, and therefore not likely to contain coal. He showed me drawings of fossil plants he had observed, which, for the first time, gave me the idea I afterwards followed up, that Gaspé might afford a fossil flora much older than that of the coal formation. [1901, p. 60]

Dawson relates that most of Logan's extensive collections from the Gaspé were lost in a shipwreck but Dawson received the remnants when he became established in Montreal. They were sufficient to incite his interest and he spent two summer vacations with some of his students exploring the coasts of Gaspé Bay and Chaleur Bay and down along the New Brunswick coast into northern Maine. He was well aware of the unique nature of these early Devonian plants and of the difficulties of dealing with them without making mistakes (1901, pp. 138–142), and in 1870 he took a large collection to England to exhibit and leave duplicates in several of the museums. His lectures and specimens were not received with enthusiasm: "to my great disappointment, the council of the [Royal] Society declined to publish my paper and illustrations, thereby losing the credit of giving to the world the largest contribution made in our time to the flora of the period before the Carboniferous age" (1901, p. 140). Not until the Rhynie plants were described by Kidston and Lang in 1917 would botanists appreciate the importance of Dawson's pioneering work.

Dawson had started to publish a few of his results before his trip to England; in 1859 he wrote a very short paper which contained the original descriptions of *Prototaxites* and *Psilophyton*, two genera of fossil plants that have since occupied the attention of many paleobotanists. The entire story could run into a small volume and I can only mention a few points along the way. *Prototaxites* is based on silicified trunks up to nearly three feet in diameter which Dawson thought had coniferous affinities. William Carruthers later showed that this was not the case, and the two became involved in a rather acrimonious dispute, which I have discussed briefly elsewhere.

It is necessary first to understand something of travel conditions and the geology of the eastern Canadian coast to fully appreciate Dawson's problems. There were no roads to amount to much in the Gaspé in the mid-1800's, travel being mostly along the coast in boats. In preparing his rather famous restoration of *Psilophyton princeps*, as it appeared in 1859, and somewhat modified in later reports, he used specimens that we now know belong to two or three different plants. The shores of Gaspé Bay and the inner (western) reaches of Chaleur Bay to the south are very rich in fossil plants of the Lower to Middle Devonian. Many of the productive localities consist of small lenses that are exposed one summer and eroded away the next. I think there are one or two places in this rather vast area of which one can say with some confidence that "Dawson collected there," but for the most part his localities have disappeared and many of his specimens have been lost or widely scattered. A tradition seems to have developed, a recurrent and erroneous one in paleobotany, that the area had been worked out. Nothing could be farther from the truth. I should like to tell something of the chain of events since Dawson's time in this area chiefly because this is *not* a unique circumstance, and one must be careful about branding a locality or area as "worked out."

There are numerous scattered and rather fragmentary reports on pre-Carboniferous plants that appeared even before Dawson's time, but they did not have any great impact on evolutionary studies or morphological philosophy until Kidston and Lang's reports on the Scottish Rhynie plants began to appear in 1917. An account of the Rhynie plants is contained in another section (see p. 120). The Rhynie plants are important because they forced botanists to consider new concepts of morphology; it was made abundantly clear that we would be better informed by searching in the early Paleozoic rocks for the real vascular plants that first evolved on the land rather than merely theorizing about them on the basis of what was known about living plants. But even the Rhynie flora did not incite the initiative in field exploration that it should have. This is perhaps partly because it seemed unlikely that deposits of such well-preserved plants would be found very often, which is true; but more important was the need for new and better techniques for extracting information from plants preserved in other ways. Most Devonian plants are pre-

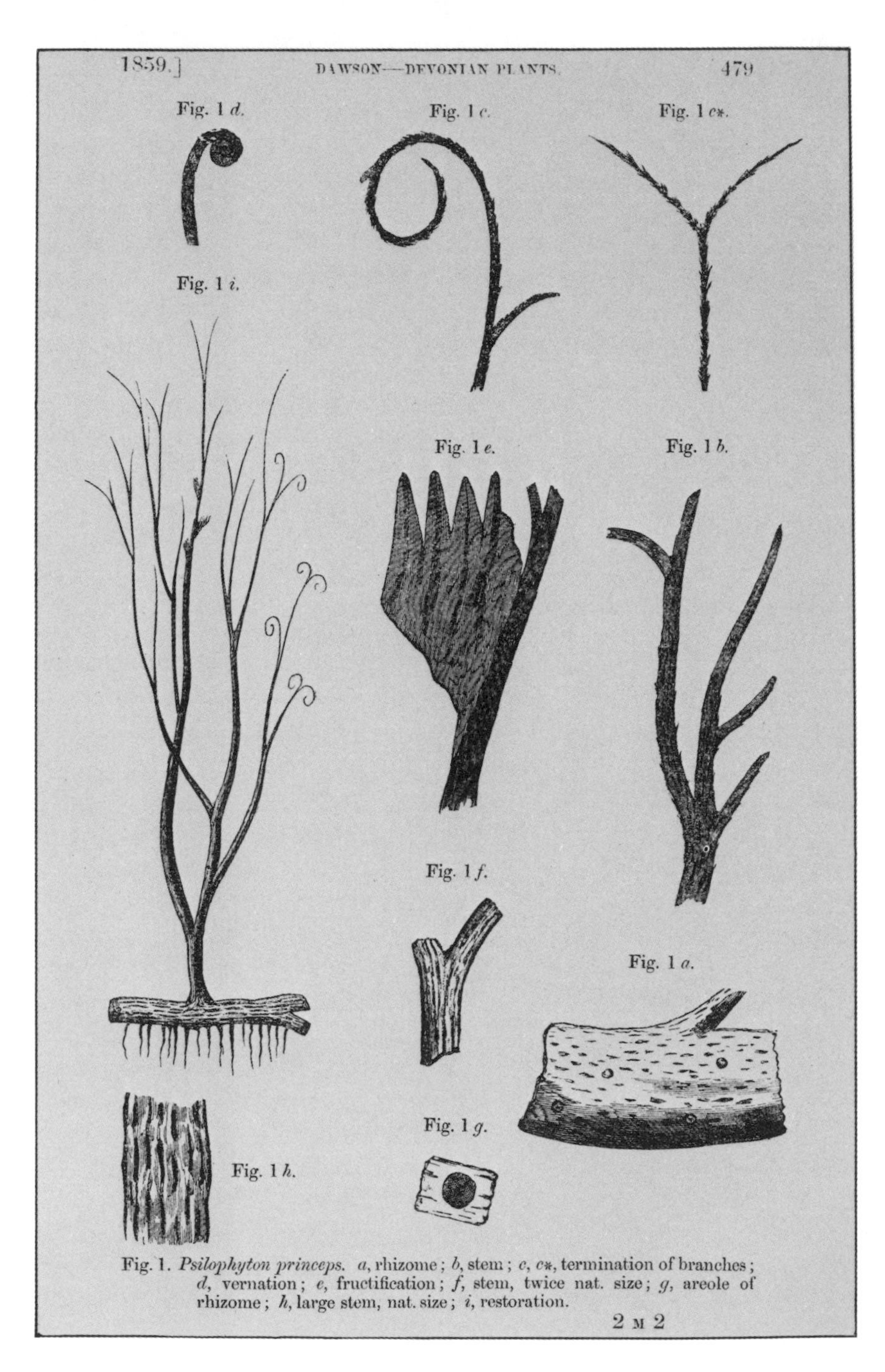

1859.] DAWSON—DEVONIAN PLANTS. 479

Fig. 1. *Psilophyton princeps.* *a*, rhizome; *b*, stem; *c*, *c**, termination of branches; *d*, vernation; *e*, fructification; *f*, stem, twice nat. size; *g*, areole of rhizome; *h*, large stem, nat. size; *i*, restoration.

2 M 2

A page from Dawson's classic little article of 1859 which appeared in the *Quarterly Journal of the Geological Society of London*.

served as impressions or compressions, or they are partially petrified. A well-preserved compression will yield gross morphological features much more readily and satisfactorily than a petrifaction, and where the vascular tissue and spore-bearing organs retain some cellular structure a good compression is superior. It has taken some time to learn how to extract the information we want and we are still learning new methods.

To the best of my knowledge, the areas along the Gaspé and northern New Brunswick coasts where Dawson did his collecting were neglected by fossil botanists until Loren Petry (1887–1970) visited the area in the 1920's. Petry was a Professor of Botany at Cornell University for many years; his written work is minimal, but he introduced several thousands of students over the years to botanical science, and helped start such notable Devonian researchers as Harlan P. Banks and Chester A. Arnold on their careers. I became acquainted with Professor Petry in the late 1930's when I was a graduate student at Washington University and enjoyed several short visits with him in Ithaca. As I had no one to guide me in my paleobotanical efforts he was helpful in several ways, among them the advice he gave me for construction of some primitive cutting and grinding equipment. I was granted the munificent sum of seventy-five dollars by the Director of the Missouri Botanical Garden for this purpose. Much of it was spent on a diamond saw, and the rest on rods and bearings and a used electric motor which I assembled to turn the saw. It worked after a fashion, tending to shake one end of the building, but I cut petrifactions and taught myself to make peels using William Darrah's formula. The peels at that time were considered quite amazing by some of my colleagues in St. Louis, but the Director found it difficult to understand why I needed seventy-five dollars for all of this if only a brush and bucket were required to make the peels!

Harlan Banks has informed me that Professor Petry made several trips to the Gaspé in the late 1920's and early 1930's; he made large collections that were deposited at Cornell. His explorations were followed by those of Banks, Arnold, and Francis M. Hueber. One of the objectives of this work was to clearly define Dawson's *Psilophyton princeps*, first described in 1859. To summarize a very complicated story, Dawson, in a study published in 1871, added a new variety, *P. princeps* var. *ornatum*, characterized by spiny axes, and he also described isolated naked branchlets bearing paired sporangia, and these were put together into a revised restoration drawing in 1870. In 1895 Solms-Laubach pointed out that the fertile branchlets were not known to be attached to the spiny axes, and in 1916 Thore Halle created the genus *Dawsonites* for the former. After a good deal of field work and study, Hueber and Banks in 1967 presented a revised description of *Psilophyton* as a genus of plants with naked or spiny axes (depending on the species) bearing terminal, paired sporangia, and the name *Sawdonia ornata* was created for the spiny axes that Dawson named *Psilophyton princeps* var. *ornatum*, which in fact bear lateral sporangia. However, Ananiev and Stepanov as a result of studying Siberian specimens that are apparently identical with the Gaspé "*P. princeps* var. *ornatum*," saw fit in 1968 to retain the generic name *Psilophyton* for these spiny axes bearing lateral sporangia.

I do not propose to judge here the relative merits of these two versions of "*Psilophyton princeps* Dawson." To me the really important aspect of the matter lies in the interest that has developed in early Devonian plants in recent years and the rather amazing assemblage of plants that have been discovered recently.

I became seriously concerned with the investigation of Devonian plants in 1958 when I had an opportunity to work in Professor Suzanne Leclercq's laboratory in Liège and to engage in a research project with her on *Calamophyton* specimens from the Middle Devonian of Belgium. However, it was not until 1964 that I devoted my research efforts entirely in the direction of the Devonian. This came as the result of an invitation from Ely Mencher and James M. Schopf to assist in paleobotanical-stratigraphical investigations in northern Maine, with a great deal of field assistance from William Forbes.

Forbes, my student Andrew Kasper, and I spent parts of several summers exploring the Lower to Middle Devonian sediments along Trout Brook in Baxter Park, Maine, and a summary account (Andrews et al., 1977) was recently published. We have also devoted parts of more recent summer seasons to extensive digging along the northern New Brunswick coast and Gaspé Bay. In both of these areas, but especially Gaspé Bay, the cliffs are erroding rapidly. The fossil plants in large part tend to be found in small lenses, and a locality that is highly productive one summer may have disappeared the next summer, other new ones having made their appearance. In a few places, though, plants are found year after year in larger deposits; a most remarkable example of this is a patch of black shale on the north shore of Gaspé Bay from which Patricia Gensel has described *Renalia hueberi*. This is a small and simple plant, beautifully preserved and very abundant in the cliff behind a delightful little beach.

We have been fortunate in finding many new plants in these Lower to Middle Devonian deposits of southeastern Canada, such as a very early heterosporous plant, new species of *Sawdonia* and *Pertica*, and others. This rich variety of early land vascular plants nicely contradicts the usual complaint about the fragmentary nature of the fossil record. Some clear evolutionary lines of development have become apparent, and these are complete enough to make it difficult to delimit species and genera in, for example, the *Psilophyton-Pertica-Trimerophyton* complex. Unstudied collections are at hand containing many more "new" plants, there are numerous fragmentary remains of other megafossils scattered through the sediments, and still others are indicated by the microfossils that have been studied by Colin McGregor; all of this evidence indicates that much remains to be found here with continued exploration and study. It will be a long time before the area is "worked out."

It may add a little variety to our story to close this chapter with an account of an incident that reveals the occasional hazards of paleobotanical field work. On one of our trips to Gaspé Bay we were prospecting the cliffs along the south shore and had left our vehicle at a small cove (l'Anse à Brillant). The cliffs immediately west of the cove are essentially vertical and perhaps seventy or eighty feet high; it was a bright and hot summer day and there was a light trickle of rock fragments falling. My companions, William Forbes and Andrew Kasper, were picking away at the cliff face with, I am afraid, more enthusiasm than discretion. I heard a voice from a hundred yards or so to the east say "Are you finding anything?"; the voice proved to be that of Colin McGregor, of the Canadian Geological Survey, whom I had not seen for some time. When he came up to us I introduced him to my two friends and, as the tide was going out, we moved down the beach a few yards and sat down to chat. Only a few seconds elapsed when we heard a sickening crash as several hundred pounds of rock landed on the beach in the exact place where Forbes and Kasper had been standing. It was the closest we have ever come to real disaster on a field trip; we abandoned our prospecting in that area.

11

North America from the Time of Lester Ward

From Ward to Wieland

IN the years that followed the great works of the men considered in the previous chapter, paleobotanical progress in the United States was centered in Washington, primarily under the aegis of the Natural History Museum and the United States Geological Survey. Again, three men stand out as preeminent in this period—Lester Ward, Frank Knowlton, and David White—while others of considerable importance came along a little later, such as E. W. Berry and Roland Brown. They were not the only important contributors to our story, but they compose a sufficiently distinct lineage as to make it fitting to deal with them as a unit.

I wish to be excused if undue space is given to Lester Frank Ward (1841–1913) who, by any standard of judgment, seems to me to have been a very great American. Chugerman (1939) calls him "the American Aristotle," a fitting title, and I acknowledge his biography, and the one by Emily Cape (1922), for much of the personal information herein. Ward was a man of two worlds—a man who lived two lives. In his biographical sketch in the *Dictionary of American Biography* (1936) he is referred to as a sociologist, although brief reference is made to his botanical work. I am sure that there are many paleobotanists who have never heard of Ward's pioneering work in sociology and I have sociologist friends who have been astonished to hear him referred to as a paleobotanist!

Lester Ward was born at Joliet, Illinois, the tenth and last child of a family "of good blood but not wealthy." His father was of a restless nature, continually moving his large family in a rather vain search for a better living. Lester obtained some schooling for a few years in Illinois, but in 1854 the family moved to Iowa; traveling in a small covered wagon and with no settled abode, he lived much of his life out of doors. His father died in January of

1858, and from this time on Lester and his favorite brother, Erastus, were on their own. Always trying to save a few dollars for their education, they worked for two years (1858–1859) for an older brother who operated a

Lester F. Ward. From *Lester F. Ward, A Personal Sketch*, by Emily Cape, 1922.

wagon-hub factory. After two years of hard work the business failed and the two brothers were left with unsalable wagon hubs as pay.

Chugerman has succinctly summarized Lester Ward's struggle for education:

> Lincoln's self-education by log fire after days of rail-splitting has grown into a world legend of the American will to succeed. Ward with far fewer opportunities than the martyred president but with greater abilities, traveled the hard road of knowledge incomparably further. In early childhood, Ward was denied even the rural education of those days, with the exception of several winters spent with the McGuffey readers in a little red schoolhouse. Since his mother, occupied with her little regiment and always on the move, could scarcely find time for tutoring, Frank and Erastus, who was his inseparable companion, started out on an adventure of self-education which is without parallel in all history, and which ten long, weary years later landed Frank in college. [1939, p. 26]

He finally saved enough to allow him to spend two terms at the Susquehanna Collegiate Institute at Towanda, Pennsylvania.

Poverty was not the only obstacle to his struggle for an education. He felt an obligation to serve the Union cause in the Civil War and enlisted as a private in August, 1862. Ward's strongly emotional character and his great urge to do something profoundly significant for humanity—which, as we will see, was clearly expressed a few years later—emerged during the war, as the following incident illustrates. "On the battle field during heavy fighting a lad carried the Flag and as he turned to speak to me, a bullet from the enemy felled the boy to the earth. It was so pitiful, so useless, so ugly, I stopped and covered the lad with his beautiful flag" (Cape, 1922, pp. 63–64).

Ward himself received several bullets from the "enemy" at Chancellorsville and was many months in recovering. Chugerman notes that, as his wounds slowly healed, "he soon recognized his enemies not in his fellowmen but in ignorance, superstition, and oppression; his weapons, no longer in the force of arms, but in knowledge—the mental dynamite which alone could change the world" (1939, p. 31).

His long struggle to obtain a formal education eventually was realized in the late 1860's; in 1869 he received the A.B. degree from Columbian College (later George Washington University) and in 1871 the degree of LL.B. Jobs at the time were not abundant, at least of the kind that Ward had hoped to obtain, but following a direct appeal to President Lincoln in 1865 he obtained a minor clerkship in the Treasury Department; several similar posts followed, and in 1881 he was appointed geologist in the U.S. Geological Survey and paleontologist in 1892, a position that he held until 1906 when he resigned to take a professorship in sociology at Brown University.

Ward's broad knowledge of fossil botany and his facility with languages are revealed in his "Sketch of Paleobotany" (1885) and "The Geographical Distribution of Fossil Plants" (1889). These are tremendous compilations of knowledge and brought together much of what was known about the plants of the past up to that time. The "Sketch" is quite readable and contains informative accounts of the important historical and contemporary figures in paleobotany. The "Geographical Distribution" does not make fireside reading but it is a great reservoir of data and indicates that Ward must have read almost everything that had been published in the field, in French, German, Spanish, Italian, and Russian, as well as English; he also had some acquaintance with several Asiatic languages. In 1862 he married Elizabeth Vought; she bore him one child who died in infancy, and she herself died after only ten years of marriage. They studied French together, and all of their love letters were written in that language.

Like many others before him, Ward was troubled by the growing volume of botanical and geological literature; unlike most of the others, Ward did something about the problem. Shortly after joining the Geological Survey in 1881 he began work on an index intended to bring together information on the sources of publication of all fossils believed to be of plant origin; the project was initiated using quarto-size notebooks, which were later abandoned in favor of slips of paper. This became known as the Compendium Index of Fossil Plant Names, and it is housed in the Paleobotanical Library of the U.S. Geological Survey; Ward was later assisted by F. H. Knowlton, David White, and Charlotte H. Schmidt, who took over the library research and preparation of the slips containing the data. The Compendium was to contain the publication sources, as well as pertinent geological and geographical information about every species of fossil plant. When I became acquainted with the Compendium in the late 1940's it included about 160,000 slips. Its existence seems to have become known, but it was not readily accessible to many paleobotanists. In 1914 Marie Stopes wrote that "there must be a complete card index of all the names ever given to fossil plants. Toward this great headway has been made in Washington, but their tens of thousands of slips are not yet complete, nor can European palaeobotanists go to Washington every time they need to use them" (1914, p. 24).

At the urging of my good friend Jim Schopf I became engaged, in about 1951, in the preparation of an index of generic names of fossil plants based largely on the information contained in the Compendium. It was published by the U.S. Geological Survey in 1955 and covered the period of 1820–1950. In 1970 I brought out a new edition for the period of 1820–1965 (Andrews, 1970) and more recently Anna M. Blazer has prepared a Supplement for 1966–1973. The *Generic Index* includes only the type species for each genus, this being the best that could be done with the funds available. I learned a great deal from compiling it but I had had enough of it when the second edition was completed. The problem of "keeping up with the literature" is still very much with us. Over the past few decades several regional efforts have been made: In 1948 and 1950 Olof H. Selling edited a *Report on European Paleobotany* that was issued from the Paleobotanical Department of the Swedish Museum of Natural History. The Birbal Sahni Institute of Palaeobotany in Lucknow periodically issues *Palaeobotany in India*. In Britain there is currently the *Report on British Palaeobotany and Palynology*, the latest issue of which (April 1976) was edited by William G. Chaloner and Alison J. Hill at Birkbeck College. In this country the *Bibliography of American Paleobotany* has been issued each year for some decades and is presently compiled under the auspices of the Paleobotanical Section of the Botanical Society of America and Arthur D. Watt of the U.S. Geological Survey.

In an effort to consolidate these regional reports the International Organization for Palaeobotany initiated a *World Report on Palaeobotany*, the first number being published in 1956; this continued under the able editorship of Edouard Boureau to the ninth number (1973), when lack of funds brought it to an end. It is time-consuming and expensive to compile and publish such reports and unfortunately their lives tend to be precarious.

But we must return to Lester Ward: in the various areas of knowledge that he probed he produced (according to Chugerman) about six hundred publications with more than eight thousand pages exclusive of his books. Among his more comprehensive works with fossil plants I believe the "Status of the Mesozoic Floras of the United States" and the *Types of the Laramie Flora* are the best known. I have "sampled" some of Ward's shorter papers which I think reveal something of both his great accumulation of knowledge and his ability to use it to critically interpret the conclusions of others, as in, for example, a short summary on the determination of dicot leaves and another on the use of fossil plants in geologic correlation (1892). In the latter paper he gives credit to David White for demonstrating by the use of fossil plants that the Gay Head sediments are Cretaceous in age and not late Tertiary, as previously supposed: "Thus has paleobotany, legitimately employed, set at rest a question which stratigraphical geology could probably never have answered. Many other illustrations of this principle might be given, but this one will suffice for all" (1892, p. 38).

To the best of my knowledge Ward gave us the word *paleobotany*, perhaps a minor achievement but a useful one. In the introduction to his "Sketch" he says, "The term *paleobotany* has the advantage of brevity over the more common expressions *vegetable paleontology* and *phytopaleontology*, while at the same time its etymologic derivation from two purely Greek words renders it equally legitimate" (1885, p. 363). This seems to have been well taken, for the "vegetables" in the literature of our science disappeared rather rapidly thereafter!

In view of his many contributions to paleontology it is a bit puzzling to read Ward's comment that "My mind has always been trimmed toward the future rather than the past" (quoted in Cape, 1922, p. 20). But the study of fossil plants was not his ultimate goal. Long before he left Washington he had been devoting much of his "spare" time to his sociological studies. His book *Dynamic Sociology* appeared in 1883 and was followed by *The Psychic Factors of Civilization* in 1893, *Outlines of Sociology* in 1897, and others.

Lester Ward left Washington in 1906 to become a Professor of Sociology at Brown University in Providence and to devote himself to being the American pioneer in sociology. I do not feel that we lost a great botanist to another science; rather, we contributed a remarkable man to a cause that he considered of the utmost importance to mankind.

Frank Hall Knowlton (1860–1926) died in 1926 before I had finished high school or indeed developed any knowledge of fossil plants, but through his publications he helped me in the early days of my career, and I feel indebted to him. Knowlton was born in Brandon, Vermont, and developed an interest in natural history at an early age. He was especially attracted to the study of birds and wrote many short articles on various aspects of ornithology, but his investigations in this direction culminated in a monumental volume of 873 pages, *Birds of the World*, published in 1909. It dealt with avian anatomy, classification, and distribution and was highly regarded by ornithologists.

He attended Middlebury College in Vermont, where he received a B.S. degree in 1884, and he then went to Washington to assist in the preparation of an exhibit by the U.S. National Museum for the Cotton Centennial Exposition at New Orleans. Shortly thereafter he obtained a job in the taxidermy shop at the Smithsonian; he had financed his way through college by doing taxidermic work for the college museum and had developed something of a reputation as a taxidermist. His interests and enthusiasm brought him to the attention of Lester Ward, who helped him obtain a position in the museum's (then) small herbarium, and in 1887 he was appointed Curator of Botany. He spent several summers in the west with Geological Survey parties and joined that organization as a geologist in 1907.

My own initial "contact" with Knowlton came from rambling through geological literature as an undergraduate searching for likely places to collect fossil plants. I combined this interest with mountain climbing in the Rockies one summer, and an article Knowlton published on the Upper Cretaceous Frontier Formation of southwestern Wyoming especially attracted me. The report was actually based on a collection that had been made in 1843 by Capt. John C. Fremont when he was exploring for a better emigrant route to the northwest. Knowlton's illustrations of the fossil plants looked interesting, and my classmate Cortland Pearsall and I found the locality in 1934, made a modest collection, and returned again in the summer of 1940. This resulted in my first significant study. The fine-grained shales contained beautifully preserved fertile fronds of *Anemia*, and using transfer and maceration methods we were able to extract considerable information about this warm-temperate fern preserved on the semidesert Wyoming hillside with its hot summer days and cold winter winds. Fronds of the subtropical fern *Gleichenia* are also present, indicating a very different climate in Upper Cretaceous times.

Knowlton's account (1899) of the petrified forests of Yellowstone National Park in Wyoming was the first comprehensive one; I still regard the area as presenting the most spectacular display of fossil plants that I know of as well as a place where one can readily appreciate the magnitude of geologic time. Early explorers in the west brought back tales of the fossil forests and the other wonders of the area that became Yellowstone Park. Probably the first scientific collections of fossil plants were made by a party under F. V. Hayden in 1871, and the fossil forests were described by W. H. Holmes in 1878. They made a great impression on him, as they have impressed many others since:

> As we ride up the trail that meanders the smooth river-bottom, we have but to turn our attention to the cliffs on the right hand to discover a multitude of the bleached trunks of the ancient forests. In the steeper middle portion of the mountain face, rows of upright trunks stand out on the ledges like the columns of a ruined temple. On the more gentle slopes, farther down, but where it is still too steep to support vegetation, save a few pines, the petrified trunks fairly cover the surface, and were at first supposed by us to be the shattered remains of a recent forest. [Holmes, 1879, p. 126]

A few years later Knowlton visited the area, and he wrote: "In the summer of 1887, Prof. Lester F. Ward and I spent about six weeks in the vicinity of the Fossil Forest, making large collections of fossil wood and leaf impressions. . . . The following season I spent two months in the same area, discovering many new beds of plants and more thoroughly exploring and collecting from beds previously known" (1899, p. 652).

Knowlton, like Ward, was concerned with ways to deal with the growing volume of literature, and his major contribution to the problem was *A Catalogue of the Mesozoic and Cenozoic Plants of North America*, which was published in 1919. Robert S. LaMotte brought out a Supplement to this in 1944 and in 1952 produced his *Catalogue of the Cenozoic Plants of North America through 1950*.

Knowlton was also a very prolific writer of popular articles on various aspects of natural history, especially ornithology. In fossil botany his most important popular piece was *Plants of the Past*, published in 1927. It is an informative volume for the nontechnical reader; Knowlton was going over the proofs of it when he died on November 22, 1926.

In the long course of preparing this volume I have read many hundreds of biographical sketches, seeking especially for personal information about the people themselves. All too often only a man's scientific publications are discussed and one learns little about his outlook on life. I would like to introduce the third member of the present trio, David White, with a few lines from an obituary written by W. C. Mendenhall (1935) which, although brief, is a concise and exemplary piece of literature:

> To his close associates in the Geological Survey and the scientific organizations of which he was a member, Dr. White was always an inspiration. His enthusiasm and industry were unflagging, his knowledge encyclopedic. His personal and professional generosity knew no limits, and to the earnest younger student who sought his counsel he would devote time without stint, pouring out for the benefit of the neophyte a wealth of information and inspiration, of suggestion and advice, which constituted both a program for a scientific career and a guide to its attainment. Always generous in his judgments, his rare displays of impatience were reserved for the slacker or the careless and particularly for lapses in ethical standards. For these he had no tolerance. But even his condemnations, although expressed picturesquely and with fervor, were couched in terms so humorous and so kindly as to convey the impression that their object was to be pitied rather than blamed.
>
> Along his pathway through life are hundreds of fellow men and women who have been helped to bear or to forget the burdens of life by his cheerful but adamant refusal to admit that there are any. In his philosophy life consisted wholly of opportunities to be made the most of, never of limitations to mourn over. How interesting and what fun it all was, and how particularly fascinating the career of science, constantly opening as it does new vistas of comprehension and understanding! [Pp. 244–45]

David White (1862–1935) was born on his father's farm in Palmyra, New York. He obtained his preparatory education at the Marion Collegiate Institute prior to entering Cornell University. For a bachelor's thesis at Cornell he made a study of the very problematical *Ptilophyton vanuxemi*. His drawings attracted the attention of Henry S. Williams, whose courses in paleontology he had taken. Thus when Lester Ward wrote to Williams inquiring whether the latter had a student who could prepare drawings of fossil leaves, which Ward needed in his study of the Laramie and Fort Union floras, White was immediately recommended. David White started for Washington on May 16, 1886, and received an appointment to the Geological Survey in October.

As I have noted briefly on a previous page, White visited the beautiful colored cliffs at Gay Head on Martha's Vineyard off the southern Massachusetts coast in 1888 and made an extensive collection of fossil plants there. The locality has been studied or visited by many notables in geology since about 1786, and the deposit was generally accepted as being of Miocene age. When White completed his study of the five barrels of fossils that he shipped to Washington, he wrote his first stratigraphic paper, "On Cretaceous Plants from Martha's Vineyard" (1890). The fine quality of the work impressed men like James Hall and John Newberry; it gave White a good start to his career, and to the best of my knowledge the Cretaceous age has not been questioned since.

In 1892 he was assigned to a study of the Upper Carboniferous plants of Missouri, which began his interest in the Paleozoic. The full account of this project appeared in his *Fossil Flora of the Lower Coal Measures of Missouri* (1893), a large volume of 467 pages and seventy-three plates. In connection with this study it is appropriate to include a few notes on the work of a physician whose collections formed much of the basis of White's monograph. Dr. John Henry Britts is typical of the many who have worked so effectively in the backstage area of paleobotanical progress. Britts was born in 1836 and began to study medicine at the age of nineteen, apprenticing first with his grandfather and later with an uncle. He served with the Confederate Army as a surgeon, losing a leg during the siege of Vicksburg. In 1865 he entered practice in Clinton, Missouri, which he continued until his death thirty years later. In the course of his frequent and extensive travels about Henry County to visit his patients he always carried his geological tools, and as time permitted he stopped at the coal mines then operating. Judging from the quantity and quality of his specimens he must have established very good working relationships with the mine operators. The major result of this activity was a collection of some eleven hundred specimens that ultimately went to the Chicago Academy of Science's Museum and another comparable collection that went to the U.S. National Museum; it is my understanding that many of White's types are included in the latter collection. Many of the Britts specimens contained varying amounts of iron pyrites and tended gradually to disintegrate, but fortunately not all of them are thus "contaminated."

In the summer of 1897, White and Charles Schuchert accompanied Peary on one of his polar expeditions, their primary objective being to recover a thirty-ton meteorite which was taken to the American Museum of Natural

History in New York. They also collected fossil plants in the classic areas on Disko Island and the Nûgssuaq Peninsula, the results being published in 1898. White contributed a study of the Permian *Glossopteris* flora of Brazil which appeared in I. C. White's report of 1908. Much later, in 1929, David White made a notable study of the Permian flora of the Hermit Shale in the Grand Canyon, which Schuchert says "is his most philosophic floral and environmental study, and is an indication of what he might have produced in much more abundance had he not been bound down by administrative work. The Supai formation, long believed to be of Upper Carboniferous age, is here shown to belong to the Permian system" (1936, p. 204).

From the viewpoint of paleobotany and geology, White's great contribution was made in revisions of Carboniferous stratigraphy. These were based initially on his study and relabeling of the many thousands of specimens in the Lacoe Collection during the summers of 1890–1893 in the U.S. National Museum, and later on extensive field observations. White had a most astute mind for correlating both kinds of evidence: "He revolutionized the general conception according to which the Pottsville, Allegheny, Conemaugh, Monongahela and Dunkard (Permian) were supposed to continue down the entire length of the Appalachian trough, proving that all the Pennsylvanian beds in Alabama, probably exceeding 10,000 feet in thickness, the entire Pennsylvanian of Tennessee, and all but a small part of the northeastern Kentucky coal field are of Pottsville age" (Schuchert, 1936, p. 190).

White's integrity and devotion to geology as a science is well illustrated by the following incident:

> In the nineties, . . . while White was doing work in the anthracite region, he noted that the lay of the rocks on one side of a valley, where coal was being abundantly mined, indicated synclinal structure. This being so, and if no faulting intervened, the coal beds should be repeated on the other side of the valley. Crossing over to test his conclusions, he saw that he was correct, and that here were buried millions of tons of anthracite unknown to the coal operators. What should he do? Resign from the Survey and turn real-estate promoter, or return to Washington and tell the Director of his discovery? He chose the latter alternative, and the facts were eventually published by the Survey. [Schuchert, 1936, pp. 197–98]

Starting in 1912 White served as Chief Geologist of the Geological Survey, and for eight years he was Home Secretary of the National Academy of Sciences; of particular importance to paleobotany, he was chairman of the National Research Council's Committee on Paleobotany from 1928 to 1934.

In his brief obituary of White, E. W. Berry wrote: "No geologist of his time had a wider influence on the scientific life of the nation, or took a more active part in that of its capital" (1935, p. 391).

Let us continue with the sequence that was started a few pages back and bring in E. W. Berry and Roland Brown; this is not strictly chronological in view of others who must be mentioned but it fits into a pattern that appeals to me and brings in people with whom I have some personal acquaintance.

In terms of quantity of publication Edward Wilber Berry (1875–1945) was the most prolific paleobotanist that the United States has had. His lack of formal education was also unique. How did he manage to produce so much under such circumstances, and were his contributions of lasting value? I may not answer these questions to everyone's satisfaction, but Berry and his record are worth examining.

Berry was born in Newark, New Jersey, on February 10, 1875. He graduated from high school but for financial reasons was unable to attend college, and entered a cotton commission house as an office boy, later traveling as a salesman through the southern states. From 1897 to 1905 he worked for the *Passiac Daily News*, eventually becoming president of the paper. It has been said that his rapid mode of writing his botanical results stemmed from this experience, but I rather suspect that it was only a contributing factor. One of his biographers says: "Berry was an extraordinary man who owed his success to inherited abilities and hard work. He is an outstanding example of what an energetic and intelligent man with motivation can achieve if given an opportunity" (Cloos, 1974, p. 57).

That Berry was energetic no one can doubt. He began to study and collect fossil plants at an early age and had produced thirty papers by the time he was thirty. During his period with the newspaper he worked part-time for the Geological Surveys of New Jersey and North Carolina. William Clark of the Maryland Survey thought well enough of his work to bring Berry in 1906 to Johns Hopkins University, where he was appointed Assistant in Paleontology at a salary of five hundred dollars a year. His task was to serve as curator of their large collections. In 1910 he was promoted to the present equivalent of Assistant Professor and became Professor of Paleontology in 1916; his administrative qualifications led to his appointment as Dean of the College of Arts and Sciences in 1929, and later to the position of Provost.

In his biography Ernst Cloos (1974) gives some piquant and concise notes on Berry's character as an academician:

> In 1916 Professor Clark urged that Mr. Berry be advanced to Professor of Paleontology because "the influence which he exerts over our students is very pronounced, probably greater than that of any other member of our staff." From then on, he was listed simply as "E. W. Berry, Professor of

Edward Wilber Berry. From a portrait by Eric Haupt in the Eisenhower Library, The Johns Hopkins University, courtesy of the National Academy of Sciences.

> Paleontology," in contrast to all other members of the faculty, who listed degrees, dates, titles. and, at times, several lines of data. His entry remained unchanged until he became dean. The contrast was striking and very typical of the man, who was no friend of pomp, glitter, and prima donnas.
>
> Berry gave a whole generation of geology students a feeling for creative research, inspired by his own example. He was never hurried or harassed and was always accessible in his room, seated at his big rolltop desk, on which was placed a board that was used for all this writing.
>
> His Saturday morning seminars were famous and are well remembered by all who ever attended. They lasted four hours and typically began with a critical review of some famous textbook. Deflation of the near great was legendary, and though he was caustic he really intended to amuse and shock his audiences. . . . One of his students who was not in his field and was afraid of his oral examination writes, "But what I remember most about him was his exceeding kindness to all graduate students. His action during that oral was that of a gentleman." [Pp. 60–61]

It is useless to attempt a general survey of Berry's written works here and I shall deal with a few, of the many hundreds, that seem representative of his best work. In 1916 he brought out one of his larger monographs, *The Lower Eocene Floras of Southeastern North America*, which dealt with plants from the clay pits of eastern Tennessee and Kentucky. Berry referred to the pit at Puryear, Tennessee, as "the most remarkable leaf-bearing clay that I have ever seen at any geologic horizon" (p. 47). I visited the Puryear clay pit with some of my students in early 1957. Two things are memorable about the visit: we met Harold Bold there with some of his students from Vanderbilt University, and it was the beginning of a long and valued friendship; and we enjoyed some fabulous collecting. The leaves and other fossils are distributed quite evenly through the clay, which is easily worked, and they make beautiful display specimens. It was the kind of digging that students dream about after sitting through many lectures and labs. On another visit one December we were caught in a sleeting rain but it was not easy to pull the students away in late afternoon, although we were all quite wet and half frozen.

There are, no doubt, mistakes of identification in this work of Berry's on the southeastern floras, and by present standards it is to some degree superficial. The recent investigations of David Dilcher and his colleagues on the clay pits of eastern Tennessee and Kentucky have opened up an entirely new and incredibly productive era of angiosperm paleobotany. The clays include well-preserved leaves, pollen, seeds, and fruits, and, perhaps most spectacular of all, flowers. Some very important publications are emerging from these studies, and they are only the beginning. I think that Berry would be delighted to see the kind of information that present-day techniques are extracting from this area. It is important to remember that he was the pioneer and laid the foundations.

The ability to locate fossils in the field is often undervalued. I have known at least three people in my career who possessed this ability to an extraordinary degree. One of them was my college colleague Cortland Pearsall. I remember one very dark night in western Wyoming when we had strayed far from our camp, wandering through rough arroyos with a thunderstorm developing and only an occasional flash of lightning to guide us. Cortland occasionally stopped and picked up pieces of petrified wood, which to me had been quite indistinguishable from the rocks of various sizes that I was trying so hard to avoid! Two others are William Forbes, a geologist at the University of Maine, and my student Andrew Kasper; the three of us worked together for nearly ten years in the Devonian horizons of Maine and southeastern Canada; their knack for locating new lenses of plant fossils as we walked along the outcrops often embarrassed but definitely pleased me. Berry apparently possessed this rare trait; as one of his biographers describes it:

> When Berry came to Baltimore in 1905, I was in residence there as a graduate student, and it was my privilege during the next five years to spend several summer field seasons with him in the southern States. By means of a canoe and camp outfit we studied the geologic conditions as they were revealed in the banks and bluffs of several of the larger southern Coastal Plain rivers. We made many trips on foot and by livery team in the interstream areas—automobiles were then only in the experimental stage of their development. Berry was in his element on these trips. He had what seemed like an uncanny ability to find fossil plants, where most geologists would overlook them. [Stephenson, 1946, p. 112]

Although much of Berry's work was with Tertiary floras, his knowledge was broad, and in 1927 he wrote a short but informative account of Devonian plants. He was familiar with important contributions such as those of Richard Kräusel and Hermann Weyland, and his conclusions were prophetic: "Although such a great advance has recently been made in our knowledge of these earliest known land plants, their chief interest still lies in the questions raised rather than in the solutions that they furnish. However, judging by the history of discovery among the Carboniferous seed ferns or the Mesozoic cycadophytes, we may now expect a rapid increase in our knowledge of Devonian floras" (1922, p. 120).

To most naturalists, and indeed to paleobotanists who are not oriented toward the angiosperms, probably Berry's best-known work is his book *Tree Ancestors* (1923). It includes chapters on geologic principles, modes of preservation, the modern forests of North America, and the geologic history of some of our more important forest trees. It was one of the first paleobotanical books that I ever read; it no doubt requires some up-dating now, but it is well written and afforded me information and inspiration.

Berry's range in subject material is extraordinary, and I can only recommend a few other of his works as good reading: "Across the Andes to the Yungas" (1921) for his ability to describe people and places, and "Far Away and Long Ago" for his vision of ancient landscapes and his ability to make them live again in one's mind—although his impressions of Patagonia are less nostalgic than those of W. H. Hudson, from whom he borrowed the title.

It is my summary opinion that American paleobotany owes much to E. W. Berry. He wrote too much and too fast, but many of his contributions were sound and well written; he knew his own faults in this direction but apparently felt compelled to move ahead at full speed. There were few active paleobotanists in this country during Berry's time, and to a considerable degree he kept the spark alive and held the fort until a new generation began to emerge in the 1930's.

In his biography of Berry, Stephenson says: "Although primarily a paleobotanist, oddly enough Berry seems not to have encouraged his students to become paleobotanists. Roland W. Brown, the only one among them who made the study of fossil plants his lifework, did so strictly on his own initiative and can in no sense be regarded as his teacher's understudy" (1945, p. 199). Whether or not they should be regarded as Berry's graduate students, both Winifred Goldring (1888–1971) and Roland W. Brown (1893–1961) received some formal instruction in paleobotany from him. It therefore seems an opportune point at which to bring them into the story.

Roland Brown, or "Brownie," as he was known to most of his friends, was born in Weatherly, Pennsylvania, and attended Johns Hopkins University, where he received his doctorate in 1926. He held positions for short periods with the Pennsylvania Geological Survey, the Pennsylvania State Forestry School, and Yale University. and then joined the U.S. Geological Survey in 1928, remaining there to the close of his career. I became quite well acquainted with him during the early 1950's on my numerous visits to Washington to work on the Generic Index project.

He was an old-fashioned type of botanist with a great fund of knowledge of natural history. Whenever the weather was at all suitable he took me on a walk around the Tidal Basin late in the afternoon before we ate dinner, and although that is not the best of places to botanize, I learned a great deal from him. Roland had a reputation for frugality that, at least in his own habits, was well earned. He had a special knack for finding small coins on the sidewalks and streets of Washington and once told a mutual friend that "in a good year he would pick up $1.25." He was a bachelor and lived a rather spartan existence, with the result that he left a considerable fortune to his heirs when he died. With others he was a generous man, and I always looked forward to my visits to the Natural History Museum for the opportunity to chat and dine with him.

Roland Brown's scholarly interests were sharply divided between fossil plants and word study. At 4:15 in the afternoon he would close his books and put away his notes and papers dealing with paleobotany and start work on his dictionary of scientific words. I believe he is generally acknowledged as a great etymologist. In 1927 he brought out a small book entitled *Materials for Word Study*; this evolved into a much more comprehensive volume, the *Composition of Scientific Words*, which he published himself in 1954. It is a unique and indispensable work designed especially to aid biologists in creating new specific and generic names. I well recall the day when he completed his manuscript, of some sixteen hundred pages, and his irritation at having to carry this mountain of paper to the appropriate office in the Geological Survey for official approval of its publication, although it was done on his own time and at his own expense.

He was a quiet and modest man who rarely attended scientific meetings, but he had a good sense of humor; his biographer Sergius Mamay recounts an anecdote that reveals these traits very effectively. As a preface it is necessary to add that one of Roland Brown's chief points of expertise was a special understanding of the Cretaceous-Tertiary boundary resulting from many summers of field work in the western states. Mamay writes:

> Sometime late in the 1940's a young geologist with only a year or so of experience with the Geological Survey was assigned to head a field party on a coal-mapping project in Montana. Inasmuch as the project centered in a remote area, the work was to be done from a field camp. A short time before leaving for the field, the party received word from Washington that they could expect a distinguished visitor, in the person of Dr. Roland W. Brown, from the Branch of Paleontology and Stratigraphy; Doctor Brown would arrive soon after the party had established their field camp. Consequently, special efforts were made to set up a neat and efficient camp because everybody—especially the chief—was anxious to make a favorable impression on Doctor Brown. Arrangements were made to hire a laborer, who would come out on a mail truck from a town some 25 miles away, to assist with the more menial chores involved in setting up camp. On the scheduled morning a meek-looking, somewhat shabbily dressed

Roland W. Brown, during his days in the West.
Photo received from Sergius H. Mamay.

individual arrived on the mail truck, and the young USGS party chief instructed him to grab a pick and shovel and start digging drainage ditches around the tents. The laborer set to work immediately and continued to work steadily. After several hours of activity on the part of the field party and the laborer, the party chief felt that the work was not progressing as fast as it should, and remarked that the pace should be stepped up a bit because Doctor Brown would probably arrive the following day. The "laborer" then laid down his ditch-digging tools and wryly announced: "I am Doctor Brown." [Mamay, 1963, p. 80]

I regard Brown's 1934 review of the Eocene Green River flora as one of his more significant studies and representative of his knowledge of Tertiary floras. He recognized some 125 species in the assemblage and concluded that it indicated a warm-temperate region of much lower elevation than the present area where the fossils are found. The Green River shales extend over a considerable part of western Colorado, eastern Utah, and southwestern Wyoming, and Brown notes, quite aptly I think, that much remains to be learned about the flora. I have prospected along the outcrops a few miles west of Kemmerer, Wyoming, where the Haddenham family has dug out so many beautiful fossil fish, and I once spent two days excavating a fine fan-palm leaf there. It represented an Eocene landscape that certainly must have been very different from the present semidesert sagebrush hills.

One of Brown's better known papers, from the standpoint of general interest, is his account of the Triassic palmlike fossil *Sanmiguelia lewisi* from southwestern Colorado. Several other paleobotanists have studied this fossil more recently. He devoted the last few years of his career to a study of Paleocene plant assemblages from the Rocky Mountains and the Great Plains. The resulting monograph was published posthumously as U.S. Geological Survey Professional Paper 375 in 1962.

He also had something of a flair for problematical fossils or presumed fossils; several drawers near his desk were filled with especially troublesome specimens and errors committed by his colleagues. He produced several papers on them, which are trivial in a sense but reveal his wide knowledge and acute power of observation and interpretation. One of these was on algal pillars from the vicinity of Rock Springs, Wyoming, that had been mistaken for geyser cones. The most interesting to me was his account of structures superficially resembling branching blue-green algae that were found in geodes.

Winifred Goldring enters into the progress of American paleobotany in connection with one very important discovery. She received her A.B. degree at Wellesley College, engaged in graduate study at Harvard and Columbia, and, as noted above, spent some time with E. W. Berry at Johns Hopkins. Much of her life was lived, and her geological studies carried on, within the borders of New York State. She is best known for her many contributions to the invertebrate paleontology and stratigraphy of New York.

It is necessary to go back in time some years to the origins of a Devonian plant story to which she made a notable contribution. In the autumn of 1869, flood waters sweeping down Schoharie Creek near the village of Gilboa, New York, exposed stump casts that are among the most interesting of American plant fossils. At a later date excavations in the area revealed more specimens as well as foliage that was probably borne by these unique trees. J. W. Dawson in 1871 was the first to study them, and he placed them in the fern genus *Psaronius*. In 1925 Miss Goldring prepared a detailed report on the stump casts and the associated foliage. The latter bore terminal ovoid bodies which she interpreted as seeds, and she gave the plants the generic name *Eospermatopteris* (the dawn seed fern). Subsequent investigations revealed that these were spore-bearing organs, and the plant is now assigned to the progymnosperm group. Although her classification proved to be in error, Miss Goldring's study stands as a fine addition to Devonian plant literature, but she did not stop with a few published papers. Largely through her efforts and under her direction a magnificent restoration was prepared by the artist Henri Marchand and his two sons. This was set up in the State Museum at Albany and was opened to the public in 1925; it is an especially successful effort to display fossils in a fashion that is attractive and informative. When I was a beginning graduate student at the University of Massachusetts I paid a visit to the Albany Museum in company with Chester Cross (presently Director of the Wareham Experiment Station and a world authority on cranberry culture); we were treated most kindly by Miss Goldring and Rudolf Ruedemann, who gave us leads to several interesting fossil localities in the Albany area.

My own serious initiation into paleobotany was during my first year of graduate study, at the University of Massachusetts in 1934–35; as a master's thesis project under the direction of Ray E. Torrey, who had studied with E. C. Jeffrey at Harvard, I studied fossil coniferous woods that I had gathered on trips to the Rocky Mountain states. This did not result in anything significant on my part but it taught me plant anatomy and introduced me to a distinctive aspect of American paleobotany, the study of coniferous woods and other plant remains in the Cretaceous clays of New York, New Jersey, and Martha's Vineyard. The most important single publication that emerged from studies of the Cretaceous in this area is Arthur Hollick and E. C. Jeffrey's *Studies of Cretaceous Coniferous Remains from Kreischerville, New York*, which appeared in 1909. The paragraphs that follow are based on the investigations included in that publication, but the research and general interests of both these men were by no means thus confined.

Arthur Hollick (1857–1933) was born at New Brighton, Staten Island, New York, and lived much of his life in that area. He graduated from Columbia College in 1879, having developed a special interest in geology and paleontology. It was there that he met N. L. Britton, a great botanist who became his lifelong friend. Hollick's first publication was *The Flora of Richmond County, New York*, written with Britton in 1879.

For about a year after graduation Hollick served as superintendent of a mine in California, but he returned to the city of New York in 1881 and was appointed Assistant Sanitary Engineer, a position he held for nearly ten years. He remained a consulting sanitary engineer for much of his life.

Hollick had once studied local botany and geology under the guidance of Professor John S. Newberry, and Newberry was to be influential in directing the course of Hollick's work. Hollick acted from time to time as private assistant to Newberry and prepared many of his drawings of fossil plants and fishes. In 1890 he was appointed a Fellow in Geology at Columbia College and carried a part of Newberry's lectures in 1891–92 when the latter was ill. He held several other positions at Columbia, and when the fossil-plant collections were transferred to the New York Botanical Garden Hollick became Assistant Curator there.

Probably Hollick's best-known study is *The Upper Cretaceous Floras of Alaska* (1930), which covers a wealth of gynmosperm fossils (cycadophyte foliage, *Ginkgo*, coniferous foliage, and cones) and many dicot leaves. Although significant in itself it is probably of more importance as an indication of what more detailed studies in the future may reveal. In fact, as I write this I have just received in the mail Jack A. Wolfe's comprehensive work, *Paleogene Floras from the Gulf of Alaska Region.*

Hollick spent four months in Alaska in 1903 under the auspices of the U.S. Geological Survey, exploring the Yukon River by canoe for a distance of more than a thousand miles, from Dawson to Anvik. He made the trip in company with Sidney Paige, a field assistant, and John Rentfro, a cook and general camp assistant. They left Seattle on June 1, 1903, and traveled by steamer to Skagway, Alaska, then by railroad to Whitehorse, Yukon Territory, then via steamboat down the Yukon to Dawson. Hollick describes his purpose as follows: "In 1903 I was detailed to make further investigations in the Yukon region, with the special object of making collections of fossil plants at all available localities and determining, if possible, their correct stratigraphic relations. One of the results of these investigations was the collection of a large amount of paleobotanic material at some 40 localities on the banks of Yukon River between Eagle and Anvik, from 24 of which Cretaceous plants were identified" (1930, p. 2). He sent nearly a ton of fossils to Washington, where he devoted six months to their study.

Hollick also had a long-standing interest in the Natural Science Association of Staten Island and a considerable knowledge of the natural history of the area. This eventually resulted in his 1909 publication with Jeffrey on the plant remains of the Cretaceous clays. This seems to me to be a real landmark in American paleobotany from the standpoint of the fossils recovered and the techniques that had to be developed to study them. The brown, nondescript remains had to be carefully washed out of the clay, kept from drying, and then subjected to maceration or celloidin-embedding methods similar to those used for studying modern woody materials. Judging from a notation in the preface of the report, the preparation and study of this material was largely the work of E. C. Jeffrey, an accomplished anatomist. Their monograph of 1909 and other shorter papers brought to light a considerable coniferous assemblage, including well-preserved woods of pine or pine-related plants and the distinctive *Prepinus* with its multi-needled branchlets, which must have looked like an oversized larch in life, as well as the foliage and cones of other conifers.

Hollick also made extensive collections of fossil plants in Puerto Rico, the results of which appeared in 1928. In the winter of 1932–33 he traveled widely over Cuba with Brother Léon, Professor in the College of La Salle at Havana; they made abundant collections which Hollick was engaged in studying at the time of his death in 1933.

Edward C. Jeffrey (1866–1952) was an influential botanist whose interests were by no means confined to fossil plants. His undergraduate studies were carried on at Toronto and his graduate work at Harvard University, where he spent the greater part of his career. His philosophy and theories of plant evolution, based largely on anatomy, were distinctive and stimulating. He is well known for his division of vascular plants into two great groups, the Lycopsida and Pteropsida, upon which he based his 1917 textbook, *The Anatomy of Woody Plants*. Although long out of print it is a unique anatomy text, strongly biased with Jeffrey's views, to which he was loath to accept much opposition! His evolutionary concepts were based on his "three R's," or canons of comparative anatomy—retention, reversion and recapitulation. I was firmly inducted into the fellowship of this philosophy by Ray Torrey, and it took me some time to recover from it and to realize that there were other points of view. Jeffrey overworked his canons, but they added zest to the study of anatomy, and the fabric he wove from those threads of information gave one a temporary picture of unity.

Professor Jeffrey was very much interested in the botanical origins and structure of coal, and he made several contributions to the literature as well as to techniques for making thin sections of coal. This is beyond the scope of my story, but his 1924 memoir on *The Origin and Organization of Coal* has seemed to me to be an especially informative account for the nonspecialist.

Aside from a few years at Harvard with Jeffrey, Ray Ethan Torrey (1887–1956) spent most of his life in the vicinity of Amherst, Massachusetts. He was born in nearby Leverett, and after receiving his B.S. degree at what became the University of Massachusetts, he made an extended tour of the United States collecting fossil coniferous woods, taught at Grove City College in Pennsylvania for a short time, and then returned to Amherst. His graduate research resulted in a memoir on American Mesozoic and Tertiary coniferous woods. This was the third part in a sequence that Jeffrey started in 1903 under the title *The Comparative Anatomy and Phylogeny of the Coniferales*; the first part dealt with *Sequoia* and the second part (1905) with the Abietineae. R. E. Torrey was one of the great botanical teachers of his time; his introductory text was a leader for many years, and some distinguished American botanists received their initial instruction and inspiration at his hands, including Chester Cross, Otto Degner, Theodore Delevoryas, John Hall, Oswald Tippo, and Taylor Steeves.

Several men who lived in the era we are now considering deserve brief mention for their interest in fossil botany. In about 1882, Dawson wrote to Asa Gray in Cambridge asking him to recommend a young man who could initiate botanical studies at McGill University. David Pearce Penhallow (1854–1910) was Gray's choice and he was appointed to the Chair of Botany at McGill where he served from 1883 to 1910.

Penhallow was born at Kittery in southern Maine, educated at Amherst College, and shortly after graduation he went to Sapporo, Japan, in company with William S. Clarke, President of Amherst College, to help the latter found the Imperial College of Agriculture, and he remained there from 1876 to 1880. Penhallow had many interests. According to Jeffrey (1911) he was the first to visit and live among the Aino group in the Yezo and Kuriles of Japan. He made several contributions to paleobotany jointly with Dawson in his early years; later he brought out several papers on *Prototaxites* and a short account on early Devonian plants from New York and Pennsylvania in 1893. He was one of the leading authorities on the structure and identification of coniferous woods, and in 1902 he produced a paper that dealt with the first fossil Osmunda (*Osmundites skidegatensis*) found in North America.

When his health became quite poor in 1910, he decided to spend a year in Cornwall, England; he sailed with Mrs. Penhallow on October 20 and died in the course of the voyage to Liverpool.

In 1880, William M. Fontaine (1835–1913) and Israel C. White (1848–1927) brought out *The Permian or Upper Carboniferous Flora of West Virginia and S.W. Pennsylvania*, a pioneer work that laid the foundations for many later studies. Fontaine was a contemporary of Lester Ward's and fought on the other side in the Civil War as an artillery officer. He was born in Virginia and studied at the University there, followed by a year at the Royal School of Mines in Freiburg, Saxony. From 1873 to 1879 he served as Professor of Chemistry and Geology at the University of West Virginia, where he was associated with I. C. White. In 1879 he accepted the Corcoran Chair of Natural History and Geology at the University of Virginia and his paleobotanical interests shifted to the Mesozoic. In 1883 he brought out his work *The Older*

Mesozoic Flora of Virginia and this was followed in 1889 by *The Potomac or Younger Mesozoic Flora.* These are now of historical interest, but F. H. Knowlton wrote of them at the time as follows:

> The full report on the older flora was issued in 1883. . . . In this he presented the full and elaborate description of the flora of the Richmond coal basin, which has furnished the most complete and best preserved of American Triassic floras. This became at once the standard work of reference on its subject and did much to stimulate investigation by others in beds of similar age, notably in Pennsylvania. . . . During the succeeding five or six years Professor Fontaine was engaged in collecting and studying what had developed into the rich and varied flora of the so-called Younger Mesozoic of Virginia and Maryland. . . . In a number of ways this was an epoch-marking work, for to Professor Fontaine belongs the honor of having described the oldest angiospermous flora known. [Quoted in Watson, 1914, p. 9]

Fontaine also contributed much to Ward's "Status of the Mesozoic Floras" (1900), which includes studies of the Oregon Jurassic flora and brief accounts of other collections from Alaska, California, and Montana.

In view of the current interest in the earliest forms of plant life on the earth I would like to include just a few lines on one of America's great geologists, Charles Doolittle Walcott (1850–1927). Among his other achievements Walcott was an authority on the trilobites and I believe the first fossil of any kind that I collected was a large *Paradoxides* from the slates of Braintree, south of Boston, Massachusetts.

Walcott served as Director of the U.S. Geological Survey from 1894 to 1907 and was Secretary of the Smithsonian Institution for twenty years. One biographical sketch (Smith, 1927) gives Walcott credit for being instrumental in the founding of the Forest Service and the Bureau of Mines. He produced two interesting papers on Cambrian algae—"Pre-Cambrian Algonkian Algal Flora" in 1914 and "Middle Cambrian Algae" in 1919—which may, I suspect, draw more attention in the future than they have in the past.

In spite of his heavy administrative duties Walcott spent a good deal of time in the west. One biographer recorded an interesting incident relative to the discovery of the Burgess shale fossils:

> One of the most striking of Walcott's faunal discoveries came at the end of the field season of 1909, when Mrs. Walcott's horse slid in going down the trail and turned up a slab that at once attracted her husband's attention. Here was a great treasure—but where in the mountain was the mother rock from which the slab had come? Snow was even then falling, and the solv-

> ing of the riddle had to be left to another season, but next year the Walcotts were back again on Mount Wapta, and eventually the slab was traced to a layer of shale—later called the Burgess shale—3000 feet above the town of Field, British Columbia, and 8000 feet above the sea. [Schuchert, 1928, pp. 283–84]

Two of the genera of algae that Walcott described, *Waputikia* and *Wahpia*, were obtained from Mt. Wapta.

G. R. Wieland's two volumes on *American Fossil Cycads* (1906, 1916) are among the great classics in paleobotanical literature and it is my impression that their importance and the wealth of information contained in them was recognized in Europe before it was in this country. They contain a very detailed historical account of the discovery of petrified cycadophyte trunks and are a great mine of information on the morphology and anatomy of the living cycads and the fossil cycadeoids. I have never read through them entirely, but they have served as a most valuable source of information in my teaching in several courses and I expect that they will do so for others for a long time to come.

George Reber Wieland (1865–1953) was born in Pennsylvania and received his undergraduate education at the Pennsylvania State College; he studied for a time at the University of Göttingen and the University of Pennsylvania and went to Yale University in 1898, where he received the Ph.D. degree two years later.

Wieland went to Yale initially with the intention of studying vertebrate paleontology with O. C. Marsh, but two events combined to direct his interests in another direction. During a summer of field work (presumably 1898) he met Lester Ward, who had made a collection of cycadeoid trunks at the Minnekahta locality in the Black Hills of South Dakota in 1893; Ward encouraged him to study the unique petrified trunks and that autumn Professor Marsh purchased a large collection of the fossils for the Peabody Museum at Yale.

For those who are not familiar with these unique fossil plants it may be noted that they consist of silicified trunks, some of which are columnar and probably attained a height of several meters; others are cone-shaped, or resemble certain cacti with bulbous branches. They differ from the true cycads most notably in the organization of the cones and the way they are borne among the leaf bases instead of at the apex of the trunk. Collecting and purchasing brought Yale hundreds of specimens, many of which are well preserved but difficult to work with because of the hardness of the petrifactions.

Wieland developed something of a mania for the acquisition of specimens

George R. Wieland, in the Black Hills, South Dakota, 1935.
Courtesy of the Peabody Museum of Natural History, Yale University.

and the preservation of the site in the Black Hills where so many had been found. Owing to his energetic efforts, a Fossil Cycad National Monument was officially proclaimed on October 21, 1922; no funds were ever appropriated for a museum or other facilities, however, and the monument was disposed of by an act of August 1, 1956, and the lands returned to the public domain. I recall that the monument appeared on road maps of the early 1940's; Mrs. Andrews and I once spent a day driving and hiking about in the general area, to the amusement of some of the ranchers; we found no fossils.

Biographical information on Wieland has not been easy to obtain. As evidence of the fact that he and his works seem to have been more appreciated outside this country than within it, he attained the status at Yale University of a mere research associate, but was held in high regard by certain European royalty. He was awarded the Archduke Rainer medal in Vienna in 1914, and, as Carl Dunbar has written: "I well remember the occasion when the Crown Prince of Sweden came to Yale to receive an honorary degree. . . . He had done some reserach on his own on the cycadeoids and was deeply impressed with Wieland's two magnificent volumes, and for this reason he wanted to meet Wieland and visit with him. There was some dismay in the administrative circle when they found that the Crown Prince was not interested in the social function they had planned for him" (personal letter).

Wieland was apparently something of a "character," and with some traits that did not endear him to all of his associates. Professor Dunbar also says: "During my early years at Yale I had only a speaking acquaintance with him. I do know, however, that he frequently came to the preparation laboratory for vertebrate paleontology to visit and talk by the hour. . . . Since this interfered with the work of the preparators, Director Lull moved to get him and his great cycad collection shifted to the Botany laboratory in Osborn Hall." When Dunbar became Director he had the collections moved back to the Museum, and Lyman Daugherty was brought in for a summer to catalogue them. Professor Daugherty has kindly supplied a few notes on this task, which must have been rather tedious:

> When I arrived at Yale Dr. Wieland's office was in the basement near the furnace. He did not allow anyone, including the janitor, to enter his office. The place was filled with boxes, crates, and fossil cycads. There was a small path to his desk and all of the debris was covered by coal dust. They gave me two Irish women and the janitor to clean the place. The janitor had been warned by Wieland to never enter his office, but I promised him I would take full responsibility for him being there. We were busy tossing out boxes and crates when Wieland arrived and I thought, here comes the fireworks. He sat down on a box and said, "Keep on with the work, you are doing a damn good job." [Personal letter]

In *American Fossil Cycads* Wieland clarified many aspects of the cycadeoid group, especially the detailed structure of the cones or "flowers," by his study of the petrified specimens from South Dakota. Perhaps their most extraordinary feature is the microsporangiate disc—the very complex pollen-producing organ that surrounded the central conical axis on which the seeds were borne. He prepared a restoration of a mature cone which appears on page 106 of Volume 1; it shows the disc spread open, as he believed it would be when discharging pollen. It is a striking illustration of a unique botanical structure and one of the most frequently copied in the entire realm of fossil botany. I think it may be said to be the focal point of his treatise, which captured the imagination of many botanists who looked to the cycadeoids at that time as angiosperm (flowering-plant) ancestors. The pollen-bearing disc was not actually found in this open state. In more recent studies of these plants Theodore Delevoryas and William Crepet have presented additional information on the reproductive biology of the cycadeoids, and they show the pollen organ in an "unexpanded" position.

Much is known about the extinct Cycadeoidales (or Bennettitales) and it is very likely that more will be learned about them in the future. Tom Harris called them "the strange Bennettitales" in the Seward Memorial Lecture that he delivered at Lucknow in 1971:

> I am sure that if Nature had been kind and had spared us just one member of the class as she did for the Ginkgoales much of the mystery would never have been thought of. This survivor would be a plant so much studied that it might seem staid, almost dull. But we have no such plant; we know the Bennettitales as flourishing in the Middle Triassic and throughout the Jurassic and into the early or Middle Cretaceous and then suddenly and dramatically vanishing. Their known span is about 100 million years and during this time they were of major importance, at least in the fraction of the world's vegetation that is known to us. [Harris, 1973, p. 3]

Second only to the *American Fossil Cycads* in size and importance is Wieland's *The Cerro Cuadrado Petrified Forest* (1935). The focal point of this work is a description of the wood and spectacular seed cones of the araucarian conifers, a group of southern-hemisphere evergreens in the Cerro de Madre y Higa of Patagonia. As in his cycad volumes, Wieland also supplies a great deal of peripheral information that is interesting to a general reader and useful in teaching. He includes a detailed discussion of petrified forests in various parts of the world, of log rafts, of the petrifying process, and an account of modern araucarian forests. The fossils were known to local people long before they were collected for botanical study. Elmer S. Riggs was probably the first trained collector to visit the area and he has the following to say about the site: "Having reached the locality in the vicinity of Sierra Madre y Higa, we found a considerable number of fossil trees, some with stumps standing, others lying prone with broken branches and cones scattered about them, revealing a forest of fossil *Arecaria* or Brazilian pines preserved on the site where it had grown" (Riggs, 1926, p. 544).

In 1909, Wieland explored the Mixteca Alta region of southern Mexico, which resulted in another of his major contributions: *La flora liásica de la Mixteca Alta*, the text appearing in 1914 and the atlas in 1916. In a preliminary account written in 1913 he notes: "The 'Mixteca Alta' or upper country of the Mixtecas is only a more or less indefinite portion of the plateau and mountain region of Oaxaca and adjoining states occupied by the original Mixtecan tribes" (p. 251). The "Mixteca Alta" flora is a large and varied one of mid-Mesozoic age and dominated by cycadophytes. Wieland's book is a fine work revealing many fascinating plants but I think even more important in being prophetic of what may be revealed in the future with continued exploration and the application of modern study methods. Indeed, Theodore Delevoryas made several trips to the area in 1966 and in collaboration with R. E. Gould described a unique fructification called *Perezlaria oaxacensis*, a rather stout branching system bearing whorls of five to eight saclike bodies which are probably pollen-bearing organs. They suggest a realtionship to some of the Mesozoic seed ferns and are associated with *Glossopteris*-like foliage. This in turn introduces a most interesting aspect of the flora, that is, the presence of leaves that seem to closely resemble those of the identifying plant of the "*Glossopteris* flora" of the southern hemisphere. Delevoryas and Person (1975) illustrated some fine examples of this foliage, to which they assign the generic name *Mexiglossa*. It seems to be a case in which a distinctive foliage type existed or evolved in two or more groups of plants that are actually not closely related on the basis of reproductive structures. This is not a rare problem in paleobotany, and Delevoryas and Person conclude that "until further collecting and investigation reveal the true affinities of the Oaxacan glossopteroids, their existence in a flora more typical of the Yorkshire, England or Rajmahal Hills, India Jurassic floras will remain an interesting dilemma" (1975, p. 119).

Professor A. C. Noé, who plays a significant role in the second part of this chapter, also visited the Mixteca Alta area, and on October 5, 1935, he wrote to Wieland:

> This is just a short note to tell you that I have returned from Mexico. Part of the time was consumed there attending the Seventh Pan-American Scientific Congress and as soon as that was over I went to the Mixteca Alta. I started from Tehuacan where I rented a Ford car which was the most

> dilapidated specimen of its kind that I ever saw and drove over an undescribably bad road to Huajuapam where I got horses and a mule and went with four Indians to the property of the former Caxaca Iron and Coal Company. This property is closed now and has become a coal reserve of the Mexican government. One of the Indians climbed on the roof of the house and got in and opened it and we slept in it. I was accompanied on this trip by a German geologist who is a member of the Instituto Geologico, Dr. Mulleried. I collected first from the dump pile of the mine and worked along the Barranca of the El Consuelo River and found a rich supply of fossil plants. After two days collecting the two boxes which I brought along were filled, and the mule laden, and the return trip taken. It was still in the rainy season but we got through the rivers all right. . . . There is an enormous amount of material still available and it would pay to make an extensive expedition there. . . . I believe much could be done with an appropriation of five hundred to six hundred dollars. [George R. Wieland papers, Yale University Library]

The Later Years

When I began to finalize my plans for this story a starting and ending point were of immediate concern. Reading through some of the late seventeenth- and early eighteenth-century literature in the Cambridge libraries led me to the decision I reached concerning the beginning. Selecting a termination point did not prove as easy; the safest and easiest procedure would probably have been to bring the account up to a few decades ago, omitting the work of all living paleobotanists. Everyone with whom I discussed the project vetoed this plan—and I think not for selfish motives. But the quantity and diversity of publications in recent decades and the number of people involved would require an encyclopedic work of several volumes to be inclusive and fair to all. I resolved the dilemma, or dispensed with it, by including here and there certain important continuing lines of research that are still very active. This did not, however, relieve me from presenting a somewhat more comprehensive account of the development of fossil botany in the United States in recent decades, for which I felt both an urge and a special obligation.

Thus the remainder of this chapter presents a brief record of what seem to me to be representative and important developments in American paleobotany of recent years. It is drawn from my own experience and from several regional histories, of which *Development of Paleobotany in the Illinois Basin* (Phillips, Pfefferkorn, and Peppers, 1973) is exemplary.

Although we got off to a rather late start as compared with our European counterparts, American "coal-ball paleobotanists" have contributed a wealth of information on the plant life of the great Coal Age, and there is presently no end in sight for this productive line of research. It is appropriate to introduce this phase of our history with some notes on the life of Adolf Carl Noé (1873–1939), although coal-ball petrifactions had been found in this country some decades before Noé demonstrated their abundance. Of real importance is the fact that he recognized the importance of this source of information and inspired others to delve into the treasure chest.

I never met Noé and my information is of necessity drawn from various sources, especially the informative biography by Croneis (1940). He was born in Gratz, Austria, and attended the University of Gratz, where he studied botany and paleobotany, thus providing a link with our story of European developments. He also served as an assistant to Constantin von Ettingshausen, who was well along in years at the time, and when he died in 1897 Noé transferred to the University of Göttingen. After two years there, and with a strong interest in a more democratic society, he decided to emigrate to this country. Thus in the fall of 1899 he entered the University of Chicago from which he received a Ph.D. in Germanic languages in 1905. After short periods of teaching at Burlington, Iowa, and at Stanford University he returned to Chicago, where, from 1903 to 1923, he served as a staff member in the German Department and an assistant librarian. In 1923 he returned to his original interest, becoming a staff member in the Botany and Geology Departments where he devoted his teaching and research efforts to paleobotany. His interest had been activated in a very real way prior to 1923. In 1921 he joined the staff of the Illinois Geological Survey, an appointment that he held for the rest of his life and which was of considerable significance to American paleobotany, as James Schopf has noted:

> Noé had a considerable interest and acquaintance with coal and coal mining. During part of the '20s he usually worked for the Illinois Survey during the summer, and with a Survey driver, for he did not drive a car himself. Noé would make the rounds of the coal company field offices in the state dispensing good will and advice, and bringing back any available collections of coal balls and plant fossils to Chicago for his students to work on. I think he was truly an ambassador of good will between the coal operators and the Illinois Survey. They were all good friends of Professor Noé. [Quoted in Phillips et al., 1973, p. 10]

Noé began to collect coal balls in the early 1920's, his first published announcement being made in *Science* in 1923. I think it is fair to say that Noé's own studies of coal-ball plants are of minimal significance. He had interests other than studying intensively the plants to be found in these petrifactions, and Croneis suggests that his work in this direction was also dampened by his mistake in identifying a medullosan petiole as an angiosperm; it was a mistake that the knowledgeable European paleobotanists of the time did not forgive. Noé's publications on coal-ball plants would not mark him as an important paleobotanist, and this was my own opinion until I came to know more about him and his work as a whole. In reference to James Schopf's comment above, I recall that my own collecting activities in the Illinois coal mines, which started in 1940, were always met with cordial cooperation from the operators and miners and Noé's name was frequently heard. This foundation of good will has contributed to the success that so many of us have enjoyed since his time.

For those who are not familiar with coal balls I strongly recommend the 1976 booklet *Fossil Peat from the Illinois Basin* by Phillips, Avcin, and Berggren. This is a detailed account of the occurrence, plant contents, and methods of study of these important petrifactions.

In the early 1920's Noé devoted much time to collecting and studying the fossils in the famous Mazon Creek nodules. A result of this is his book *The Pennsylvanian Flora of Northern Illinois*, which I found very useful as a student and later in teaching. The nodules, or concretions, contain a wide variety of plant and animal remains and were originally found in Mazon Creek in northern Illinois, and later in great abundance as a result of strip-mining operations. I will engage in a short digression here to mention two amateur collectors who were intimately involved with this phase of American paleobotany.

John McLuckie of Coal City, Illinois, was a shovel operator for the Northern Illinois Coal Company who began to notice and collect the nodules in the late 1920's. I became acquainted with John and his wife Lucy in the early 1940's, and on several occasions John led my students over the vast dumps to good collecting areas. This was some time before "rock hounds" became numerous and one was assured of finding some good specimens. On one visit late in the fall we ended a rainy day very wet and cold; the collecting had been good but we faced a fifty-mile drive to our lodgings. The McLuckies took us into their home, where everyone was able to change to dry clothes after a hot bath, and we learned more about the fossil flora from the "museum" in the McLuckies' basement.

Frederick Oliver Thompson (1883–1953) was another amateur who amassed a vast collection of the Mazon Creek nodules which were deposited at Harvard University where he had graduated in 1907. Fred became interested in fossils in 1930 and employed several boys during the summer months to gather and select the nodules from the Coal City mine dumps. He later became interested in coal balls and collected some especially well preserved material in the vicinity of his home city of Des Moines, Iowa. I was introduced to Fred by a former classmate of his and made two visits in 1944 to his home. The basement of his large house was pretty well occupied by his coal balls, which he enjoyed cutting himself. Much of this material also went to Harvard but he kindly turned over some fine specimens to me, including a new medullosan seed fern which I described in 1945.

Fred Thompson and John McLuckie were keen and enthusiastic collectors and extraordinarily generous; they wanted their fossils placed where they would be of the greatest service. Not all amateur collectors are as altruistic.

Noé was a man of many facets. Croneis notes that he "was an enthusiastic horseman, a redoubtable fencer, and an expert marksman," He coached the Chicago fencing team and was the instructor for the rifle club. Ralph Chaney records that he first met Noé when the latter was drilling a University of Chicago infantry unit in the early months of our participation in World War I. Noé was a staunch supporter of his adopted country, while remembering his homeland, and in 1920 he served as treasurer of the American Commission for Vienna Relief for which Austria awarded him a high honor.

Paleozoic plant studies have advanced rapidly since Noé's time, and I will try to present a very brief summary of some of the highlights. Noé was particularly concerned with turning out students who would delve into the coal-ball plants. Fredda Reed was one of his early students: among her contributions was one on *Arthroxylon* in 1952, which correlated with some work we were doing on the calamites at Washington University. Fredda served for many years on the faculty of Mount Holyoke College in Massachusetts. J. Hobart Hoskins (1896–1957) was also an early worker in the Chicago group, and I can hardly separate him here from his own student, Aureal T. Cross. Hoskins and Cross collaborated in the production of numerous papers on Coal Age plants, perhaps most notably a 1943 monograph on the cone genus *Bowmanites* and two 1946 papers on the pteridosperm seed genus *Pachytesta*. A few years later they started a series of studies on the Devonian-Mississippian petrifactions of the New Albany shale. This is a tantalizing aspect of American paleobotany that many people have dipped into. D. H. Scott and E. C. Jeffrey issued a rather lengthy report in 1914 describing plants such as *Calamopitys*, *Archaeopitys*, and *Lepidostrobus*. Charles B. Read wrote at least two papers on the flora in the 1930's, and some years later Sergius Mamay turned over to me an intriguing specimen that had been collected by Read from a locality near Boston, Kentucky—certainly the most problematical one that I have ever worked with. It was described in collaboration with Karen Alt Grant, who was then an undergraduate student. We called the plant *Crocalophyton readi*; our article received a somewhat mixed reception from paleobotanists at the time and I have alternated since between trying to forget it and wondering just what the fossil really is.

Maxine L. Abbott also studied with Hoskins and has brought out several important works on Carboniferous plants, including a revision of the fern genus *Oligocarpia* in 1954 and a paper on sphenopsid fructifications in 1968.

Roy Graham was a very promising young geologist and paleobotanist who received his introduction to fossil plants from Noé. Graham also studied at Cambridge for a year and he produced three papers on coal-ball plants, the most important of which was an excellent and useful anatomical study of the leaves of the arborescent lycopods in 1935. After his year in England, Graham became an instructor in geology at the University of British Columbia. He worked summers as a mine geologist in British Columbia and while thus employed he was killed by a rock fall in 1939. I do not know that he intended to go on in paleobotany, but his start was impressive and his early death a tragic loss.

Raymond E. Janssen, another student of Noé's, became head of the Geology Department at Marshall College in West Virginia, and wrote a nicely illustrated popular book on paleobotany, with particular emphasis on the Coal Age.

In 1939, William C. Darrah initiated a series of studies on fossil plants found in Iowa coal balls which I believe were for the most part specimens collected by Fred Thompson. Darrah's articles include studies of the huge and strange sporangial aggregate of *Botryopteris* and an account of *Cordaianthus*. He brought out a short article on a petrified female gametophyte of a *Selaginella* found in a Mazon Creek nodule; his illustrations show what are, to the best of my knowledge, the most likely examples of fossil chromosomes—although recent studies of certain pre-Cambrian plants have led to a controversy concerning the real nature of fossil "nuclei." Darrah's description of the peel method (1936) was my first introduction to this technique, and I used his solution formula very successfully for some time prior to the development of the sheet-film modification.

Members of the Botany Department at the University of Illinois and the State Geological Survey have played a prominent role in coal-ball studies, and in Paleozoic botany in general. One of the leaders of this group, and indeed a world leader in our science for several decades, was James M. Schopf. His contributions have been numerous and varied and few men have been so helpful to their colleagues. He was a close personal friend of mine for forty years, and as President of the International Organization for Palaeobotany he was a friend and servant to all of us. To me it brought a great era to a close when I received a telephone call from his wife, Esther, on the morning of September 15, 1978, telling me that Jim had passed away. Thus it seemed appropriate to deal a little more fully with his life and work than I have done with most of our other contemporaries.

I first met Jim Schopf shortly after my return from England in 1938. He had received the doctorate the year before from the University of Illinois and remained in Urbana as a member of the State Geological Survey until 1943. He then joined the Bureau of Mines in Pittsburgh and in 1947 transferred to the U.S. Geological Survey. He established a special coal-geology laboratory at Ohio State University in Columbus in 1950, where he also held a professorship, and remained there for the rest of his career.

When I visited him in 1938 in his laboratory in Urbana he was just beginning his work on American coal balls and I was anxious to do the same. Knowing of my graduate work with the Paleozoic pteridosperms he turned over to me a small collection of *Heterangium* specimens which served as the basis of one of my first investigations. It was the first of many such favors on his part. He was always generous in sharing what he had and always anxious to place fossil-plant collections in the hands of those who he thought could make the most of them. He was known for two other characteristics as well: he was a keen and tireless explorer in the field and could find fossil plants when all others had given up; and he could extract significant information from even the seemingly poorest specimens that most others would ignore. He discovered an Upper Devonian horizon in West Virginia that contained superbly preserved *Archaeopteris* and *Rhacophyton* material and turned this over to Tom Phillips and myself. Very few men would have "let go" their claim to such a find, but he had other work to do at the time and was anxious to see the deposit studied.

As to the second characteristic mentioned above, my work in northern Maine began with an invitation to visit the area by both Ely Mencher and Jim Schopf. In the course of our explorations there Professor Mencher arranged one summer to have some heavy equipment brought in to an early Silurian site where there seemed to be some hope of finding vascular fossil plants. It did not turn out to be a very productive locality, but by simply refusing to give up Schopf found some small fossil plant remains that were described in a joint paper of 1966, although Schopf was clearly the senior author.

Jim Schopf was born in Cheyenne, Wyoming, on June 2, 1911. He received his undergraduate education at the University of Wyoming and then went to Illinois, as I mentioned. He was trained or influenced by such highly competent botanists as Aven Nelson at Wyoming and John T. Buchholz at Illinois. His graduate studies were strong in both botany and geology, and he developed a particular interest in coal. As a result of continued studies based on these interests of his graduate days Jim Schopf became a scientist who probably combined a greater knowledge of *both* botany and geology than any person of his generation.

I can only recount here a few representative samples of his many contri-

James M. Schopf, with a *Glossopteris* specimen at Coalsack Bluff, Antarctica, 1969. Photo by James Collinson.

butions to paleobotany in the broadest sense. His 1943 doctoral thesis consisted of a fine study of the embryology of the larch (*Larix*). In 1944 he brought out, with L. R. Wilson and Ray Bentall, a classic memoir on Paleozoic spores; he was one of the real founders of Paleozoic palynology. He produced several studies of coal-ball plants, of which one in 1948 dealing with *Dolerotheca* is exemplary. He wrote scores of studies dealing with the composition of coal; and in his later years he spent several field seasons in the Antarctic where I believe his greatest contribution was the discovery of petrified plant material comparable to the northern hemisphere coal balls. He had an intense interest in nomenclatorial matters and contributed much in this direction at many of the national and international meetings.

Probably the greatest formal honor that he received was the Mary Clark Thompson gold medal, awarded by the National Academy of Sciences in 1976 "for most important services to geology and paleontology." When he was first informed of the award, his response was "Oh, but my best work hasn't been published yet!"

There was, I am sure, much left undone, but few have advanced the cause of paleobotany so well in so many different areas, and the high regard that his colleagues, worldwide, felt for him as a man and as a scientist was reflected in his election to the Presidency of the International Organization for Palaeobotany in 1975.

It is opportune here to mention briefly another paleobotanist, Robert M. Kosanke, who was also associated with the Illinois Geological Survey for some years and who later joined the U.S. Geological Survey. Most of Kosanke's studies have been in palynology, which falls outside my main theme, but in 1955 he brought out a study of a new genus of calamite cones, *Mazostachys*. Based on a partially petrified specimen from an ironstone concretion from the Mazon Creek area, it is an exemplary study of a compression-petrifaction in which a maximum amount of information has been extracted and documented with excellent illustrations.

In about 1947, Wilson N. Stewart initiated at the University of Illinois a productive coal-ball study program which has been continued there and at other universities by his students and colleagues. Among Stewart's own publications are several which have added much to our knowledge of pteridosperm seeds. His student Thomas N. Taylor produced a monograph on the *Pachytesta* seeds that was published in 1965; Taylor in turn, after several years at Chicago, has established a very vigorous program at Ohio State University. Theodore Delevoryas is one of Stewart's earlier students who brought out several fine studies of Carboniferous petrified plants. I have noted elsewhere his later work at Yale University with Mesozoic plants. And among Delevoryas's students who have organized their own programs in coal-ball research I wish to mention Donald Eggert at Chicago; especially significant are his comprehensive studies of the ontogeny of the arborescent lycopods (1961) and the arborescent sphenopsids.

John Hall, although not a student of Stewart's, was influenced by him and in turn initiated fossil-plant studies at the University of Minnesota. And from Hall's laboratory came Gilbert A. Leisman, who developed paleobotany at Kansas State College in Emporia. Among Leisman's numerous papers is one that strikes me as especially interesting—his study of the small lycopod *Paurodendron* done in collaboration with Tom L. Phillips. Leisman also wrote a short account of the history of paleobotany in Kansas (1968).

For my own initiation to coal-ball collecting it is a pleasure to credit Eloise Pannell, who came to Washington University to study for a master's degree. One of her professors at Carbondale, Illinois, had told her of a coal mine near

Pinckneyville where the petrifactions were to be found. We made many trips there beginning in 1940, from which several papers resulted, including what is certainly one of the most informative studies of a single species of *Lepidodendron* stems written by Pannell in 1942.

I have enjoyed a very pleasant liaison with two students who came to work with me at Washington University immediately after the war, Robert W. Baxter and Sergius H. Mamay. Baxter came from service in the Navy; Mamay had mastered the Japanese language and served as an interpreter in Japan. They were almost my contemporaries, a relationship that has both advantages and disadvantages for a teacher, but the former have prevailed.

Baxter has carried on with coal-ball studies at the University of Kansas and developed mass cutting methods to investigate the rather vast quantities of petrifactions that occur in the mines in the eastern part of the state. In 1950 he brought out a short but significant paper on a new sphenopsid cone that was found in a collection of coal balls that we made in Indiana. The preservation is remarkable in that the contained spores display what seem to be well-preserved nuclei. I leave it to colleagues of the future to solve the present controversy over the true nature of such structures. Baxter's earlier studies include a description of a well-preserved *Palaeostachya* cone and others on the vegetative structure of the zygopterids. More recently he described a fine specimen of *Litostrobus*, a sphenopsid cone genus that had been established previously by Mamay.

Sergius Mamay's 1950 doctoral thesis dealt with several Upper Carboniferous fern fructifications, and others have continued to build on his work. In 1954 he brought out a paper on a distinctive cupulate pteridosperm seed, *Tyliosperma*. His research then shifted to the Permian of Texas, and I had the pleasure some years ago of spending a few days out in the desert digging with him in what seemed to me to be a very rich Permian flora. Several memoirs have developed from his investigations of this flora, the latest being a very informative and readable one, *Paleozoic Origin of the Cycads* (1976), which brings together old ideas and presents new ones which inject new life into this ever-fascinating group of gymnosperms.

I will conclude this subsection on Carboniferous plants with a few comments on another student who has continued to work with me over the years—I have recounted some of our adventures elsewhere in my story—Tom L. Phillips. As a graduate student Phillips began a monumental study of both American and European species of *Botryopteris* of which a portion, illustrated in part with his own fine drawings, has been published. He contributed a much-needed article, written with Joan Courvoisier in 1975, on the correlation of the spores in coal-ball fructifications with dispersed spores. His memoir *The Evolution of Vegetative Morphology in Coenopterid Ferns* is an especially worthy example of his summary studies.

Through his own extensive work in this country and recent visits to Europe to consult with paleobotanists and study the collections there I believe that Phillips has an especially keen understanding of the overall status of our knowledge of coal-ball petrifactions. I can think of no better conclusion than to quote from the summary statement he prepared at my request.

> Since the first reported discovery of coal balls by J. D. Hooker and E. W. Binney in 1855, in the Lower Coal Measures of England, such permineralized peats have been found at more than 200 reported localities in the Upper Carboniferous of Europe and North America as well as in some Lower Carboniferous and Permian coals. . . . With some exceptions most of the coal balls from Europe are stratigraphically older than those from North America with some overlap across the Westphalian B-C boundary. . . . Histological and ontogenetic studies have expanded alongside classical morphological and taxonomic approaches; attention is given to cytological preservation, gametophyte generations and other life cycle aspects which constitute reproductive biology. . . . In some coal basins, particularly in Lancashire, England, with its many coal ball localities and in both the Illinois and Donets Basins with their many coals (17–29) yielding coal balls, floristic and vegetational studies of swamp communities have begun which unite paleoecology with the traditional evolutionary thrust of coal ball studies.

Phillips is now the chief mentor at Urbana, Illinois, and is sending out students of his own, such as Benton Stidd, who has produced several splendid studies of the *Psaronius* tree ferns.

And so the "coal-ball genealogy" goes. From Noé's start, nearly a half century ago, several "generations" of students have evolved. Much has been added to our understanding of the plant life of the past.

It is a particular pleasure to recount something of the life and varied contributions of Chester A. Arnold (1901–1977), who, with Ralph Chaney, unites the past with the very recent present in the chrononogy of my story. When I was a young graduate student Arnold was the first person I ever wrote to for "reprints." I was informed by one of my professors that authors had private copies of their publications available and these might be obtained by request. Chester responded generously not only with the reprints but with a cordial note encouraging me in my studies.

Chester Arnold was born in Leeton, Missouri, the son of parents who were farmers and whose forebears had been farmers. Most of my information on this aspect of his life comes from a biographical sketch by Kenneth Jones and Charles Beck (1977) who have this to say on the economy of the times and life on the farm:

> Elmer Arnold [Chester's father] had a cutting machine which he used not only on his own acres but hired it out to other farmers. Also he had a cider press where apples were pressed for cider at 2¢ a gallon which could be payed in cider if the farmer was short of cash.
>
> Once, the day's milk overturned in the bucket being lowered into a well where it would have been cooled. Chester being the only one small enough and sufficiently agile had the job of being lowered into the well and bailing it. It took moons before the water was clear for drinking.

The Arnold family later moved to Ludlowville, New York, and since this was not many miles from Ithaca, Chester enrolled in agriculture at Cornell University. There he met Professor Loren Petry, a fine teacher and enthusiastic student of fossil plants, as I have noted elsewhere. Petry himself produced very little in the way of original research in fossil botany but his varied educational achievements include the guidance of both Chester Arnold and Harlan Banks in the early days of graduate study. Arnold attracted the attention of Harley H. Bartlett, Chairman of the Botany Department at the University of Michigan, and as he was interested in adding paleobotany to their program Chester was duly employed in the fall of 1928. He remained there throughout his career except for a year as Visiting Scientist at the Sahni Institute in Lucknow, India.

Chester Arnold's 1930 doctoral thesis dealt with the *Callixylon* woods from the Upper Devonian of central and western New York. It includes descriptions of several new species as well as a historical summary of Devonian plant studies in that general area. As noted elsewhere here, *Callixylon* is now known to be the stem that bore the *Archaeopteris* foliage and the combination constitutes a plant that presently stands as the best-known member of the progymnosperm group which partially bridges the gap between pteridophytes and seed plants. He added another important link to the progymnosperm story in 1939 by demonstrating that a species of *Archaeopteris*, *A. latifolia*, was almost certainly heterosporous. Since then several other species have been shown to be heterosporous. In a short paper of 1935 he described some "seed-like" compression fossils from the Upper Devonian of northern Pennsylvania; later these were investigated more fully by John Pettitt and Charles Beck who in 1968 gave them the name *Archaeosperma arnoldi* and demonstrated that they consist of a pair of seeds enclosed by a primitive cupule. This is the earliest evidence that we have of seeds in the fossil record.

A later paper describes a remarkably well preserved species of *Prototaxites* from Ontario (1952). This curious Devonian plant, consisting of trunks up to nearly three feet in diameter, is probably the stem of an alga, and the Ontario specimens reveal the unique anatomy much better than any previously described species. He also included a very readable summary of the rather bitter controversy that J. W. Dawson and William Carruthers engaged in over the presumed affinities of *Prototaxites*. Chester was generous with his fossils,

Chester A. Arnold. Courtesy of Charles Beck.

and peel preparations that I made from a specimen served me well in my class work.

On a visit to the famous Berryville, Illinois, coal-ball locality in the mid-1950's he found several specimens of a beautifully preserved heterosporous *Calamostachys* cone. In 1940 he described a *Lepidodendron* (*L. johnsonii*) which added much to our knowledge of the bark anatomy, and some years later he described another *Lepidodendron* stem, this one from Kansas, which shed some interesting if somewhat controversial light on the nature of cambial activity in the arborescent lycopods.

These are but a few of Chester Arnold's original contributions. His broad knowledge of botany and fossil plants in particular is revealed in his review articles (1968, 1969, among others) and especially in his textbook *An Introduction to Paleobotany*, which was for many years the standard work in English. Arnold was physically a large man, somewhat shy, but forthright with his opinions of others and their work. He was hardly an extrovert; it required a little time with him to see what was below the surface, and I came to like and admire him more each time we met.

It seems fitting to end this sketch with a comment that Charles Beck has made concerning the origin of the progymnosperm concept:

> Working in my laboratory one cold Sunday afternoon in the winter of 1959, I suddenly realized that I had accumulated evidence that proved beyond the shadow of a doubt that *Callixylon* and *Archaeopteris* represented the same plant. These were genera on which Chester had done much excellent and significant work early in his career. Although commonly occurring in geologic horizons of the same or similar ages, they had been classified in totally different taxonomic categories, one a gymnosperm, the other as a fern. I immediately called Chester and told him what I had concluded. His response was: "Well, I had always thought that might be the case." His words do not indicate the true nature of his response. There was excitement, pleasure, even drama in his voice, and within several minutes he was in my laboratory to see for himself and to discuss the evidence. [Jones and Beck, 1977]

I have mentioned the work of Harlan P. Banks in a previous chapter as one who revived the investigations of Devonian plants in southeastern Canada that J. W. Dawson initiated and left off many years ago. For more than two decades now Banks has been the leader in this country in investigations of plants of Devonian age. His contributions and those of his several students have ranged through the Devonian from botton to top and have been concerned chiefly with the New York State area. Banks received his bachelor's degree at Dartmouth College and his doctorate at Cornell, and after serving on the staffs of Acadia University and the University of Minnesota he returned to Cornell in 1949 and from 1952 to 1961 was head of the Botany Department. I am indebted to him for a good deal of aid over the years in relation to our studies of Devonian plants, and for various bits of information in the preparation of the present volume.

Harlan Banks's research contributions are numerous and I will simply cite a few representatives. His study, with Suzanne Leclercq, of the Middle Devonian plant *Pseudosporochnus nodosus* placed on record one of the best-known plants of that age; it is typical of the beautifully preserved compression fossils from the quarry at Goé in Belgium, and the partially petrified axes convey some information on its anatomy. In 1967 he brought out with Patricia Bonamo a study of the fertile parts of the Upper Devonian progymnosperm *Tetraxylopteris schmidtii*. And to complete this Devonian sequence, in 1975 Banks, Suzanne Leclercq, and F. M. Hueber produced a fine memoir on an exceptionally well preserved species of *Psilophyton* (*P. dawsoni*) that was found on the south shore of Gaspé Bay in Quebec.

These are landmarks in our rapidly expanding knowledge of plant evolution in the very critical Devonian age. I think Harlan Banks would agree that the more recent work is founded on the earlier studies of such fine paleobotanists as Richard Kräusel and Suzanne Leclercq, but Banks clearly deserves the credit for the resurgence of "Devonian activity" in this country. And his contributions have extended to other aspects of botany and paleobotany. By the middle of the present century our knowledge of early land vascular plants and evolutionary lines in the pteriodophytic groups had become both voluminous and somewhat chaotic. At the occasion of the hundredth anniversary of the foundation of the Peabody Museum at Yale University in 1968 Banks contributed a summary memoir on Devonian plants entitled *The Early History of Land Plants*. He brought togehter the abundant and scattered information and gave us a revised classification that is a tremendous aid to anyone concerned with research or the teaching of Devonian paleobotany.

His unique text *Evolution and Plants of the Past*, which came out in 1970, presents paleobotany in a distinctive light, emphasizing the aspects of plant evolution and stratigraphic correlations that characterize much of his work. He has also served us as a President of the International Organization for Palaeobotany.

A few American paleobotanists of recent decades have been especially adept at dealing with stratigraphic problems and thus aid in bridging the interests of botany and geology. I think that the work of Erling Dorf falls in this category; his contributions range from the Devonian to the Cretaceous and for personal reasons that I will note below I may add that he has also inspired others to develop leads that he has opened.

In 1938 and 1942, Dorf brought out two important memoirs on late Cretaceous angiosperm floras from Wyoming and Colorado. The late Cretaceous is a difficult age in paleobotanical research because most of the fossils cannot be compared closely with modern plants, but he was able to conclude that "the dicotyledonous leaves in the Lance flora indicate lowland, humid, warm temperate conditions of growth, approaching subtropical" (1942, p. 123). This finding has been confirmed by many studies of early Tertiary floras in the mountain states.

He spent considerable time in the Yellowstone Park area and has given us a very readable and beautifully illustrated account (1964) of the fossil forest succession. On a steep bluff overlooking the Lamar River he was able to identify at least twenty-seven successive forest layers through a total thickness of twelve hundred feet, representing a span of about twenty thousand years. In the spring of 1945, Lee W. Lenz and I spent a few days in the northwest (Gallatin) corner of the Park observing the succession of forests on Big Horn Peak, where we were able to identify some sixteen forest levels. I believe there are few places in the world where paleobotanical and stratigraphic phenomena combine in so spectacular a fashion to reveal to any observer something of the immensity of geologic time. Dorf also discovered an early Devonian flora in nearby Beartooth Butte, Wyoming, which includes a distinctive species of *Psilophyton* and *Bucheria ovata* Dorf, with terminally clustered sporangia.

Quite a few years ago I read his 1943 article, written with John R. Cooper, "Early Devonian Plants from Newfoundland." It meant more to me when my own interests turned from the Carboniferous to the earlier forms of vegetation in the Devonian. These interests were sharpened when I had an opportunity to spend several months studying with Suzanne Leclercq in her laboratory in Liège in 1958. We collaborated in a study of *Calamophyton*, one of the many beautifully preserved plants in the Middle Devonian rocks of Goé in eastern Belgium. However it was not until the mid-1960's that I was able to devote my research interests exclusively to Devonian paleobotany in northern Maine and southeastern Canada. I then remembered the Dorf-Cooper paper on Newfoundland Devonian plants. It seemed a good excuse to visit an intriguing part of the world.

In the late summer of 1967, Francis M. Hueber, Andrew Kasper, and I were digging along the New Brunswick and Gaspé coasts. We had arranged to explore the Long Range Mountains of southwestern Newfoundland to investigate the locality described briefly by Dorf and Cooper. Thus, after considerable planning, we found ourselves at noon of a late August day on the shore of Codroy Pond, about forty miles north of Port aux Basques, waiting for a helicopter to fly us in over the mountains. All of this had been arranged through the very kind aid of the Newfoundland Forest Service and a colleague of mine at the University of Connecticut, Antoni Damman, a former staff member of the Forest Service. After an initial attempt early in the afternoon that was turned back because of heavy fog, we managed to fly in some forty miles to a place that was central to the area we wanted to examine. Two trips were required to transport the three of us and our equipment. After the pilot left us for a scheduled four days and we set about making our camp, it was discomfiting to find that the box containing most of our staple foods had been left in the station wagon on the shore of Codroy Pond! I had packed two full weeks' supply in case of emergency, but we now found that we had a modest supply of condiments, some sugar and bacon, and little else. In the course of the next two days we located the fossiliferous horizon, and although plants were present they were highly metamorphosed and of little value. I would not discourage future explorations in the area, but what we found simply was not promising. We also visited the upland area above our camp; it was unique in the abrupt transition from forest to alpine zone, the abundance of moose, and a generally fascinating flora. As it proved later, the particular region we saw was somewhat misleading as to the ease of foot travel generally in that part of Newfoundland.

Our pilot did not return on the scheduled day due to heavy fog, and three more days went by with the same result. I had been informed that if the helicopter was required for forest-fire duty we should be prepared to walk out over the mountains. After the fourth day of waiting, and with little to eat other than fish, it was decided to break camp, pack most of our gear for the ultimate arrival of the helicopter, and start walking; it was a bad decision, for which I was responsible.

The first day of hiking went rather well, but the second proved more arduous, as the dense and almost impenetrable "krummholz" or spruce-fir growth some one to two meters high could be traversed only by following game trails and these coincided only vaguely with the path of our planned route over the Long Range Mountains. The second night brought a torrential rain and in spite of all our efforts our sleeping bags were thoroughly soaked by morning. We tore them open and left them spread out on a high point where we hoped they might guide the pilot along our route. A driving rain continued for two days. Late in the fourth day we dropped down into the forest where it was possible to build a fire and feel dry and warm, although definitely hungry. We had a half pound of sugar, some orange flavoring, and a few nuts remaining.

The fifth day came bright and clear, revealing the beauty of the rolling mountains with myriads of ponds of all sizes and rushing streams. We descended through the forest to the valley floor and then walked up a long open slope which appeared to be about two miles long. But four days of hard hiking with almost no food now began to tell; we simply did not move very fast uphill. We mixed the remainder of the sugar and orange flavoring with some water at the next pool, drank it, and were somewhat revived. We reached the top of the mountain and looked out over the next fifteen miles of rough country that had to be traversed. It seemed best now to try to follow a pattern of streams to our destination, going downhill instead of taking the more direct westward route. At this point two things happened: a caribou with incredibly huge antlers approached us to within thirty or forty feet, certainly more curious than afraid; and off to the south we heard our helicopter, now more than a week overdue as a result of the persistent fog and rain. We waved our one blanket and presently the helicopter turned and headed toward us, and in a few minutes it settled onto the meadow beside us. It was the end of our trek, but we had the caribou to thank: our pilot and a friend with him said they had not seen us but had noticed the caribou with his great antlers and came to have better look at him.

For a very long time philosophers and biologists have wondered about the origin of life on the Earth. How did it come here, if from another planet? There have been several reports in the literature of presumed micro forms of life contained in meteorites. Articles by Frank Staplin and Martine Rossignol-Strick are significant and thought-provoking, but it seems to me that the present evidence for life having come to this Earth in cellular form is not very convincing.

A more likely question is, how did life originate from the various elements that compose the rocks and waters of the surface of the globe? A considerable amount of literature has appeared on the subject in recent years; some of it is interesting and much of it is speculative, and I do not propose to review it in detail. Although paleontology probably cannot inform us about the very earliest forms of life, recent studies have made some very sound contributions to our understanding of the records of life at a primitive cellular stage. E. C. Jeffrey called paleontology "the last court of appeals" in evolutionary matters; I am not sure that he coined the phrase but I think it still holds good. One may weave all kinds of theories about life of the past and perhaps create strange forms of life in the laboratory, but the fossil record is the only direct source of information about the plants and animals that actually lived in former times.

Paleontologists have been involved recently in a distinct spurt of activity directed at exploring some of the older rocks for evidence of the earliest forms of life, and I think the work of Elso S. Barghoorn and Stanley Tyler initiated the most productive line of research that has yet been developed.

Stanley A. Tyler (1906–1963) was a Professor of Geology at the University of Wisconsin who also served as a consulting geologist for an iron and steel company. In the course of exploring old iron-ore prospect pits in upper Michigan he encountered masses of a black rock with the general appearance of a very hard coal. Suspecting that biological activity might have been responsible for its formation, he discussed the matter with Robert Shrock, a noted Massachusetts Institute of Technology paleontologist, who in turn referred Tyler to Elso Barghoorn at Harvard. Adding to their interest in the material was its age, which was determined as approximately 1.9 billion years.

Upon examining thin sections Barghoorn immediately recognized a considerable assemblage of filamentous and unicellular organisms, and Tyler and he published a report on them in 1954. In 1965, in a more detailed account (*Microorganisms from the Gunflint Chert*), Barghoorn and Tyler described and illustrated numerous new species and genera of thallophytic plants—spheroidal bodies, multicellular and tubular filaments. This seems to me to be one of the great landmarks in paleobotany. Others had suspected the existence of primitive plants in the older pre-Cambrian rocks, but the evidence from Tyler's "coal" left no doubts. The report is important in itself and also for the interest that it initiated in further explorations of very ancient rocks.

Shortly thereafter several studies appeared by Barghoorn and one of his students, J. William Schopf, a son of my old friend James M. Schopf. In 1966 they reported bacteriumlike organisms, *Eobacterium isolatum*, from South African rocks dated at 3.1 billion years, and in 1967 they described "spheroidal microfossils" believed to be the remains of unicellular algae of a similar age and also from South Africa.

In 1971 J. W. Schopf and Jan Blacic reported a considerable assemblage of microplants from the 900-million-year-old Bitter Springs Formation of Australia. This is much younger than the formations examined previously, and the plants are somewhat more complex, as might be expected. and the assemblage more diverse, but a particularly interesting aspect of this report is that some of the plant cells appear to contain distinct nuclei, a great advance over bacterial and blue-green algal forms in which nuclear material is scattered through the cell.

I have mentioned several controversies in the history of paleontology; paleontologists are human beings with intellectual limitations and passions like those of other people. I think that progress is generally upward, but it is not always directly so. The validity, for example, of the correct identification of nuclei in the Bitter Springs fossils has been questioned, most recently by Andrew Knoll and Barghoorn, who feel that the presumed nuclei are the result of decay or deterioration of the cell contents. This brings my story perhaps a little too close to the present, and I leave the solution of this matter to a future historian.

I first became acquainted with Elso Barghoorn some thirty or more years ago when he was serving as a consultant at the Saugus, Massachusetts, Iron Works Restoration project. I spent several days one summer studying the fossil-plant collections in the Botanical Museum building at Harvard. In the laboratory where I worked Barghoorn had a huge vat of melted paraffin bubbling away with samples of the buried three-hundred-year-old timbers of the old iron works, the object being to determine the best way in which to preserve them after removal from the entombing sediments. He has had a long-standing interest in the anatomical and chemical aspects of plant decay and preservation. Of numerous articles that have resulted from this research a very recent one, "Silicification of Wood," written with Richard Leo in 1976, is particularly significant and informative.

To date nature has not been especially generous to American paleobotanists seeking information about plant life in the Triassic period. During my ten years in Connecticut I did some exploration in the sediments of that age there and students occasionally brought in fragments. But very little of significance has been found thus far, although one should not give up hope. Fossil plants are known in some abundance in the Triassic of North Carolina, and short accounts by Theodore Delevoryas indicate that we will hear more about that area in the future.

The most interesting and productive horizon that has been reported thus far in this country is the Upper Triassic Chinle Formation of the southwest, the fossils having been found chiefly in the Petrified Forest National Park of Arizona and in northwestern New Mexico. The classic study of the flora is Lyman H. Daugherty's *The Upper Triassic Flora of Arizona* (1941). Significant additions have been made since then by Sidney R. Ash, who has also written a very readable and effectively illustrated account of the history of paleobotanical exploration in the area. Ash's account (1972) is the kind that makes the preparation of this kind of book fascinating; I recommend it and hope that there may be more of the same type for other areas in the future.

The Chinle flora includes about fifty species that contrast strikingly with the vegetation of northern Arizona today. Several fern families are represented, including the Osmundaceae (cinnamon ferns), as well as the Matoniaceae and Dipteridaceae, which are found living today in the Indo-Malayan region; there are articulates referred to the Calamitaceae and Equisetaceae; foliage and petrified stems of the Cycadales and Bennettitales indicate that these groups were important parts of the vegetation; and there is the abundance of petrified wood, much of which is referred to the Araucariaceae, a present-day family of conifers found in the southern hemisphere.

In his summary Daugherty says: "The Upper Triassic of Arizona was characterized by subtropical to tropical temperatures, as indicated by the fern element. Although both organic and inorganic evidence points toward a dry season, there appears to have been sufficient precipitation to insure permanent streams and swampy conditions in the lowlands. The growth layers of the fossil wood indicate that most of the rain fell during a growing season, which included intense dry spells between periods of rain" (p. 35). Ash's studies also confirm the opinion that the climate was a humid tropical one.

Two great series of floristic studies began to appear in the early 1920's, written by two men who almost exactly coincide in time—Ralph W. Chaney (1890–1971) in this country, and Walter A. Bell (1889–1969) in Canada. Chaney's studies dealt largely with the sequence of Tertiary floras in the western states while Bell's many publications cover a much wider stratigraphic range of Canadian localities.

Ralph Chaney was born in Brainerd, a suburb of Chicago, and developed an early interest in natural history, particularly in birds; his first scientific paper (in 1910) was concerned with the birds of the Hamlin Lake region, Michigan. At the University of Chicago he started in ornithology but soon shifted to botany and geology. He was fortunate in being able to receive his instruction from a galaxy of great men of the time, including the botanists John M. Coulter, Henry Cowles, and J. M. Greenman, and such geologists as T. C. Chamberlin, Rollin Salisbury, and the paleontologist Stuart Weller. I would like to add a word or two on Jesse Greenman, to emphasize the great range in teaching methods and abilities of different men. I sat in on one of Greenman's taxonomy classes during his later years at the Missouri Botanical Garden. His notes were yellow with age; he was old, slow, and methodical, and some of his evolutionary ideas were hardly up to date, but it was a time when I needed basic information about the classification and morphology of plants, and what I learned from his lectures served me well through my entire academic career. He rather reminded me of Hervey Shimer, the great invertebrate paleontologist who first introduced me formally to the world of fossils. Both of these men fell far short of being spellbinding lecturers, but they were among the most effective and influential teachers that I ever had.

In their biographical sketch Gray and Axelrod (1971) record the following relative to Chaney's start in paleobotany: "In the summer of 1916, using funds largely acquired from his teaching [at a private high school], Chaney went, at the encouragement of J. Harlen Bretz, to the Columbia River gorge where fossil plant localities had been discovered in the Eagle Creek Formation. He was later to write, ' . . . from that time on I knew I was primarily interested in Tertiary paleobotany'" (p. 2).

In 1916 Chaney visited the U.S. National Museum to study the Tertiary collections of Knowlton, Hollick, and Berry. Presumably he met Knowlton in Washington, and he visited Berry, Hollick, and Wieland at their respective residences. By training and acquaintance with leading men of the time he was off to a good start. In 1922, after teaching geology for five years at the

University of Iowa, he went to the University of California, at Berkeley, as a staff member and a Carnegie Institution research associate. He was appointed Professor of Paleontology in 1931 and also assumed the chairmanship of that department.

Ralph W. Chaney. Courtesy of Jane Gray.

Chaney and his associates produced a monumental series of memoirs on the Tertiary floras of the western states which documents in floristic detail the change from the warm, humid conditions of the Late Eocene in the northwest to the much cooler climate of the present, and the desert to semidesert environment of the intermontane area. It would take a book of significant size to review the series briefly. I suggest his study of the Eocene Goshen flora of west-central Oregon, coauthored with Ethel Sanborn (1933), as representative and especially interesting.

Although Chaney's work developed naturally from the research and methods of men such as Knowlton and Berry, he carried the study of fossil plants to an appreciably higher state of refinement. Axelrod (1971) points out that whereas his predecessors were concerned with the "what" and the "when" of fossil plants, Chaney was more concerned with the "why." His bent was ecological, acquired partly from his acquaintance with Frederic E. Clements and developed by detailed field studies in Central and South America during the 1930's.

He was interested in floras as a whole and developed methods for studying leaf morphology that were a distinct advance over most of the work of his predecessors. His biographers note that "in the method he developed of using plant fossils to date sedimentary strata, two ideas are of prime importance: (1) wide differences existed in vegetation at different latitudes at any one time, and (2) similar or essentially similar vegetational units have lived at different times over a wide range of latitude" (Gray and Axelrod, 1971, p. 4).

In 1925 Chaney was a member of Roy Chapman Andrews's expedition to Mongolia and Manchuria, and in 1937 he became involved in a study of the Shangwang flora of Shantung Province in China. At this point Chaney emerged as a central character in one of paleobotany's most interesting chapters, the discovery of the conifer *Metasequoia*. In 1944 the botanist Tsang Wang, of China's Central Bureau of Forest Research, first collected, in Szechwan Province, specimens of a living and previously unknown (to the scientific world) coniferous tree. It had been recognized in the fossil state in 1941 by the Japanese paleobotanist Shigeru Miki and given the name *Metasequoia*. It is of special interest because of its close relationship with two other living conifers, *Sequoia* and *Taxodium* (the swamp cypress). In 1948 Chaney hiked into the remote hills of Szechwan to see a stand of *Metasequoia* and in 1951 he wrote a detailed account of the discovery which includes a revision of fossils previously assigned to other genera.

Finally, I would like to pay high tribute to Ralph Chaney's great interest in conservation and especially his long service with the Save-the-Redwoods League, of which he was president from 1961 to 1971. He was one of the leaders responsible for the preservation of a large tract of these great forest trees so that future generations may enjoy one of the most glorious achievements of the plant world.

Ralph Chaney initiated a new era in the study of (predominantly) angiosperm floras of the American Tertiary, but of equal or perhaps greater importance was the direct influence he exerted through his teaching and guidance of graduate students, as well as the inspiration that he provided to others who developed their own research programs. The resultant voluminous literature has contributed much to our knowledge of the floral and climatic successions that have taken place in the western United States during the past fifty million years. It is of necessity a branch of paleobotany in which the investigator must have a profound knowledge of the general biology and distribution of modern flowering plants. For this reason, and because of the time limits I have set, it falls largely outside my account. I would like, however, to mention two of the leaders, and I trust that others after me may deal with this special area of fossil botany in their own recollections and reminiscences. Daniel I. Axelrod has been the most prolific worker in this area in recent decades. His written work includes a great many regional floristic memoirs as well as numerous useful review papers and philosophical studies dealing with angiosperm origins and distribution. Herman F. Becker, at the New York Botanical Garden, has been very active during the past two decades; his studies, dealing chiefly with certain Tertiary floras of Montana, have appeared in recent issues of the journal *Palaeontographica*.

I will close my discussion of the "Chaney era" with a few personal notations which, if they do not fit perfectly into the sequence of my story, may perhaps be of some general interest. The Rocky Mountain states attracted me in my student days for both mountain climbing and the great variety of fascinating botanical and geological phenomena that they contain. We made two ascents of the Grand Teton in northwestern Wyoming, one in the summer of 1932 and the second two years later. The latter was a night climb and we reached the top just as the sun came up over the eastern horizon. On later trips my efforts were directed into paleobotanical channels, chiefly into explorations of certain Cretaceous horizons, and one of these resulted in a study of some nicely preserved *Anemia* ferns from a locality south of Kemmerer, Wyoming. I made several trips in search of the curious petrified trunks of *Tempskya* which brought me into contact with some very helpful mineral collectors in Idaho. Had I not been introduced to the world of coal balls in about 1940 I believe I would have gone on with these explorations in the western states and might possibly have become an angiosperm paleobotanist.

Two localities where my friend Cortland Pearsall and I collected petrified wood and leaf impressions will always be recalled with special pleasure: the fossil forests of Yellowstone Park and the fossil deposits at Florissant, Colorado. For this reason I have been interested in some of the works of Harry D. MacGinitie, especially his fine monograph entitled *Fossil Plants of the Florissant Beds, Colorado* (1953). Florissant is located in the center of the state, about thirty-five miles west of Colorado Springs. The shales there are largely of volcanic origin and contain a great abundance of plants and insects as well as petrified stumps, many of which are *Sequoia*. The area has attracted many paleontologists since the middle of the last century, and MacGinitie notes in his introduction that "the number of species of fossil plants reported by Lesquereux, Kirchner, Knowlton, Cockerell, and others is about 258, of which only about one-half are valid" (1953, p. 1). In 1911–1913, T. D. A. Cockerell, a noted entomologist who was also interested in fossil plants, had written a series of short papers in *Torreya* entitled "Fossil Flowers and Fruits." In view of the present interest in fossil flowers, as noted in Chapter 16, it is likely that paleobotanists will be taking another look at the Florissant beds in the near future.

Also, following the previous work of men such as Newberry, Lesquereux, Knowlton, Cockerell, Brown, and Chaney, MacGinitie brought out a thorough revision of the Eocene Green River flora in 1969. The Green River Formation outcrops over a vast area of some 25,000 square miles in Colorado, Wyoming, and Utah, and has long been known for the fossil fish that it contains in certain layers. MacGinitie also produced a memoir on the Eocene age Yellowstone-Absaroka region of the Wind River Basin in Wyoming, and he brings these three areas together in the following words: "The fossil flora showing the most significant correlation with the Green River flora is the Early Oligocene flora from Florissant, Colorado, with 29 identical or closely similar species. There are also significant similarities between the Green River and the Kisinger Lakes and Rate Homestead floras of the Wind River Basin, Wyoming, and the upper West Branch Clarno flora of eastern Oregon" (1969, p. 2).

I noted in Chapter 1 that the emphasis here would be on studies that are evolutionary in nature rather than stratigraphic. This is not to imply that the stratigraphic use of plant fossils is unimportant—quite the contrary. This emphasis is due in part to my own limitations of knowledge and also in part to the fact that I cannot cover the entire waterfront. But I use an author's privilege of making exceptions, and the work of Walter A. Bell seems to fully justify such an exception. Like Ralph Chaney he produced many large floristic memoirs based on considerable field experience and a vast knowledge of plants. His objective, to use fossil plants as a service agent in stratigraphic geology, was rather different from Chaney's ecological-climatic approach, and Bell's publications cover a much greater time range. I have chosen to look at just two of them as examples.

Bell initiated field studies of the Nova Scotia and New Brunswick coalfields as early as 1911, and several monographs resulted. The *Fossil Flora of Sydney Coalfield, Nova Scotia*, of 1938, is profusely illustrated and reveals a great knowledge of Carboniferous plants. When, in 1944, Bell was awarded the Professional Institute medal, the citation noted, in part:

His stratigraphic investigations and faunal and floral studies furnished new and exact correlations of strata in the various coal-fields which have been of tremendous economic importance in appraising the value of coal deposits in the various Nova Scotia basins. . . . The previous uniform classification of the strata of the coal-fields implied that all the fields had similar histories. Dr. Bell's studies of the geological history of Nova Scotia indicated that each coal-field had a distinct and separate history. [Anon., 1944, pp. 93–94)

Walter Bell. Courtesy of Colin McGregor, Geological Survey of Canada.

On the other side of Canada and in quite a different age Bell's 1956 study, *Lower Cretaceous Floras of Western Canada*, reveals something of the diversity of the Chalk Age floras there. He notes that this report is based on collections from 365 localities. I have not had an opportunity to examine any of these collections but a perusal of the many plates, showing unique and fascinating fossils, some of which appear to be fertile specimens, suggests that much additional information may still be expected from them.

Walter Bell was born at St. Thomas, Ontario, enrolled in Queen's University in 1909, and started his long association with the Geological Survey of Canada as a junior assistant in the Yukon. Gradually rising through the ranks, he was ultimately appointed Director of the Survey in 1949. His studies were interrupted by World War I, when he served with the Canadian Field Artillery in Europe from 1916 to 1919. Shortly thereafter he spent a few months in Cambridge with Sir Albert Seward and then returned to North America and received his Ph.D. from Yale in 1920. His chief academic mentor was Professor Charles Schuchert, who described Bell as "by all odds the best man I ever had" and prophesied that he would be a worthy follower of Sir William Dawson, a prediction that seems to have been well justified.

I never met Walter Bell and I am very grateful to Mrs. Bell for sending me some personal information about her husband—whom she says she first met when he was tenting with a survey party on her mother's potato patch. Walter's interest in geology stemmed, at least in part, from his family background. His father was a noted civil engineer with a strong interest in conservation and instilled in his sons a love and respect for nature. As might be surmised from his travels and published record, Walter was a vigorous field worker, and he was by no means a large man. In this connection Mrs. Bell remembers the following incident:

> Walter was a tireless walker. One summer he had a young assistant in his survey party, six foot six, who said to me before going out the first morning: "Mrs. Bell, I am worried about going out with Dr. Bell in the field. I take such a long stride." I alerted Walter and that evening, after they returned, the young man limped into the cook tent and prostrated himself at my feet. I asked innocently "What's wrong Heath?" whereupon he gasped, "My God, that husband of yours is like a moose in the woods. I've been running after him all day with my tongue hanging out." [Personal letter]

Walter Bell was a modest man, somewhat withdrawn, and not at his best in large gatherings; he was also partially deaf as a result of his artillery service. Although a few words can hardly do full justice to a man's personality, the following incident, which Mrs. Bell has described, tells a good deal:

> We owned a much loved cairn terrier, Brigand, that Walter used to exercise. After he died, I was walking Brigand early one morning in a nearby

Henry N. Andrews, at a fossil plant locality near Stockholm, Maine, 1965.
Photo by Richard W. Sprague, *Maine Line* magazine, Bangor and Aroostook Railroad.

cemetery when an elderly man approached, obviously the caretaker. "Are you Mrs. Bell? I recognized the dog." Upon being assured that I was, he continued: "You know, I often met your husband in the morning and we would stop and have a chat. Until I read about him in the newspaper, I never realized he was a great man. I thought he was just a simple man, like myself." That was Walter.

12

Scandinavia and the Arctic

THE subject of this chapter has especially vexed me in trying to sort out people, places, and fossil plants in the most appropriate fashion. There has been a special correlation between the interests of Scandinavian paleobotanists and the fossil floras of the North, but of course men of many other countries have also found a fascination in the Arctic, as well as the Antarctic. Much has transpired in the Soviet laboratories in recent decades, but it has seemed best to deal with this research as a unit in another chapter. Oswald Heer also appears in this chapter, in view of the vast quantity of literature on Arctic fossil floras that he produced.

As to a starting point, there are a few brief reports on fossil plants from the northern climes dating back to the early part of the nineteenth century, and studies in general up to 1889 have been reviewed by Ward in his "Geographical Distribution of Fossil Plants" (1889). But of the scientific explorers who were probing into the northern lands I believe that Nordenskiöld was the real pioneer in fossil botany, and I have therefore chosen him to introduce this section.

Adolf Erik Nordenskiöld was born in Helsingfors, Finland, in 1832, and after some travel in various parts of Europe he received his doctorate and left Finland for Sweden in 1857. Finland had become a part of the Swedish "Baltic lake area" during Sweden's rather short-lived empire, but Finland came under Russian domination in the early nineteenth century and with this came a lack of freedom that was largely responsible for Nordenskiöld's departure.

In Stockholm, Professor Sven Lovén was greatly impressed with Nordenskiöld's knowledge of geology and mineralogy and enabled him to accompany Otto Torell on the latter's expedition to Spitsbergen in 1858. A second voyage was made with Torell in 1861 to Spitsbergen and Bear Island, and they returned to both places in 1865 where, apparently for the first time, a serious interest was taken in fossil plants and large collections were made. And in 1870 he visited Disko Island and other localities in Greenland.

Nordenskiöld is probably best remembered for his long voyage in the *Vega* which started at Tromsö, Norway, on July 21, 1878. The *Vega* made the long and difficult passage east along the north coast of Siberia, eventually reaching Bering Strait, and then went south to Japan and Egypt. Nordenskiöld made a special and successful effort to do some collecting in Japan and the fossils were later described by A. G. Nathorst. Nordenskiöld wrote from a fishing town near Nagasaki in southeastern Japan:

> When my arrival became known I was visited by the principal men of the village. We were soon good friends by the help of a friendly reception, cigars and red wine. Among them the physician of the village was especially of great use to me. As soon as he became aware of the occasion of my visit he stated that such fossils as I was in search of did occur in the region, but that they were only accessible at low water. I immediately visited the place with the physician and my companions from Nagasaki, and soon discovered several strata containing the finest fossil plants one could desire. During this and the following day I made a rich collection, partly with the assistance of a numerous crowd of children who zealously helped me in collecting. They were partly boys and partly girls, the latter always having a little one on their backs. These little children were generally quite bare-headed. Notwithstanding this they slept with the crown of the head exposed to the hotest sun-bath on the backs of their bustling sisters, who jumped lightly and securely over stocks and stones, and never appeared to have any idea that the burdens on their backs were at all unpleasant or troublesome. [1882, p. 690]

At Suez he visited the Cairo fossil forest that is mentioned by several travelers before and after Nordenskiöld's time:

> A day was . . . devoted by some of us, in company with M. Guiseppe Haimann, to a short excursion to the Mokattam Mountains, famous for the silicified tree-stems found there. I had hoped along with the petrified wood to find some strata of clay-slate or schist with leaf impressions. I was however unsuccessful in this, but I loaded heavily a carriage drawn by a pair of horses with large and small tree-stems converted into hard flint. These lie spread about in the desert in incredible masses, partly broken up into small pieces, partly as long fallen stems, without root or branches, but in a wonderfully good state of preservation. Probably they had originally lain imbedded in a layer of sand above the present surface of the desert. This layer has afterwards been carried away by storms, leaving the

heavy masses of stone as a peculiar stratum upon the desert sand, which is not covered by any grassy sward. [1882, p. 725]

Although he was an intrepid explorer, Nordenskiöld gave particular attention to matters of scientific importance and often gathered the fossil plants himself. This interest was not confined to the Arctic regions, for in 1873 he made a large collection at Pålsjö in southern Sweden (Scania) which were described by Nathorst in the first of a series of papers on the Swedish Rhaeto-

Alfred G. Nathorst, wearing the emblem of the Swedish Yachting Club. Courtesy of Britta Lundblad.

Liassic floras. Nathorst accompanied Nordenskiöld on his last northern expedition, to western Greenland, in 1883; and Halle (1950) notes that Nathorst had been asked to go on the *Vega* expedition but was unable to do so. We thus meet at this point one of the greatest figures in the history of fossil botany.

Alfred Gabriel Nathorst (1850–1921) was born at Väderbrunn, Södermanland, Sweden; he attended Lund University in 1868 and later Uppsala

University, where he graduated in 1873. His early interests were botanical and geological; the paleobotanical studies of Mesozoic deposits for which he is best known developed a little later. His polar travels actually began some years before the 1883 trip with Nordenskiöld; in the summer of 1870 he accompanied H. Wilander, an engineer, to Spitsbergen, the primary objective being to examine phosphor beds, but he spent some time studying the living vegetation and also gathered fossil plants. In these early days he developed a particular interest in the Pleistocene migrations of Arctic plants; in 1871 in company with Japetus Steenstrup he found glacial plant remains at the bottom of Danish peat bogs; in 1872 he traveled to Germany and Switzerland, finding *Betula nana*, the Arctic birch, in peat at Mecklenburg, and in the same year he visited England, where he found *Salix polaris* under a moraine on the Norfolk coast. These travels initiated contacts with British botanists which continued throughout his life; Seward has recorded that Nathorst's English visits were divided between conferences and excursions with Clement Reid in the Tertiary localities, and collecting plants from the Jurassic beds of Yorkshire. A good many paleobotanists have either researched, or studied for their own edification, as I have myself, the great plant deposits of Yorkshire. And in Switzerland, Nathorst was introduced to Oswald Heer, who apparently encouraged his interest in the plants of the past.

These early studies were important in themselves and laid the foundations for his three major contributions: the development at the Natural History Museum in Stockholm of a great paleobotanical facility, many important studies of Mesozoic floras, and the development of techniques which enabled him, and many others after him, to extract much more information from fossil plants.

He was encouraged by Nordenskiöld to study the Rhaetic floras of Sweden, and the results are among the classic studies of Mesozoic plants; papers appeared on the flora of Pålsjö in 1876, the flora of Höganäs in 1878, and the Bjuv flora in three sections between 1878 and 1886. Halle notes that Nathorst had a very narrow view of a species, in contrast, for example, to the broad view of men like Seward. One gains the impression that Nathorst was a man of great precision in his work and rather expected the same of his colleagues. Seward, in his short biographical sketch, notes that

> Nathorst had the true scientific spirit. His work was based on a firm foundation of accurate and wide knowledge of botany and geology; he recognized the limitations of his material and never ventured to deal with matters on which he was not competent to speak with the authority of a specialist. In 1895 he wrote in one of a long succession of most helpful letters, "The chief rule in dealing with fossil plants is that one ought to say

> precisely as much as the material allows, neither more nor less. This is the ideal, but one cannot help sometimes saying a little too much in consequence of what one besides (that is beyond the evidence) does believe!" [1921, p. 464]

Although precise and demanding, Nathorst was also evidently very human.

Halle (1921) says that Nathorst seldom received a paleobotanical paper without noticing some negligence on the part of the author, and that he usually took occasion to write to the author in question. One may wonder whether this apparently well-intentioned criticism was *always* welcomed. Seward, who had much correspondence with Nathorst, indicates that such comments were, however, always well meant: "He took delight in helping others with kindly encouragement and frank criticism. For him, purely destructive criticism had no charm; he always took pains to be stimulating and constructive. His sincerity and generosity inspired confidence and affection" (1921, pp. 464–65).

Paleobotanists have rarely been abundant in any one place and Nathorst suffered from the same loneliness that many of us have experienced at times. He once wrote Seward: "You can hardly imagine how isolated I am here. My correspondence with friends and fellow-workers has been a great source of joy and satisfaction" (Seward, 1921, p. 465). Also, Nathorst was deaf, and that may partially explain his voluminous correspondence.

Nathorst's first great study resulting from a polar expedition was actually his work with the late Tertiary plants that had been collected by Nordenskiöld in Japan near Nagasaki. This was published in 1882 in both Swedish and French editions. A second work on the Japanese Tertiary came out in 1888. Nathorst's reputation had increased considerably by this time, and the 1888 study was based on over thirty different collections that had been sent to him directly or through universities in Berlin and Uppsala. He regarded these floras as indicating a more northerly climate than existed in Japan in 1888 and leaned in favor of a displacement of the North Pole toward the Pacific Ocean. Halle says that Nathorst felt somewhat less certain about this in later years and showed no resentment when he (Halle) strongly opposed it.

In 1898, fifteen years after his Greenland trip with Nordenskiöld, Nathorst engaged in a well-planned expedition to Bear Island and Spitsbergen. Although much of his earlier writing is in Swedish, he wrote an account in English of this trip for the *Geographical Journal* in 1899, and it makes rather good reading. This trip yielded, among other fossils, the unique *Pseudobornia ursina* from the Upper Devonian of Bear Island; in recent years more extensive collections have been made by Hans-Joachim Schweitzer, who spent the summer of 1964 on Bear Island and published additional information on *Pseudobornia* in 1967.

A short excerpt from Nathorst's account of the 1898 expedition, relative to fossils found in the vicinity of Bell Sound, Spitsbergen, cannot fail to excite any paleobotanist.

> I must confine myself to stating that we discovered the presence of the same fossil Rhaetian flora, which exists in Scoresby fjord on the east coast of Greenland; and proved that the Tertiary strata so rich in leaf-impressions, which I discovered in Ice Fjord in 1882, were also extensively represented in the branches of Bell sound. In the places where these strata occur, you literally walk in fossils up to the knees, amongst them being magnificent specimens of leaf-impressions of marsh cypresses, alders, limes, magnolias, and other trees. A very noticeable feature about these leaf-impressions is that they are of an extraordinarily large size; and yet it vexed us with the tortures of Tantalus, seeing that we had to leave behind us so much that we desired to take away. [1899, p. 62]

Among other important and well-known genera that Nathorst described may be mentioned *Lycostrobus*, *Cephalotheca*, and *Wielandiella*. Certainly one of the most intriguing fossils that he described was an *Artocarpus* (bread fruit) from western Greenland; I have looked at the specimens, fruit and leaf, several times in the Museum in Stockholm; they seem to be correctly identified but remain something of a climatic puzzle. Halle notes that the whole generation of Swedish geologists studying in the 1890's and immediately after 1900 are grateful to Nathorst for his comprehensive books *Jordens Historia* (History of the Earth) and *Sveriges Geologi* (The Geology of Sweden).

Nathorst deserves much of the credit for the fine facilities and fossil-plant collections that the Paleobotanical Department enjoys in the Natural History Museum in Stockholm. Nathorst's original museum quarters were in an old two-story building in Wallengatan in Stockholm, and several who visited him there have left brief comments on the cramped and decrepit nature of the place; but one does develop nostalgic feelings for such "facilities," and Seward says: "It was my good fortune on two occasions to spend several days in the old museum with Nathorst, and it will always be the small and crowded rooms in the heart of Stockholm that some of us will remember with feelings of admiration, gratitude, and affection" (1921, p. 463).

More and better space was essential, however, and efforts to achieve this end were initiated by Nordenskiöld as early as 1875. In 1884 the government established a professorship for a new Department of Paleobotany and Archegoniate Botany in the Swedish Museum, but it was not until 1910 that this seems to have been placed on a firm footing. And it was only a few years prior to Nathorst's death in 1921 that the move was made from the former quarters in the city to the present wing of the Natural History Museum at Frescati outside of Stockholm. At first the new department included living bryophytes, pteriodphytes, and gymnosperms as well as fossil plants, owing to the small size of the paleobotanical collections and overcrowding in the Botanical Department.

Although Nathorst was considerably before my own time I developed what might be called a personal feeling for him after reading a short paper that he wrote on *Archaeopteris* in 1904. The specimens were collected by Per Schei, the geologist on Otto Sverdrup's expedition of 1898–1902 to Ellesmere Island, and were turned over to Nathorst for study and description. I read the article many years ago never expecting to see where the fossils came from. In 1960, however, I obtained some substantial (for that time) research funds that made it possible to do some exploring without having to submit a clear-cut and convincing statement of just what I expected to find. It seemed a good time to have a look at the Arctic. Aside from the *Archaeopteris* locality I thought that there were large areas that needed investigation, and my love of alpine terrain, developed from many years of mountain hiking in this country and a brief venture in the Himalayas, had given me a long-standing desire to visit the northern latitudes.

Thus, after a good deal of planning and aid from several organizations and Canadian officials, N. W. Radforth, Tom L. Phillips, and I took off in the summer of 1962 for Goose Fjord in southwestern Ellesmere Island. Our primary objective was to explore the vast deposits of Upper Devonian rocks at the head of the fjord. Even with the advent of air transport one can still encounter problems traveling in the Arctic, but this first trip went quite well. We reached Churchill in Hudson Bay in late June via a commercial flight, having dispatched three-quarters of a ton of supplies and equipment some weeks previously through Winnipeg. On arriving in Churchill we found that our baggage was being held in Winnepeg for some unexplained reason; a few frantic phone calls cleared the misunderstanding, and everything arrived a few hours before our arranged departure. A Canadian Air Force plane took us to Resolute Bay. From there we employed a DeHaviland Otter with a very good pilot at the controls and took off for Ellesmere. The plane was equipped with huge tires and we were fortunate in spotting a site at the head of Goose Fjord that allowed a reasonably safe landing.

We camped there for several weeks and spent another week at a locality some forty miles to the north. After a couple days of searching we located the place from which Per Schei probably made his collection. We then continued to explore the area rather carefully for a radius of a few miles around the head of the fjord and found *Archaeopteris* specimens at a few other places. The most spectacular sight we encountered, from the standpoint of the life of the past, was an exposure of ledges which were covered with impressions of plant stems up to ten inches wide. These almost certainly were the impressions of *Callixylon* trunks which bore the *Archaeopteris* foliage, the latter also being present in the near vicinity. Only a few hundred yards from these ledges there lay a great snowfield that capped the mountain. The contrast between the present and the past was easily visualized: the barren windswept mountains and snowfields of the present with the forest of Upper Devonian times.

On the west side of the valley we found numerous coal fragments which apparently formed the basis of a reported coal "seam" on the geologic map. It seemed reasonable to conclude that these fragments were only isolated specimens of coalified trunks of *Callixylon*.

We made a modest contribution to the progymnosperm story and learned a good deal about Arctic botany and geology, and Radforth was able to add to his great fund of knowledge about peat deposits. Accordingly we decided to try to go back the next summer, this time with considerable aid from the Canadian government, and our headquarters were to be at the Lake Hazen camp usually occupied by a group of entomologists.

We left Edmonton on July 15, 1963, in a Canadian Air Force Hercules, a so-called flying boxcar which is more famous for its freight capacity than for passenger comfort. After an intermediate stop at Resolute Bay we proceeded to Alert at the northernmost tip of Ellesmere Island. But this summer, certainly through no fault of the various people who were trying to help us, everything seemed to conspire against our plans. We waited about a week in Alert for a plane to take us the ninety miles south to Lake Hazen. Two helicopters were due there sometime later in the summer, and we were assured transport to several areas we wished to investigate. However, an accident and mechanical trouble eliminated them from our plans; they never arrived. A boat was available at the camp, but that summer Lake Hazen never thawed sufficiently to permit its use. Foot travel on Ellesmere can be quite fascinating but because of the innumerable bodies of water, rushing streams, and rough terrain it is a very time-consuming way to explore for fossil plants. We located a few plant beds, including a deposit of petrified wood some forty miles from the camp, very little of which could be taken back because of the weight.

I think one loves the Arctic very much or not at all. For me it was the former, and I believe we would have returned again, hoping for a better transport system, but in the summer of 1964 I became involved in digging in northern Maine and later in southeastern Canada, and these areas proved so highly productive that they occupied my time for the next ten years—and there is still much to be done there.

Returning to Nathorst after this little diversion: he also had a vast knowledge of Tertiary floras, much of which died with him. He had made large collections in Spitsbergen and had arranged a set of plate illustrations which I believe are stored in the library of the Paleobotanical Department at the Natural History Museum in Stockholm, but he never finished the text.

Certainly one of Nathorst's greatest contributions was the improvement of maceration methods in dealing with compression fossils. He became concerned with this rather late in his career, and although he was by no means the originator of the general method, his use of it marks a distinct turning point in the productive study of compression fossils and guided many paleobotanists to investigate "significant" specimens in contrast to simply collecting and illustrating showy or "pretty" museum pieces. This, therefore, seems an opportune point at which to make a few notes on the history of this important development.

The earliest illustration I have encountered of a leaf cuticle showing cellular structure is in the second volume of Lindley and Hutton's *Fossil Flora* (1834). They show leaves described as *Solenites murrayana* (probably a ginkgo relative, *Czekanowskia*) which could be removed quite easily from the rock and could be cleared by plunging them into boiling nitric acid. They said "the sides were evidently composed of prismatical cellular tissue," which is shown in their plate 121. They did not follow up their little experiment with any other plant material and apparently were not aware of its having any special significance. In several works of the 1840's and 1850's brief reference is made to cuticular structure, but the figures are crude and contribute rather little.

Credit is usually given to Johann Georg Bornemann's "Uber organische Reste der Lettenkohlengruppe Thüringens" (1856) as the first significant study of fossil cuticles. However, another half century would elapse before Nathorst gave this aspect of paleobotanical investigation the push that inspired and directed others to the store of information that was latent in well-preserved compressions.

Bornemann (1831–1896) studied natural science in Leipzig, Göttingen, and Berlin, where he graduated in 1854, presenting a thesis on a study of Liassic plants. He intended to continue with his academic studies, but an excursion to Sardinia led him into mining interests, which apparently proved too rewarding financially to give up, and he entered into a mining and real-estate business with his father in Eisenach. But he had sufficient time to continue to devote some attention to fossil botany, although the 1856 paper was his most important work. Bornemann's preparations are preserved today in the paleobotanical collections of the University of Halle.

Rather few paleobotanists took advantage of the method in Bornemann's time. August von Schenk in his *Die fossile Flora der Grenzschichten des Keupers und Lias Frankens* (1867) made a considerable series of maceration preparations and showed the advantage of the technique in studying fern spores, and Gothan (1950) states that Schenk introduced the use of the Schulz reaction (HNO_3 + $KClO_3$) followed by alkali treatment. In 1882 René Zeiller, in his article "Observations sur quelques cuticles fossiles," figured and described the cuticles of cycadophyte leaves from Italy and some French Cretaceous conifers. This seems to me an extraordinarily fine study that might have been followed more vigorously by other workers. Zeiller also described the cuticles of the famous paper coal of Tovarkovo from the central Russian province of Toula; Scott mentions in his *Studies in Fossil Botany* that Zeiller described the cellular structure and discussed the effects of various oxidizing agents in clearing the cuticles.

In the first decade of the present century Nathorst began to experiment with maceration methods and demonstrated their potential very effectively in a series of papers. One of these, dealing with *Lycostrobus scotti* in 1908, includes two very impressive plates showing macerated spore masses and individual spores. I have mentioned previously that Hamshaw Thomas spent part of a summer in Nathorst's laboratory; he applied the technique he learned there to cycadophyte compressions, and he later made significant developments of his own in studying the fruits of *Caytonia*. Tom Harris has probably devoted more attention than anyone else to extracting information from Mesozoic compression fossils; in 1926 he introduced the bulk-maceration method, which, he notes, reveals plant fragments that are either lost or never observed when one examines only the surface of a piece of shale.

It is my intent to deal only with the historical aspect of cuticle studies up to the time of Nathorst and his immediate "disciples." Since that time many others have used and developed the method to suit various modes of preservation and plant materials.

I will return to the paleobotanical succession after Nathorst at the Stockholm Museum, but at this point I wish to bring in Oswald Heer (1809–1883) who, along with Nordenskiöld and Nathorst, makes the great trio that initiated paleobotanical studies in the Arctic. In paleobotany Heer is best known for his *Flora fossilis Arctica* but he was a man of astonishing productivity in two fields, botany and entomology. A somewhat fragile state of health during much of his life prevented him from engaging in the rigorous field work that the other two conducted, but this was more than compensated for by the long hours he devoted to the numerous and huge collections that were sent to him from all quarters.

Oswald Heer was born at Niederutzwyl in the Canton of St. Gall in Switzerland, but his family soon moved to the village of Matt, where he spent much of his early life. Like Leo Lesquereux, who left Switzerland to make a comparable name for himself in the United States, Heer loved and was influenced by the alpine regions of his homeland. He began very early to gather his own voluminous collections of plants and insects; one of his biographers (Scott, 1883) says that he bribed his schoolfellows to help him with his collecting by offering to give them singing lessons on Sundays—an offer that I suspect would draw meager results today!

At his father's urging he entered the University of Halle to study for the clergy and was ordained in 1831. But when he had to choose between accepting a pastorship at Schwanclen or a curatorship of the insect collections in Zurich, he chose the latter. He soon became associated with the University of Zurich and served as Professor of Botany for many years.

Although he was unable to carry on extensive field work he did do some traveling. He was in Madeira in the winter of 1850–51, and it is evident that all of his time was not spent basking in the sun, for on his return he produced papers on the fossil plants of the island, as well as the probable origin of the living flora and fauna of the Azores. This visit also seems to have incited in him an interest in Atlantis, based on the presumed identity of Miocene species found in Switzerland and in North America. In the spring of 1856 he went to Italy, and he visited southern England in 1861 and spent some time digging in the Bovey Tracey deposits in Devon.

It is impossible here even to list the innumerable works that continually dripped from his pen; I will therefore scarcely touch on his entomological work and consider what seem to be his most important studies with fossil plants. He very early found himself the recipient of literally floods of collections, many of which came from the early Arctic explorations. The explorers themselves were anxious to have the results of their efforts appreciated and brought to the attention of the public and the scientific world. Heer produced published results at a rate that few other naturalists have equaled, and as his reputation grew, collections poured in at an ever increasing rate.

Heer's greatest paleobotanical work is his *Flora fossilis Arctica*, which was produced between 1868 and 1883. This fifteen-hundred-page work may be found in many libraries bound in seven volumes under this title, although much of it originally appeared in a wide range of journals. The first volume deals with fossil plants from northern Greenland, northern Canada, Iceland, and Spitsbergen; the second with Bear Island, Alaska, Spitsbergen, and Greenland; the following volumes include studies of fossil plants from Siberia, Grinnell Land, Sakhalin Island, Novaya Zemlya, and other parts of the Arctic. There are nearly four hundred plates, and he included discussions of the modern vegetation as well as insects.

Heer was very much interested in the general problem of the origin of the past and present Arctic floras and the climatic changes that have ensued. Although he did not visit the polar latitudes himself he did the next best thing by keeping in close touch with the explorers who brought back to him great quantities of information and specimens. His account (1869) of Whymper's Greenland expedition of 1867 and Nordenskiöld's expedition to Bear Island, Spitsbergen, and Greenland makes good reading and includes some of Heer's basic ideas about the development of the plant kingdom, former Arctic climates, and plant distribution. He concludes that "a whole series of new facts, established by the recent discoveries, confirm the opinion that the glacial zone must formerly have enjoyed a climate much warmer than that which it has in our days. This fact springs from the study of all the geological formations from the Carboniferous Epoch to the Miocene period" (1869, p. 99).

Heer also began to look upon the Arctic as one of the major centers of origin of the higher seed plants: "the plants of the Miocene . . . have probably been propagated starting from several centres; but their diffusion on the surface of the globe must have been slower, on account of the weight of their seeds, which are generally larger [the comparison here is with pteriodophytic plants of the Carboniferous]. *One of these centres of diffusion was evidently in the polar zone, whence plants and animals have spread in radiating directions*" (1869, p. 100).

In 1865 Heer brought out a two-volume work, *The Primaeval World of Switzerland*, in German, followed by a French edition and an English one in 1876. This is almost an encyclopedia of paleobotany and certain aspects of geology, although the emphasis is on Switzerland. The first volume begins with Carboniferous horizons and works up through the Tertiary, including very detailed discussions of the various plant and animal groups. The second volume starts with the Swiss Miocene localities and includes a long discussion of the Miocene, a period that Heer was especially fond of.

He did not take kindly to Darwinian evolution and was especially bothered by the point that man had not actually observed the origin of new species.

> Hence, it may be affirmed that no new species has had its origin since the drift-period. A certain number of species have disappeared, and great changes have taken place in the intermixture of forms. . . . Although a species may deviate into various forms, it nevertheless moves within a definitely appointed circle, and preserves its character with wonderful tenacity during thousands of years and innumerable generations, and under the most varied external conditions. [1876, pp. 283–84]

He of course recognized great changes from one geologic period to another, but to me his vagueness in explaining them is reminiscent of Brongniart:

Oswald Heer. Engraving by W. Meyer, from *Oswald Heer—Bibliographie* by Godefroy Malloizel, and *Notice Biographique* by R. Zeiller, Stockholm, 1887.

> Times of creation occurred during which was accomplished a remoulding of organic types, and there was a primaeval epoch during which the first species were brought into being. Even if the first species were extremely simple, for them an act of creation must be admitted. . . . Great creative renewals are indicated within the limits of the principal geological periods; and during those periods important transformations also took place, the

significance of which cannot at present be satisfactorily estimated. [1876, pp. 291–93]

How does one judge the tremendous literary output of men like Heer? Along with a few others in the history of paleobotany he was deluged with collections and he had to choose between taking them as they came to him and doing the best he could or working more carefully with a few and placing the rest in storage—always a precarious move. Although he made mistakes I think he made the correct choice. It is better to move ahead with work that is respectable but contains errors, perhaps many errors, than to do nothing. It is of interest to note what others have thought of his work, both his contemporaries and writers of the present day.

Of Heer's work on the Tertiary floras of Switzerland, the three-volume *Flora tertiaria Helvetiae*, Ward says: "The exceedingly great care, accuracy, and thoroughness with which this *chef d'oeuvre* of science was executed, especially in the matter of illustration, is a marvel to contemplate. Nothing comparable to it had appeared before, and nothing equal to it has appeared since" (1885, p. 378). Lester Ward was a kindly critic and tended to see the best side of a man's work, but Nathorst also thought highly of Heer's studies, in spite of some obvious errors. Charles Lyell was one of Heer's ardent defenders. And of a much more recent vintage is the opinion of L. Y. Budantsev, who says that "in spite of the fact that many of the determinations of extinct plants made by Heer are hopelessly antiquated, as are his impressions about the age of several Arctic floras, *Flora fossilis Arctica* is still the most valuable classical study to which paleobotanists of the whole world turn" (1969, p. 485; translated).

It is said that Heer was confined to bed for much of the last ten years of his life, but poor health and old age did not hinder him. One of his biographers says: "Much of his work . . . was produced under conditions the reverse of favourable for exact determination and comparison: a friend relates that when calling to convey one of the numerous awards made to him by English scientific bodies, he found the Professor lying down with a small table arranged to cross the bed, upon it being specimens which he named while an assistant made drawings" (*Nature*, 1883, p. 613).

In 1913 an assistantship was created in Nathorst's domain in the Natural History Museum in Stockholm; it was filled by Thore Gustav Halle (1884–1964), who in turn became Professor and Director of the Paleobotanical Department when Nathorst retired in 1918.

Halle was born at Mullsjö in the county of Skaraborg, Sweden. He took part in a Swedish expedition to South America under the leadership of Karl

Skottsberg in 1907–1909; his thesis for the doctorate (received in 1911) at Uppsala, based on this trip was a geological history of the Falkland Islands which includes a description of a *Glossopteris* flora of Permian age.

Also among his earlier papers is "The Mesozoic Flora of Graham Land" written in 1913 and based on plant collections made by J. G. Andersson during the Swedish South Polar Expedition of 1901–1903. In his summary notations on the development of the Department of Palaeobotany in 1950 Halle comments on the source of the fossils:

> This expedition brought back the first collections of determinable fossil plants ever made in the Antarctic. The largest material came from the Jurassic of Hope Bay, where J. G. Andersson, in addition to other activities, once more proved himself an unsurpassed collector of fossil plants. (The work partly carried out under severe hardships. With two companions he had to pass the whole Antarctic winter in a stone hut built by themselves, living almost entirely on penguins, the fat of which also provided almost the only fuel.) [P. 10]

Halle's paleobotanical publications are not numerous, but several are of great and lasting significance. In 1916 he produced an important work on the early land vascular (Lower Devonian) plants of Röragen in western Norway: this flora includes *Psilophyton* species and some other interesting primitive land plants. Mrs. Andrews and I made a short visit to Röragen in 1964; it is in the general vicinity of Röros, and we had a little trouble finding the locality as no one had mentioned that "Röragen" is simply a cluster of farm buildings! The surrounding country is a rather lovely and wild bit of landscape, with moss-covered ledges and rocks, partly bordering the end of a small lake. We obtained some plant fossils and it was my impression from a one-day visit that the area might well yield new information with more extensive searching. Harlan Banks has shown that some of the plant axes are petrified, revealing good preservation of the woody tissues. Halle also found the curious plant *Sporogonites* at Röragen.

Halle's greatest single publication is his monumental volume on the Paleozoic plants from Central Shansi, China (1927). The flora of Shansi is one of the most important Permian floras and includes a great assemblage of articulates, pteridosperms, cycadophytes, and other groups. In the introduction Halle says: "I had long wished for an opportunity to study the plant-bearing deposits of China in the field, and in the autumn of 1916 I was able to realize my plans. . . . I arrived at Peking in November 1916 and devoted the next two months to the study of the fossil plants in the Museum of the Geological Survey in Peking and to collecting new material, partly in the hills west of

Thore G. Halle and Rudolf Florin, at Stockholm, 1950.

Peking and partly in the Kaiping mining district" (p. 5). He also spent several months collecting in the early part of 1917, but illness prevented him from doing all that he had hoped. He goes on to relate the ultimately tragic result of this field work: "Because of the shipping difficulties caused by the war the collections had to remain in China till 1919, when they were sent to Stockholm by the Swedish steamer 'Peking.' This ship went down with all hands in a typhoon in September 1919, and so all the collections were lost" (p. 6). A second collection, obtained by the Geological Survey of China, served as the basis of Halle's great account. One may despair of what went down in the China Sea but it is perhaps better to wonder what may still lie hidden in the rocks of Shansi. There are many drawers containing the collection in the Stockholm Museum, and other museums have benefited from it by exchanges.

As a follow-up Halle brought out a much shorter study in 1929 entitled "Some Seed-bearing Pteridosperms from the Permian of China." This is the most informative article on fernlike foliage bearing seeds up to that time and includes descriptions and illustrations of *Sphenopteris*, *Pecopteris*, *Emplectopteris*, and *Nystroemia* fronds. It contributed much to our understanding of the morphological diversity of the pteridosperm group.

Halle was a careful and meticulous worker who carried on Nathorst's tradition of developing new techniques for extracting information from fossil-plant remains, the most important result of this being his article "The Structure of Certain Fossil Spore-Bearing Organs Believed To Belong to Pteridosperms," which appeared in 1933. "In it," says one of his biographers,

> he skilfully adapted the method used by E. C. Jeffrey and H. Hamshaw Thomas, to swell and soften the coaly remains of Mesozoic fossil plants so that they could be embedded in collodion and sectioned by microtome, to his investigation of this harder, more resistant Carboniferous material. In one method he enclosed the carbonaceous material with the softening reagents in small but thick-walled gunmetal cylinders so that the contents could be subjected to much higher pressure and temperatures than those used by Jeffrey and Thomas. The results were so successful that he was able to distinguish the position of the pollen grain masses in the complex organs which produced them and in more than one instance proved that some fossils which had been thought to be seeds were in fact complex synangial pollen-bearing organs. [Walton, 1966, pp. 21–22]

Halle also brought out a fine paleobotanical text, *De Utdöda Växterna*, in 1940 which might have been advantageously translated into other languages.

To continue with the great Swedish sequence at the Stockholm Museum: when Halle became head of the Paleobotanical Department in 1918 his own post was filled by Rudolf Florin (1894–1965). There are many paths to greatness and many kinds of greatness and it is not always easy to draw fair comparisons among our colleagues of the past, but I think there is no doubt that Florin has given us the greatest unified story in plant evolution, based on fossil evidence—that of the origin of the seed-bearing organ in the evergreens (conifers).

Florin was a botanist all of his life; he was born at Solna, near Stockholm, and from his earliest days learned to love and study plants, his father having been the head gardener at the Hortus Bergianus where Rudolf eventually became Director. He attended the well-known "Beskowska skolan" where Gustav V and his brothers had studied, and received his advanced education at the University of Stockholm. His earliest works were in horticulture and reveal a thorough practical understanding of plants. One of his first botanical papers was a joint publication with his father describing a new kind of apple, "P. J. Bergius." In the decade following World War I he and his wife, Elsa, carried out important research on the fertility and partial sterility of pollen from apple and pear varieties, cherries, and other fruits in order to eludicate the problem of compatibility. He also had an early interest in bryophytes.

Florin's publications, like those of Halle, were not abundant in numbers.

By most people he would probably have been considered slow; in any event he was extraordinarily meticulous and unhurried, and allowed no significant detail to escape him. He gathered together a vast amount of data before putting anything into print—a characteristic that some present-day workers might be encouraged to adopt. This is well illustrated by his first outstanding work, a 1931 volume of nearly six hundred pages on the epidermal structure of recent conifers. It represented years of work and formed the basis of much that followed. One of his biographers notes, "I first met Florin in 1924 I think, both in the Museum and at his home where I saw his three young sons and his wife Elsa. Here I went to study cuticles and I found that he had an enormous body of work beneath the surface, like an iceberg!" (Harris, 1968, p. 175).

I have mentioned elsewhere the work of Thomas and Bancroft on the cycadophytes, which resulted from Thomas's period of study in Stockholm. Harris points out that Florin's

> study of the cuticles of the fossil Cycads and Bennettitales (1933) was masterly. Thomas and Bancroft had made the distinction 20 years earlier and indeed pioneering suggestions are to be found in Nathorst's work, but in Florin's paper matters are worked out fully. The characteristic arrangement of the subsidiary cells of the Bennettitales was related to their presumed course of development from a single embryonic cell which finally formed the two guard cells, a course of division exactly parallelled in certain dicotyledons but not seen in other gymnosperms (apart from *Gnetum* and *Welwitschia*). For this form of subsidiary cell group he gave the new term "syndetocheilic," the alternative where the subsidiary cells and guard cells have no immediate common ancestor being "haplocheilic." [Harris, 1968, p. 175]

In 1936, Florin brought out two comprehensive studies on the fossil ginkgophytes of Franz Josef Land. The material on which this study was based is a kind of lignite petrified with silica and beautifully preserved. In studying it he used ground thin sections in part, and also extracted the leaves and twigs from the matrix after treatment with hydrofluoric acid, and some of these he embedded in wax and sliced on a microtome. The result was the establishment of six new genera of Ginkgoales and a great advancement in our knowledge of the diversity of this group in the Mesozoic. I recall that, on one of my visits to the Botanical Garden, not long before Florin died, he showed me some similar material of the conifers that also came from Franz Josef Land. I believe that he did not finish this study and I have wondered whether any efforts have been made in recent years to locate the source of this unique fossil material from the Arctic island. The specimens that I saw in his laboratory, although dark brown inside, had a white weathered surface and might well be ignored in the field unless one knew exactly what he was looking for.

Florin's greatest work, possibly the most noteworthy single study ever carried out with fossil plants, is his *Die Koniferen des Oberkarbons und des unteren Perm* (The Conifers of the Upper Carboniferous and Lower Permian). This was issued in eight parts (with a total of 729 pages and 186 plates) in *Palaeontographica* from 1938 to 1945. It resulted from several decades of study of at least seventy-five collections stored in European and American museums. Very briefly, Florin explained in large part one of the great morphological problems in botany: the nature and evolution of the seed cone of the evergreens (conifers). He showed that in plants such as the pines the ovuliferous scale (the scale of the cone, on the surface of which two seeds are borne) was derived from a short shoot with radially arranged appendages, some of which bore terminal seeds. The number of appendages composing the short shoots was progressively reduced, the seed-bearing appendages became recurved so that the seed micropyle pointed toward the axis, and the whole structure became flattened.

There is still work to be done; it seems to me that we still have a big gap from the open inflorescence of the cordaites to the first distinct cone of the conifers, but the basic evolutionary plan is clear. In his own unique and delightful use of the English language, Tom Harris sums up Florin's work as follows: "His theory of conifer morphology does indeed consist of numerous separate but connected hypotheses and I dare say some will wear better than others, but even if wholly different views are later to prevail (and I do not believe they will), this at least can be said—he gave conifer morphologists peace for a time and I suspect it will be for a long time" (1968, pp. 176–77).

Probably rather few botanists have read Florin's monumental work in full, and fortunately he brought out a much condensed summary in English, "Evolution in Cordaites and Conifers" (1951), which resulted from the Prather Lectures he delivered at Harvard University in 1948–49. Those who are fortunate enough to have acquired a copy of this own one of the greatest classics in scientific literature.

Following the appearance of this work Florin devoted his attention chiefly to the taxads, separating *Taxus* (the yews) and some closely related genera from the true conifers. These works are listed in a biographical sketch of Florin that was written by Britta Lundblad in 1966.

It seems especially important to look in some detail into the character of a man whose researches were so significant. It was my pleasure to have been able to visit with him several times and become quite well acquainted with him and his friendly and hospitable wife Elsa. I will offer a few comments but first let's look at those of two others who knew Florin quite well. Tom Harris writes:

> How did he achieve so much? He gave me the impression of having a slow mind; probably it was quick enough but he mistrusted the quick answer. I often wondered. It seemed to me he achieved his success not at all through what is called brilliance but by ordinary virtues, virtues which others could emulate but in fact do not. He had enormous industry and the physical and mental strength to support it and also sustained drive. He was thorough and orderly. He had the courage to accept an outstanding fact and to use it logically, however awkward the conclusions might seem. But these and an undeviating devotion to scientific truth do not explain all. He did have the gift of scientific imagination which makes all the difference between a pedestrian work and a great paper. But beyond all a great man's virtues and qualities I think there often stands another; a serene home provided by the sensible and loyal woman he made his wife. [1968, pp. 177–178]

Britta Lundblad says: "The impression remains with me that Florin worked in a spirit of calm conscientiousness, yet, strangely enough he never seemed hurried. This is all the more remarkable in view of his great accomplishments, and he must have had an extraordinary power of concentration and imagination" (1966, p. 91).

Florin once visited St. Louis, Missouri, and spent several days in my home and at the Botanical Garden. He was something of a disappointment to my students in that he seemed to have little interest in any fossils other than the conifers, and unfortunately my collections were rather sparse in that direction. But his knowledge of and devotion to the investigation of a great group of plants was so great and so apparent that it was certainly a unique experience for a student to come into contact with him.

He was personally delightful and had a fine sense of humor, and he and his wife were the most cordial of hosts. I spent a pleasant week with them in their house at the Botanical Garden once, shortly after the appearance of Roland Brown's report on the problematical *Sanmiguelia* which suggested the possibility of angiosperms in the Triassic and created considerable interest. (More recent studies of *Sanmiguelia* indicate that it is not an angiosperm.) I asked Florin what he thought about it and his reply was quick and final: "I have no concern or interest in these sensational reports!" Actually Roland Brown was a very meticulous worker himself, a real scholar and hardly given to the sensational, but I thought that this expressed very clearly Florin's method of study.

Florin received the title of Professor in 1942 and was appointed Professor Bergianis in 1944; he was then living in the director's home at the Hortus Bergianus, which is across the street from the Natural History Museum. A few words on this garden may be of interest. According to Britta Lundblad the Bergius Foundation owes its name to one of Linnaeus' best pupils, P. J. Bergius (1730–1790), who, after a life devoted to the medical profession, bequeathed all of his property, including a garden, to the Swedish Academy of Sciences. The present garden owes much to the care that Florin and his father gave it. It is a delightful place to wander about in on a summer day. I have not been there since Florin left; it is probably still as lovely but there would be something missing for one who had known this great botanist.

Although micropaleontology generally falls outside the scope of my account, I wish to include a brief note on another talented botanist and paleobotanist of Stockholm, Henning Horn af Rantzien. I had occasion to meet him briefly on one of my visits and thought him a promising scientist and a personable young man. He was a knowledgeable taxonomist with a special interest in the living charophytes which had led him into studies of the fossil forms. After graduating from the University of Stockholm, Horn held research positions at both the Natural History Museum and the Geology Department of the University, where he taught. Thus he was on the way to a very promising career when, at the age of thirty-eight, he was killed in an accident while searching for fossils in a quarry at Montmorency, France, on September 14, 1960.

The more recent paleobotanical developments in Stockholm are also beyond the scope of my account and I can only include a few brief notations that may serve as a lead to any readers seeking additional information. Olof H. Selling took over the headship of the Paleobotanical Department at the Natural History Museum in 1951 relinquishing the post in 1966, and shortly thereafter Britta Lundblad became Professor of Paleobotany. Selling is probably best known for his comprehensive studies in the late 1940's of Hawaiian palynology and the history of Hawaiian vegetation in late Quaternary times. Among his studies of older plant fossils his investigation of a Carboniferous articulate cone, *Palaeostachya schimperiana* Weiss, is noteworthy; he employed a serial-grinding method to extract information from a type of fossil that in the past had often been either discarded or simply filed away in a museum drawer.

Britta Lundblad's early studies deal with a revision of the Rhaeto-Liassic floras of the mining district of northwestern Scania in Sweden. They amount to more than a revision, for they bring in much additional information using modern techniques and are well illustrated. The first one describes the pteridophytes, seed ferns, and cycadophytes. The second describes and gives fine illustrations of the ginkgophytes *Ginkgoites* and *Sphenobaiera*. Somewhat later she brought out two papers on Mesozoic liverworts. The first describes especially well-preserved fossils from Scania, and the second deals with Lower Cretaceous plants from Patagonia that are probably referable to the Marchantiineae. And also very worthy of mention is a short article of 1950 describing a fertile *Selaginella* from Scania in which both micro- and megaspores are well preserved.

I regret having to close this summary of activities in the distinguished Stockholm Museum on another tragic note but such is the case. A few years before Britta Lundblad took over the Professorship a very talented young man, Dr. Hans Tralau, was appointed Assistant. In 1966 he produced a fine floristic study of a Mesozoic flora from Scania which compares closely with the flora of Yorkshire, and in 1968 he wrote a very useful summary of the fossil history of the genus *Ginkgo*. Tralau had a strong interest in bibliographic matters and devoted a great deal of time to a compilation of all the pertinent paleobotanical literature from the early nineteenth century to the present. The initial result of this work, *Bibliography and Index to Palaeobotany and Palynology, 1950–1970*, appeared in 1974. This was a tremendous undertaking and I believe it was carried on in large part at his own expense. But it is sad to have to relate that Hans Tralau's other activities, including the Secretaryship of the International Organization for Palaeobotany, were brought to an end by his death at the age of forty-five in March of 1977.

I am personally most grateful to all of these Swedish paleobotanists and other workers in the Swedish Museum for making my several visits there pleasant and profitable.

One of the most significant and revealing studies of the early Arctic floras is Ove Arbo Høeg's (1942) detailed account of the late Silurian and Devonian plants of Spitsbergen. Devonian rocks occupy much of the north-central part of Spitsbergen, consisting of thick series of sandstones and shales with occasionally abundant, although not especially well preserved, plant and animal remains. I had an opportunity in 1964 to study part of the collections in Oslo and was greatly impressed with the amount of information that Høeg had been able to extract from them.

The collections were made largely by Norwegian expeditions but some specimens obtained earlier by Nathorst were included in the study. Of his explorations in Spitsbergen in 1942 Professor Høeg says: "as a member of Hoel's expedition, I had the opportunity to search for fossil plants in the Devonian of Dicksonfjorden, also making a journey overland to the inner part of Wijdefjorden (Vestfjorden, Gråakammen). Although a number of fossil fishes were collected, and plant fragments . . . were observed in many places, no plant fossils of any value were found" (1942, p. 10).

Fossil plants were collected in some abundance by Dr. A. Heintz, a member of the Vogt expedition of 1925, and again in 1928 on a trip in which Høeg also participated: "The paleobotanical results of this expedition were a fossil flora from the Downtonian of Raudfjorden, discovered in Fraenkelryggen (Fraenkel Ridge) by Strand and Störmer, plant remains from various localities at Wijdefjorden, and a large material from Mimerdalen. These collections, together with those from 1925, form the main basis of the present paper" (1942, p. 10).

The plants described cover a considerable time span from the Upper Silurian to the base of the Upper Devonian—and the sixty-two accompanying plates reveal a considerable number of species. There are numerous fine specimens of *Psilophyton*. There are also several new genera, including the unique *Enigmophyton* with its fan-shaped leaves and associated heterosporous sporangiate organs; *Psilodendron*, consisting of a spiny branch system, which more recent evidence suggests may be a distinct species of *Psilophyton*; *Svalbardia*, which seems to be not distantly related to some species of *Archaeopteris*; and *Germanophyton*, a new generic name for *Prototaxites psygmophylloides* of Kräusel and Weyland. The latter is described as a rather large plant with irregularly branched axes bearing large fan-shaped leaves and containing tubes of the *Prototaxites* type. It is to be hoped that future explorations in the area will bring to light new localities and more information.

In 1937, Ove Høeg prepared a very useful summary of Devonian floras that was published in the *Botanical Review*, and in 1967 there appeared his fine treatment of the Psilophyta in the *Traité de paléobotanique*. While the coverage in these two publications is not quite identical it is a great tribute to paleobotany to see the tremendous strides that have been made, during thirty years, in our knowledge of the early groups of vascular plants.

It was my pleasure to have been invited to teach a course in paleobotany at the Geology Institute of Aarhus University in Denmark in the autumn of 1976. This enabled me to renew an old acquaintance with Eske Koch, who is known for his work with the Cretaceous-Tertiary floras of western Greenland, and to see at first hand some of the work that his vigorous group of paleobotanists is doing with the fossil plants of the Tertiary lignites in Jutland. The earliest comprehensive and systematic studies of the plants in these (Miocene) lignite beds are those of Friedrich Jens Mathiesen (1886–1976) which were published between 1965 and 1975, although he began to work on them many years before.

Ove Arbo Høeg, at a fossil plant locality on the Gaspé, Canada, 1959.

The story of Mathiesen's "paleobotanical life" stands as an extraordinary example of fortitude. He was born in Svendborg, Denmark, on May 29, 1886, and studied pharmacy there, graduating in 1910. He obtained a position as assistant at the Pharmacological High School in Copenhagen in 1912, and, having some time available, he started botanical studies at the University, where he took his M.Sc. degree in 1918. He worked with Professor E. Warming, who was in charge of Arctic plant studies, and prepared accounts of the Primulaceae and Scrophulariaceae which were published in 1916 and 1921. His paleobotanical interests seem to have originated in about 1918, when he joined the Mineralogical Museum as a temporary assistant. There he became acquainted with Professor Ravn, whom he accompanied in 1917 on a trip to Silkeborg and Moselund in Jutland, where lignite deposits containing well-preserved fossil plants had been found. Much later, in 1965. he wrote, "Tertiary strata containing such an abundance of well-preserved vegetable remains had hitherto been unknown in this country and so the find claimed the greatest interest."

In the summer of 1919 he spent about a month exploring and collecting in the various lignite pits near Moselund, Fasterholt, Silkeborg, and other places in the general area. He prepared a manuscript on the stratigraphy and flora of the lignite deposits that was awarded a prize by the Royal Danish Academy in 1923. Mathiesen's chief interest was clearly tending toward the study of fossil plants but he was unable to obtain a position that would enable him to pursue it professionally. Instead of obtaining the Chair of Botany at the Pharmacological High School which he had aspired to he received, in 1927, the Chair of Pharmacognosy, a position that he held for nearly thirty years. He devoted himself with great energy to this post, and although his interest in fossil plants never waned he was able to devote but little time to research in that direction. He did spend some time in the summers of 1926 and 1927 at the Stockholm Museum, where he came in contact with Florin and Halle.

Then after a full career in which his time had of necessity been devoted to pharmacology, Mathiesen retired in 1956 at the age of seventy and, when most others would have turned to a less arduous life, took up once again his study of fossil plants, a pursuit that he maintained until he died at the age of ninety in July of 1976.

Mathiesen's botanical knowledge, especially in morphology and anatomy, is quite evident in a comprehensive article he wrote in 1921 on the Arctic Scrophulariaceae, which is nicely illustrated with drawings of floral morphology as well as stem and leaf anatomy. Among the Arctic plants illustrated in a 1961 study of some especially well preserved fossil coniferous and dicot woods from the Nûgssuaq Peninsula of western Greenland is a specimen of the wood and bark of a plant related to *Aphloia* (Flacourtiaceae). This is one of the most detailed and careful descriptions of a fossil wood that I know of and should stand as a model for others working in this difficult area of paleobotany.

His greatest work, the "Palaeobotanical Investigations into Some Cormophytic Macrofossils from the Neogene Tertiary Lignites of Central Jutland," appeared in three parts from 1965 to 1975. Part 1 contains a general introduction to the area with a rather detailed historical account of mining operations beginning in about 1917. It deals specifically with the pteridophytes and describes several ferns, including species of *Osmunda* (cinnamon fern), *Lygodium* (climbing fern), several members of the Polypodiaceae, and a species of the 'water fern,' *Salvinia bjerringii*. Part 2 is devoted to gymnosperm remains and includes descriptions of the wood, foliage, and cones of trees closely related to the modern *Taxus* (yew), *Sequoia*, and *Glyptostrobus*. Part 3 describes leaves and seeds of numerous angiosperms that are predominantly temperate-climate plants, such as *Salix* (willow), *Fagus* (beech), *Alnus* (alder), *Acer* (maple), and *Liquidambar* (sweet gum), as well as some warm-climate plants including *Ficus* (fig) leaves and fragmentary fossils that are probably referable to the palm *Nypa*. The descriptions are accompanied by Mathiesen's excellent illustrations showing anatomical details of the leaves and seeds.

In recent years the paleobotanical group at Aarhus University initiated a new series of studies of the fossil plants from the lignite deposits in central Jutland. The lignite had been mined during World War II as an emergency source of fuel. I had an opportunity to see this work in progress during my visit at Aarhus during the fall semester of 1976, and it is a very promising development in Tertiary paleobotany. The fossils come chiefly from two localities a few kilometers south of the town of Herning and are referred to as the Søby and Fasterholt floras.

The first contributions dealing with one of the plants from the Fasterholt flora, written by Eske Koch and Walter Friedrich, describe the exceptionally well-preserved fruits of *Spirematospermum*, a plant that compares closely with the living *Cenolophon oxymitrum* (Zingiberaceae) of Thailand. And based on an examination of some ten thousand specimens of the smaller plant remains (seeds, fruits, and other parts under two millimeters long), Marie Friis had identified about 125 species by 1975. The Fasterholt flora consists predominantly of wetland and aquatic plants of warm temperate affinities and may be compared with some of the Atlantic Coastal Plain communities of North America. Other macrofossils, including the cones and fruits of *Taxodium*, *Sequoia*, *Liquidambar*, *Alnus*, *Nyssa*, *Myrica*, *Vitis*, and *Potamogeton*, were described by Koch and others in 1973.

In two recent studies Erik Christensen described a number of well-preserved leaves from the nearby Søby flora (including *Smilax*, *Comptonia*, *Juglans*, *Alnus*, *Castanea*, and *Liquidambar*), as well as improved techniques developed to extract information on the epidermal structure.

This project impresses me as being an exemplary investigation of a Tertiary flora involving the interests and talents of several people and the most up-to-date techniques. Enough is already known to reveal the striking difference between the open, rolling farmlands of present-day Denmark and the varied, warm-temperate lowland forest of the Miocene.

Danish studies also continue on the fossil floras of Greenland. In 1963, Eske Koch brought out a comprehensive account of the predominantly angiospermous Lower Paleocene flora of the Nûgssuaq Peninsula of western Greenland. In 1964, in a very useful review, he examined the errors in previous studies of the Cretaceious and Tertiary floras of western Greenland, such as insufficiently recorded information on the geographical and stratigraphical origin of specimens, identifications based on poorly preserved specimens, and the need for correlating palynological studies. And finally, Raunsgaard Pedersen's nicely illustrated 1976 summary, *Fossil Floras of Greenland*, is a readable account of the paleobotany of the great island, and his palynological studies that are in progress will contribute much additional information.

I have chosen to conclude this chapter by going back in time to the beginnings of scientific geology and paleontology to recognize one of Denmark's greatest scholars and a most interesting personality, Nicolaus Steno (1638–1686). Steno's work provides a significant link with the early development of paleontology in Britain, and he is of particular interest to us here for the rational and sound ideas he experessed on fossils and related aspects of geology. He was a contemporary of Robert Hooke's; their concepts were similar, but were probably developed independently of each other. There is abundant biographical material available on Steno and I have drawn largely from the recent accounts of Scherz (1958), Winter (1968), Garboe (1958), and Eyles (1958).

Steno was born in Copenhagen on January 10, 1638; because of poor health in his early years and a precocious mind, as a child he much preferred the company of older people to what he called the "frivolous chatter of younger companions." But judging from some of his subsequent rigorous travels it seems evident that he was not a weakling in later life, and when he was a student at the University of Copenhagen (1656–1659) he helped in the defense of the city against the attacking forces of Carl Gustav of Sweden.

Through much of his life, and especially the earlier years, Steno was fortunate to have been associated with some of the finest scientists, craftsmen, and philosophers of the time. His stepfather was an accomplished goldsmith; he was influenced by two early teachers, Ole Borch and Jörgen Eilersen; in France he met Martin Lister and John Ray; he became acquainted with men such as Francesco Redi and Marcello Malpighi in Italy, and was befriended by the royalty of Tuscany and Denmark.

He probably left Copenhagen in late 1659. "At any rate," says a biographer, "on 7th April 1660 he was in Amsterdam, for on this day, about three weeks after his arrival in the Netherlands, he dissected a sheep's head in the house of Professor Gerard Blasius (Blaes), his host, and discovered the parotid salivary duct, which is still known as the Ductus Stenonianus" (Scherz, 1958, p. 15).

He returned to Copenhagen in 1664, partly on account of the death of his stepfather and apparently because he expected to obtain an academic post by virtue of his achievements in anatomy. His hopes were not fulfilled, and in the same year he departed for Paris. In late 1664 or 1665 Steno visited Montpellier, in the south of France, a center of intellectual life which included a medical school. John Ray was there at the time, and "Lister, too, met Steno in Montpellier, for he has recorded that he had 'ye honour to assist att an Anatomie Lecteur on some particular dissections and demonstrations made by Mr. Steno ye Dane himself in my Lord of Ailesburys cabinet'" (Eyles, 1958, p. 168).

Steno went from France to Italy and met Francesco Redi in 1666 at Pisa, the winter residence of the Grand Duke of Tuscany, Ferdinand II. The latter had a strong interest in science, especially the earth sciences—he operated several mines and quarries—and he recognized the Dane as an exceptional scholar.

Steno's interest in geology seems to stem from about this time. He had already established a reputation for his discoveries in anatomy, and an event took place which most of his biographers have placed much weight on. In October of 1666 an enormous shark had been landed near Livorno, and the head was taken to Florence and turned over to Steno for dissection. This gave him an opportunity to compare the teeth of a modern shark with the glossopetrae or "tongue stones" (fossil shark's teeth) found in the Cretacous cliffs of Malta and the origin of which had long been a source of speculation by naturalists and collectors of curiosities.

In 1667 he produced an account stemming from the shark study and his geological observations, *Canis Carchariae dissectum caput* (The Dissected Head of the Dog Shark). Scherz says that, in addition to dealing with the question of glossopetrae, Steno gives an explanation of the layers of the earth and the fossils to be found in these, and then in six conjectures logically presents the conclusions of these, the last of which ends: "Nothing seems to contradict the theory that the bodies excavated from the earth and which resemble the parts of animals, must also be regarded as parts of animals." On account of this treatise, Steno is regarded as the founder of scientific palaeontology, especially perhaps because of the careful, modest and hesitating manner in which he presents the results he has reached. [1958, p. 28]

During 1667 Steno traveled extensively through Italy. and especially Tuscany. As a direct result there appeared in 1669 the curiously titled treatise for which he is best known in historical geology: *De solido intra solidum naturaliter contento dissertationis prodromus*. As the last word indicates, it was intended as a forerunner to a more comprehensive treatise on (translated freely) "a solid body enclosed by process of nature within a solid"—which refers to a body such as a shell enclosed in a sedimentary rock.

In the *De solido* Steno discusses in considerable detail the formation of sedimentary rocks and the way in which animals and plants were deposited and transformed into what we now call fossils. It hardly does justice to such a treatise to extract only a few lines, but the following comments on the plants of the past are of particular interest here.

> If in a certain stratum we find a great abundance of rush, grass, pine cones, trunks and branches of trees, and similar objects, we rightly surmise that this matter was swept thither by the flooding of a river, or the inflowing of a torrent. . . .
>
> What has been said regarding animals and their parts holds equally true of plants and the parts of plants, whether they are dug from earthy strata or lie hidden within rocky substances; for they either completely resemble actual plants and parts of plants (this kind is found rather rarely), or they differ from actual plants only in color and in weight (this kind occurs more frequently, sometimes burnt in charcoal, sometimes impregnated in a petrifying fluid), or they correspond to actual plants in form only; of this kind there is a great abundance in various places. [Winter, 1968, pp. 228–29, 260–61]

Steno's conclusions concerning the true nature of animal fossils undoubtedly stemmed in considerable part from his vast knowledge of anatomy, which, combined with his high level of intelligence and geological field observations, rendered it impossible for him to interpret fossils other than as the remains of formerly living animals and plants.

He discusses several classes of plant fossils, partly on the basis of their mode of preservation, some of which are more readily identified as having been formerly living plants than others. I think that for the most part he recognized certain pseudofossils, such as dendritic formations (mineral deposits superficially resembling plants such as mosses), as being inorganic, but in part he was caught in the dilemma that Martin Rudwick three hundred years later would characterize as a distinction between "easy" and "difficult" fossils.

The full account of *De solido* was never completed. In February of 1670, King Frederick III of Denmark requested that Steno return to Copenhagen; having been converted to Catholicism he was reluctant to go back to Lutheran Denmark, but he answered the call of his monarch. Religious controversies developed, as he had expected, and he departed again, in 1674 returning to Florence; in 1675 he took Holy Orders, was later appointed Bishop of Titopolis, and from that time on was fully occupied with a religious life which I will not go into here. He died on November 26, 1686.

It is difficult to avoid drawing a comparison between Nicolaus Steno and Robert Hooke. From such literature as I have consulted I conclude that the two men had very similar notions of the importance of the geological phenomena they had observed as having created the world around them, and both arrived at the same conclusions concerning the nature of fossils; both were extraordinarily accurate and far ahead of most of their contemporaries. I suspect that Steno had engaged in more extensive field observations than had Hooke. Steno's two principal works were issued in 1667 and 1669; some of Hooke's observations appeared in his *Micrographia* in 1665, but the bulk of them were published posthumously in 1705, although he had been presenting them in lectures to the members of the Royal Society as early as 1663. Lister and Ray, had been very favorably impressed with Steno's knowledge when they met him and probably had communicated accordingly with their London friends. Eyles (1958) states that a copy of the *De solido* must have reached England shortly after it was published, and an English translation was made in 1671. Lister must have been one of the first to see this, for on August 25 of that year he wrote to the Royal Society criticizing one of Steno's arguments.

In summary it seems clear to me that Hooke and Steno were two of the greatest intellects as well as two of the most honorable men that the seventeenth century produced and that there is no reason to doubt that their observations and conclusions were made quite independent of each other.

13

Paleobotany in Central Europe: From Goeppert to Kräusel

A long and important series of productive paleobotanists began to appear on the scene in about 1830 in what I think is best referred to geographically as "central Europe." The work of some of them has been reviewed in Gothan's (1950) article "Die Paleobotanik in Deutschland in der letzten 100 Jahren" (Paleobotany in Germany in the Last 100 years), which covers approximately the period 1848–1948. I have drawn some information from his account but this chapter is more inclusive as to both time and place.

The principal central European paleobotanist of the mid-nineteenth century, Heinrich Robert Goeppert (1800–1884), was enormously productive and must be ranked with such outstanding figures as Brongniart and Seward. Goeppert was born at Sprottau in Lower Silesia and obtained his early education at the Mathias Gymnasium in Breslau from 1813 to 1816, where he was introduced to plant lore by the vicar, Dr. Kaluza. He then worked for a year in his father's apothecary shop and for another apothecary at Neisse for a year. We next find him studying medicine at the University of Breslau, where he developed a lasting friendship with Christian L. Treviranus, Professor of Botany and Director of the Botanical Garden. Goeppert's thesis on plant nutrition was presented in 1825 and he was awarded the doctorate in medicine. He set up a medical practice the following year in Breslau as physician, surgeon, and eye-doctor that he maintained until he was seventy. In 1827 he became Treviranus's assistant at the Botanical Garden and eventually succeeded to the directorship in 1852; he also served for many years as Professor of General Pathology, Therapy, and Pharmacology. It is difficult to understand today how some of the great figures in biology in the early and mid-nineteenth century accomplished so much in both the breadth of their interests and the depth to which some of them were pursued. There were, of course, many differences attending their working conditions but, nevertheless, some of them were extraordinary people. One of Geoppert's biographers (Conwentz, 1885) lists over a hundred articles and books that he produced on various aspects of botanical science other than paleobotany, including works on pathology, pharmacology, plant anatomy, and forestry.

Heinrich R. Goeppert. From *Naturforschenden Gesellshaft zu Danzig*, 1885.

Goeppert had an especially strong interest in the historical aspect of botany, and, indeed, his first publication on fossil plants, which came out in 1835, is a historical sketch of the development of paleobotany in his native Silesia. Its breadth indicates that Goeppert had devoted considerable time to the literature before he started his own researches. He notes especially the work of G. A. Volkmann and Scheuchzer, to whom he gives credit for stimulating many of his contemporaries to paleontological inquires.

In the following year Goeppert's first great palcobotanical book appeared, his *Systema Filicum Fossilium*. It includes forty-four plates, chiefly of Silesian fossils, and insofar as possible draws comparisons between the fossil and living ferns. This work is prefaced by a seventy-six-page historical summary of fossil-plant studies, which remained the most detailed account available until Lester Ward's "Sketch of Paleobotany" was published, in 1885. Gothan says of the book: "Despite the generally erroneous assumptions concerning the systematic arrangement of the Carboniferous ferns, which was expressed in generic names like *Cheilanthites*, *Cyatheites*, *Aspidites*, *Adiantites*, etc., he correctly recognized, for example, the genus *Oligocarpia* on the basis of the sporangia, which he had correctly observed" (1950, p. 94; translation). It is evident that some years of study must have gone into the production of this volume.

Perhaps the depth and thoroughness of Goeppert's interests and observations are best revealed in an interesting account published in 1837 entitled *On the Condition of Fossil Plants, and on the Process of Petrifaction*. He notes the loose way in which the term "petrifaction" had been used by previous authors and distinguishes between impressions, compressions, and petrifactions. He describes experiments he conducted with fern and dicot leaves, placing them between plates of clay, and impregnating plant materials with a wide range of chemicals in order to produce petrifactions of a sort. And he notes that some plant materials may be preserved without being petrified: "The vegetation preserved in brown coal often hardly deserves any other term than that of dried vegetable matter; and in fact fossil wood often differs but little externally from wood which has lain for a long period in water" (p. 76). His studies concerning the chemistry of living plants relative to fossilization processes led him to far-reaching suggestions: "If we proceed in this way we shall in future possess in chemistry an important and serviceable assistance for the determination of fossil plants" (p. 81).

Somewhat along the same lines, using a technique that was far ahead of his time, he described, in his *Les Genres des plantes fossiles* (1841–1846), a fossil alder (*Alnites kefersteinii* Goeppert) with well-preserved catkins from which he extracted and illustrated the pollen; the same plate illustrations were used earlier in a thesis he presented in 1838. This is, to the best of my knowledge, the earliest instance in which spores were obtained from a fossil-plant compression and used as an aid in identification.

In 1849, Goeppert published, with Beinert, a prize essay entitled *Ueber die Beschaffenheit und Verhältnisse der fossilen Flora in den verschiedenen Steinkohlen-Abagerungen eines und desselben Reviers*, and Gothan cites this as the first instance in which Carboniferous plants were used effectively for stratigraphic purposes.

His extensive knowledge of wood anatomy is demonstrated in his *Mono-*

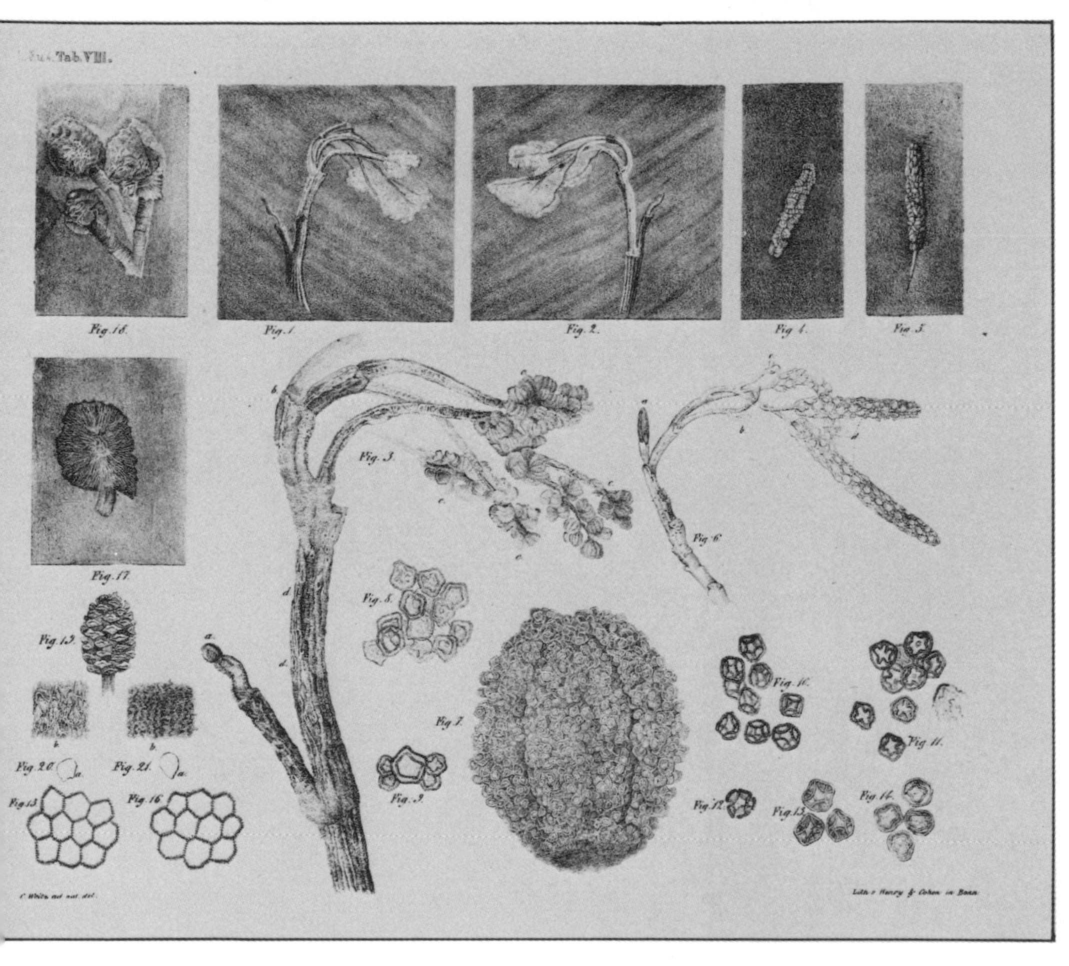

A plate from Goeppert's *Les Genres des plantes fossiles*, 1841. Probably the first time that pollen, extracted from a macrofossil, was illustrated.

graphie der fossilen Coniferen, which came out in 1850. This is a difficult phase of paleobotany, as I well know from my own initial efforts. His historical bent is again shown here as this great work includes a lengthy introduction describing the study of fossil woods from very early times.

In 1845, in collaboration with Georg Carl Berendt, Goeppert published a widely read monograph on the Prussian ambers, expressing the opinion that they were derived from coniferous trees, and in a paper read before the Silesian Society in 1853 he inferred that the ambers were of Tertiary age. He seems to have been concerned again with amber studies in his later years, but researches in this area were left for his biographer and student, Hugo Conwentz (1855–1922), to carry out.

For centuries amber has had a distinctive fascination for many people as an item of semiprecious jewelry, and judging from its abundance and rising price in shops today, its charm continues to grow. References to it appear frequently in the literature, but it seems to me that the name of Conwentz is almost synonymous with the botanical study of amber, at least in a historical sense. Hugo Conwentz was born in Danzig; his education at Breslau and Göttingen included considerable botany, and in Breslau he served as Goeppert's assistant at the Botanical Garden. Shortly before he was twenty-five he returned to Danzig to found and develop a natural-history museum where he served as Director for thirty years. Conwentz's paleobotanical works were few in number but I think are best described as magnificent. His book *Die Angiospermen des Bernsteins* (The Flowering Plants of Amber), which appeared in 1886, has long impressed me as one of the great achievements in paleobotany. Its thirteen beautiful plates illustrate pieces of amber (shown in yellow), apparently all natural size and accompanied by excellent drawings of the flowers, leaves, and other plant fragments. There are, for example, flowers of *Smilax baltica* Conwentz, flowers and leaves of several species of *Quercus* (oak) and a staminate inflorescence, a *Cinnamomum* flower, a rosaceous flower with the petals intact, and several branch fragments referred to the Loranthaceae.

Conwentz is probably more important as a pioneering conservationist than as a paleobotanist, and in view of present trends a few words in this context seem relevant. He traveled widely through Germany in connection with his museum position and became especially concerned with problems that are very much with us now, such as the destruction of wilderness areas and the protection of endangered species of plants and animals. He was, after much effort, successful in bringing about the establishment of a conservation department in the Prussian government, first in Danzig and then in Berlin, and he served as its director for twelve years. His interests were not confined to his own country, for he wrote to Lester Ward in 1899 asking him for information on conservation activities in the United States, mentioning specifically the Arizona petrified forest and the giant Sequoias.

Another Prussian with an interest in amber is Robert Caspary (1818–1887), who contributed a few short articles on amber plants and fossil wood during the period 1881–1887. I include him partly because he reminds me a little of John Ray. Caspary was born in Königsberg and attended the university there as a theology student. His interest in the subject seems to have been of an intellectual nature and he aimed at teaching rather than preaching. He spent his free time collecting insects, and since he was unable to obtain the kind of teaching position in theology that he had hoped for, he abandoned it at the age of twenty-five in favor of the natural sciences and returned to the university to start his education anew.

His earlier interest in zoology shifted to plants, and after several years of serving as a private tutor he obtained a university position in Berlin and became associated with Alexander Braun, who had made a study of the Tertiary plants of Oeningen in Switzerland (and he married Braun's oldest daughter, Marie). Caspary returned to Königsberg in 1859, where he served as Professor of Botany and Director of the Botanic Garden for twenty years. His botanical interests were varied, the most important ones relating to studies of the living members of the Nymphaeaceae (water lilies).

In 1864 Goeppert's *Die Fossile Flora der Permischen Formation* appeared, written in collaboration with Karl Gustav Stenzel (1826–1905). Stenzel was born in Breslau and spent much of his life in that area. He attended the university there, obtaining his botanical training from Nees von Esenbeck and his paleontology from Goeppert. Stenzel seems to have had broad academic interests in literature and history which he acquired from his father, and he earned his living for many years as a chemistry teacher, indulging in his botanical interests, especially of a morphological nature, in such spare time as he could find. His doctoral dissertation was a study of the trunks of palms and his more important contributions in paleobotany were studies of fossil palms and the medullosan seed ferns.

Two of Goeppert's other contemporaries, Gutbier and Geinitz, should be mentioned briefly here. Christian August von Gutbier (1798–1866) was born in Rosswein, Saxony, and entered military service at the age of eighteen. He apparently served as a full-time army officer most of his life, twenty-six years of which were with a regiment at Zwickau. He found time for other pursuits, including a study of the Westphalian coal seams in that region. He is noted for his works on Carboniferous stratigraphy, especially his 1835 *Abdrucke und Versteinerungen des Zwickauer Steinkohlengebirges und seiner Umgebungen*, a work accompanied by eleven plates of somewhat mixed quality that were prepared by Gutbier himself. In 1848–1849 he brought out his work on the petrifactions of the Zechstein Range and the Lower Red Sandstone in Saxony with H. B. Geinitz. Gutbier seems to have been something of a general naturalist with an interest in ferns, Pleistocene vertebrates, and geomorphological problems, with fossil plants as a side issue.

Hans Bruno Geinitz (1814–1900), an important geologist of the time, served as Professor of Mineralogy and Geology at Dresden. His contributions to fossil botany include studies of the Quadersandstein, as well as Permian and Carboniferous floras. He investigated some fossil plants collected by

Charles Darwin in Argentina and he produced a paper on presumed Permian plants from Nebraska.

In his *History of Geology and Paleontology* (1901) Zittel gives Geinitz and Gutbier partial credit for the establishment of the name Permian. The term had been proposed by Murchison for a series of rocks in the province of Perm in Russia, and Geinitz and Gutbier furthered its recognition in their studies of the German Zechstein.

Paleobotany has lost through death at an early age a number of highly promising people, and I believe one of the most important is another contemporary of Goeppert's, August Joseph Corda (1809–1849). I have mentioned previously his contribution to the last part of Sternberg's great work, and this led to Corda's own *Flora Protogaea* of 1845, one of the most important studies of petrified plants of the first half of the century and apparently well enough regarded at the time to justify a second printing in 1867. It is a folio volume of 128 pages and sixty plates, most of the stone engravings having been executed by Corda himself. Although from diverse geologic horizons the plants are predominantly Carboniferous. The work is a starting point for several important genera, including *Heterangium*, *Anachoropteris*, *Chorionopteris*, *Senftenbergia*, and *Tempskya*. It seems to me the definitive work of the time for the *Psaronius* tree ferns, there being numerous plates illustrating in detail the distinctive aspects of the stem anatomy. There are also several plates showing palm woods.

Senftenbergia has had a checkered career and is to some degree representative of the way our knowledge of plants of the past evolves—sometimes along a tortuous route. In Volume 2 of his *Fossil Plants*, Seward says:

> An examination of the suspiciously diagrammatic drawings published by Corda of the small fertile pinnules of a Carboniferous fern from Bohemia, which he named *Senftenbergia elegans*, leads us to conclude that the sporangia are almost certainly those of a Schizaeaceous species. . . . It has already been pointed out that the apical annulus of recent Schizaeaceae, though normally one row deep, may consist in part at least of two rows. Zeiller examined specimens of Corda's species and decided in favour of a Schizaeaceous affinity. . . . Zeiller's figures confirm the impression that Corda's drawings are more beautiful than accurate. [Pp. 346–47]

I have a personal interest in *Senftenbergia*, having described some quite well-preserved fertile fronds from the Upper Carboniferous of Illinois; I became acquainted with the genus as a graduate student when I visited Glasgow, where N. W. Radforth was studying some nicely preserved material in Walton's laboratory for his Ph.D. thesis. Radforth's study appeared in two papers during the late 1930's and provides significant information on the early evolution of the Schizaeaceae. There is, however, some striking divergence in the leaf morphology between the Carboniferous members of the family and the living genera such as *Anemia*, *Schizaea* ("curly grass"), and *Lygodium* (climbing fern). These three are, themselves, something of a "mixed bag," and in his morphology textbook David Bierhorst gives them individual family status.

Very little has been known about the anatomy of the presumed fossil members of the Schizaeaceae; a very recent contribution in this direction is thus significant: in 1977 James Jennings and Donald Eggert described petrified fronds bearing the typical *Senftenbergia* sporangia; they indicated that there is a structural similarity with the coenopterid ferns *Ankyropteris* and *Clepsydropsis*, and also noted that the apical annulus, presumably typical of the Schizaeaceae, is found in other fern families. There is clearly much to be learned about these plants and their proper relationships.

Corda was born on October 22, 1809, in Reinchenberg, Bohemia. He lost both parents at an early age and drifted along in his younger years with little guidance in his education. In 1829 he enrolled in the University of Prague in medical studies and obtained a diploma. He apparently devoted little time to practicing medicine, but in 1832 he helped for some months to suppress a cholera epidemic which had penetrated into the capital of Bohemia. He also studied at Dresden, Breslau, and Berlin, where he became acquainted with Alexander von Humboldt, who in turn introduced him to Kaspar von Sternberg. He engaged in studies of insect anatomy, liverworts, and the structure of palms and cycads. In 1837–1842 he brought out a five-volume work on the fungi, *Icones fungorum hucusque cognitorum*.

Much of Corda's life seems to have been plagued with frustrations—the loss of his parents, little guidance in his early years, difficulty in getting many of his works published, failure to obtain positions that he applied for. He did eventually obtain a curatorial post in the Bohemian Museum in Prague, the exact nature of which I have not ascertained. In 1848 he was granted a year's leave of absence from his museum duties to go on a collecting trip to Texas, under the aegis of Prince Colloredo. Corda's biographer, Weitenweber, does not elaborate on his experiences there. It is known that he corresponded with several people in Prague during this year, and if such letters are still extant they might offer an interesting and significant record. He made large natural-history collections in Texas, some of which he sent back separately, but the bulk of them traveled with him. He started his return journey in September, 1849, from New Orleans in the steamer *Victoria*, which encountered a storm, probably an early-autumn hurricane, and Corda and all of his collections were lost with the ship.

In 1832, the year the first volume of Brongniart's *Histoire* was completed, Bernard Cotta brought out his book *Die Dendrolithen, in Bieziehung auf ihren inneren Bau* (Petrified Trees with Reference to Their Inner Structure), and a second edition appeared in 1850. Cotta (1808–1879) was born in Klein-Zillbach near Meiningen, Germany, and studied mining and natural sciences at Freiberg and Heidelberg. He would be classified today, I believe, as an economic geologist and is best known for his 1839 text, an introduction to the study of geognosy and geology; he also did some important work in preparing geologic maps of Saxony.

Die Dendrolithen is based on a study of some five hundred ground sections from his father's collections and was done as a "spare-time" project. His classification is admittedly rudimentary, using three divisions: Rhizomata, Stipites, and Radiati. The Rhizomata include *Tubicaulis* (Cotta), *Psaronius* (Cotta), and *Tempskya*, although the latter was described under the name *Porosus*.

The name "Starling stones" is frequently encountered in this early literature, and the following from Scott is explanatory: "The name 'Starling-stones' for these ornamental fossils [*Psaronius*] is familiar; it may not be so generally known that this name properly applies only to the specimens showing the *roots*; those in which the long, curved sections of the vascular bundles of the stem are visible used to be called 'Maggot-stones,' 'Madensteine,' '*Psaronius helmintholithus*.' In earlier days these fossils had been regarded as Corals or Encrinites" (Scott, 1911, p. 24).

Cotta's "Stipites" include fossil palms, and the "Radiati" include *Medullosa* and *Calamitea*. The figures illustrating three species of *Medullosa* are quite good, and the *Calamitea* include fossils that were later placed in *Calamodendron* and *Arthropitys*. In his "Supplementary Remarks" Cotta considers correlations between petrifactions and compressions, tending to identify *Tubicaulis* with *Lepidodendron* and the ribbed sigillarias with cacti.

Scott says: "On the whole Cotta's book is not to be taken too seriously from a scientific point of view. He was only a beginner at the time, and evidently no great botanist. His observations, however, were good, and sometimes his natural instinct led him right when more learned authorities went wrong" (1911, p. 26). I am inclined to be a little more generous in appraising Cotta's book. It is one of the pioneering efforts in describing the internal structure of plants based on ground thin sections. The illustrations do not compare favorably with those of Witham, but he placed on record for the first time several very important genera.

By the early 1840's the number of fossil plants that had been described had become so great that paleobotanists began to see the need for a central source of information. I believe the first one to deal effectively with the problem was Franz Unger (1800–1870), who brought out his *Synopsis Plantarum Fossilium* in 1845. This was a small volume of 328 pages, prefaced by a good bibliography, that listed all of the fossil plants known to Unger, and it was dedicated to Brongniart, Goeppert, and Lindley. References are given to the place of publication of every work describing a new species, and the locality from which each species originated. In 1850 Unger prepared an expanded version under the title *Genera et Species Plantarum Fossilium*, this time dedicated to G. P. Schimper, Alexander Braun, and Heer. These two compilations list 1,648 and 2,421 species, respectively, according to Ward; I should add that I have not taken the time myself to check these figures.

From 1841 to 1847 Unger issued the ten parts of his *Chloris Protogaea*, of which Ward says:

> The body of the work is strictly descriptive, and here we find 120 species characterized, all new to science or consisting of corrected determinations of other authors. What specially distinguishes this work, however, . . . is the very large percentage of dicotyledonous species, mostly from Parschlug, embraced in these descriptions. Considerably over one half of the number belong to this subclass and to such genera as Ulmus, Alnus, Betula, Quercus, Acer, Rhus, Platanus, Ceanothus, Rhamnus, etc. He seems to have reached his determinations of these genera by an intuitive perception of the general and special resemblance of the fossil to the living leaves, with which, as a botanist, he was perfectly familiar. [1885, pp. 419–20]

It may be added here that in 1833 Jonathan Carl Zenker published his *Beiträge zur Naturgeschichte der Urwelt* (Contributions to the Natural History of the Former World), of which half of the sixty-seven pages and six plates are devoted to Cretaceous dicotyledons of Blankenburg in the Harz district. I believe this is the first attempt to treat dicot fossils systematically. Thus these two works of Unger and Zenker usher in a period in which numerous studies of dicot leaves form a conspicuous element in paleotobtanical literature.

Franz Unger was born in southern Steiermark; his father held a position with the tax commission, having found it necessary to depart from a religious life in order to support his large family. He apparently urged his sons to follow the kind of life that he had had to give up, with the result that Franz was torn between a sense of obligation to his father and a religious career on one hand and a strong interest in natural history on the other, with the latter more or less predominating. He decided in favor of medicine in 1821, and his studies for several years were a confused mixture of philosophy, religion, botany, entomology and medicine. He traveled extensively, apparently paying little attention to his personal appearance; he was locked up by the police for seven months, being freed in July, 1824, since no reason could be found for detaining him other than his suspicious appearance.

His botanical interests continued to grow; he made significant studies of the alga *Vaucheria*, leaf mildews, and diseases of trees; in 1835 he was appointed Professor of Botany and Zoology and Director of the Botanical Garden at Graz, and later Professor of Botany in Vienna.

Like many before and after him, Unger was attracted by the Atlantis legend, and in 1860 he delivered a lecture at the University in Vienna entitled "The Sunken Island of Atlantis" (1865). The presence of presumed North American species of plants in the German Tertiary deposits led him

> to the conclusion that there must have been a continental connection. In the Tertiary period, or at the time when the lignite was formed, Europe must have been connected with North America, and the Atlantic Ocean must have been divided at one place or another by a continent. . . .
> At present we must be content to know that during the Tertiary period an intermediate continent, which we shall call Atlantis, really existed, and that it extended northwards as far as Iceland, and southwards beyond the present Atlantic Islands. But any attempt at tracing its exact configuration must be regarded as ideal. [1865, pp. 18, 22]

Among his other travels, in 1852 Unger made a journey through Denmark, Norway, and Sweden, and he was shown some of the Icelandic fossils, which must have had some influence on his Atlantis ideas.

Perhaps inspired by the works of such men as Unger and Zenker, researchers in the following decades brought forth a considerable volume of studies on angiosperm floras. Among the more notable contributors was Constantin von Ettingshausen (1826–1897), whose works appeared in the period of about 1850 to 1870. His name is linked with two distinct facets of botany and paleobotany: ideas concerning the Australian character of certain European Tertiary floras, and his application of a leaf "nature-printing" technique. In about 1850, a process for showing the outline and venation of leaves had been invented in the Austrian state printing office; Ettingshausen seized upon it with great enthusiasm and made use of it in his numerous botanical and paleobotanical publications. In 1854 he produced one on the Euphorbiaceae and another on the Papilionaceae, and in 1855 he and A. Pokorny were asked to prepare a comprehensive work for the Paris Exposition of 1867 showing the botanical use of the method. The result was, in Ward's words, "that immense and astonishing production entitled 'Physiotypia plantarum Austriacarum,' with its six enormous volumes of most exquisite plates, not only illustrating the leaves of the trees and shrubs, the flowers with their petals, sepals, stamens, and pistils, but the entire plants wherever within the ample limits of size, and these stand forth from the plates in actual relief like a vertiable *hortus siccus*" (1885, p. 381).

I have thumbed through these large volumes on several occasions and concluded, I hope not unkindly, that their rightful place is in an art library rather than a botanical one.

Ettingshausen also produced a number of memoirs on European Tertiary and Cretaceous floras in which he interpreted a strong Australian relationship. One may question many of his identifications; Nathorst characterized most of them as "completely worthless" (see Halle, 1921), and he was not the only one to find fault. Ettingshausen spent four months in London in the winter of 1878–79 examining the Bowerbank collection of seeds and fruits from Sheppey, and his list of species was published by the Royal Society of London in the latter year. In their account, based on many years of study, Reid and Chandler (1933) point out that most of Ettingshausen's identifications, unaccompanied by descriptions or illustrations, are in effect *nomina nuda* (that is, names without any accompanying description).

I have discussed in other chapters the great difficulty in identifying Cretaceous leaves and wood of flowering plants, and this is also true with respect to some of the early Tertiary fossils. But in decades past paleobotanists simply have not been able to resist giving this a try. It is perhaps to their credit that they had the courage to do so even though the results are of questionable value. There is an understandable urge, and indeed a necessity, to attach a name to a fossil—to give it some kind of identification. I was greatly disturbed in my earlier days, when I did some work with Cretaceous plants in the Rocky Mountain area, to observe the casual way in which some amateur but well-meaning collectors compared petrified woods with present-day trees on such superficial characteristics as the color of the petrifaction.

It was an opportune time for a man who was both competent and outspoken to appear on the scene, and August von Schenk (1815–1891) filled the need. Schenk was born in Salzburg, Austria, and began his botanical studies with Martius at Munich. His university education seems to have been primarily in medicine, but I have not encountered evidence that he became a practicing physician.

When he was thirty he was appointed to a professorship at the University of Würzburg, and a few years later he also became Director of the Botanical

Garden; he remained there for twenty years, later taking a professorship at Leipzig. Schenk is known for his study of the paleobotanical collections made by Baron von Richthofen, which was published in the fourth volume of the latter's work on China in 1883. Schenk also contributed several papers on European Triassic floras and on petrified woods. His broad knowledge of fossil botany is indicated by his preparation of the greater part of the fossil-plant section of Karl A. von Zittel's handbook of paleontology; this section was started by Schimper, who died after completing only the cryptogams and cycads.

Schenk strongly criticized the work of several of his contemporaries, among them Unger, Heer, and Ettingshausen; he made it clear that he regarded many of their species and genera as worthless. His biographer, Drude, says that, although he was friendly and warm-hearted, he was known for his sarcasm. He apparently did not significantly inhibit the three noted above, but his words of caution may have had some effect on others coming along at the time.

I do not pretend to have any special competence myself in the matter of judging much of the Cretaceous and Tertiary floristic work that has been done since Schenk's time. Here as elsewhere in my story the sequence is far from a complete one—I bring in men whose work has attracted my attention and who seem important, interesting, and I hope representative of this aspect in the development of fossil botany.

Fridolin Krasser (1863–1922) seems to have been either a direct disciple of Ettingshausen's or was favorably impressed with his floristic monographs. Krasser produced several accounts on Cretaceous fossils during the late 1800's and the first decade of this century, one of them being a treatise on the Chalk Flora of Kunstat in Hahren. He also studied the interesting Lunz Triassic flora and described several Asiatic collections.

Krasser served for some years at the German Technological College in Prague and later in Vienna, where he felt much more at home. He was a qualified botanist, well known for his work with algae and lichens, and he had a strong interest in applied botany generally. Although he did considerable lecturing he seems to have bordered on being a recluse. His biographer says that "he avoided large societies whenever possible. If a situation arose where a meeting was unavoidable, he remained laconic, taciturn, alone with his thoughts. However, among his small circle of friends he enjoyed opening up in a charming, often whimsical way. His distinguished, sensitive nature allowed him to grant the pleasure of being on close terms with him to only a few people" (Greger, 1922, p. 114; translation). He had a keen love of music and was a grand-nephew of Franz Schubert.

We have had all kinds of people join the Fellowship of Fossil Botany.

On one of my visits to the Natural History Museum in Stockholm, after I had decided on this historical venture, I occasionally took some time from my other research to browse through the very pleasant and well-stocked paleobotanical library, and it was there that I first encountered the works of Josef Velenovský. Again, I do not pretend to judge his work critically, but he was a well-known figure in European botany at his time—and it was a rather long time. He brought out a series of large memoirs on Cretaceous floras in the 1880's, gave this up for a considerable period, and returned to it again in the 1920's when he collaborated in another series with Ladislav Viniklář.

There is a fine set of biographical sketches on Professor Velenovský (1858–1949) in the journal *Preslia* in 1958. They were written on various facets of his long career, the section on his paleobotanical studies having been prepared by František Němejc who says, in part:

> His work on the plant life of the Czech Cretaceous formation presents an exceptionally detailed picture of the overall composition of the Cretaceous flora in Central Europe and of the morphological nature of the different types, so that since his day it has been possible to compare our Cretaceous flora with those of various other areas. Velenovský also contributed not only to clarifying the local palaeontological conditions in our Czech Cretaceous, but also in inestimable measure to resolving the floristic and palaeogeographical questions relating to the middle phase of the Cretaceous on a world scale. [1958; translation]

Velenovský was born in Čekanice, near Blatná, where his father was a farmer; he entered the University of Prague in 1877 and became an assistant in paleobotany in the museum two years later. He succeeded Professor L. Celakovsky as Director of the Botanical Institute and Garden, where he remained until he retired. His many publications were based in part on fossil plants obtained by previous collectors and in part on much field work of his own. He carried on the fine Czech tradition in fossil botany that had been established by predecessors such as Sternberg, Stur, and Corda. A. N. Kryshtofovich, in his 1956 *History of Paleobotany in the U.S.S.R.*, comments: "Speaking about the large increase in our knowledge of the Cretaceous flora, it is noteworthy that its study in Western Europe in the twentieth century almost ceased. This is a serious impediment for generalized works because the old works are already very much obsolete. The only exceptions to this are the works of Velenovský and Viniklář on the Cretaceous flora of Czechoslovakia, published in 1926–1931, and of Teixeira on the flora of Portugal, published in 1950" (p. 72; translated).

It is fitting to continue here with a few words on František Němejc

(1901–1976), who received his botanical education in part from Josef Velenovský. Němejc was born near Ostrava, Czechoslovakia, a mining district, which probably had some influence on his scientific pursuits. He received his doctorate from Charles University in Prague and was then appointed to a position at the National Museum there. Němejc was one of a distinctive group including Paul Bertrand, Walther Gothan, W. J. Jongmans, and Richard Kräusel, all recently departed, who made many important contributions to morphological and stratigraphical paleobotany. Němejc's interests fell in both areas, and he was a great compiler of information as well.

He initially engaged in a series of studies of the Quaternary but soon shifted to the Carboniferous where most of his work was carried out. He brought out numerous small papers as well as several comprehensive monographs dealing with the stratigraphy of the Bohemian coal basins. Thumbing through some of his many short studies of a morphological nature reveals some very interesting items: an article on *Discinites* in 1937 in which he demonstrates the probable heterosporous nature of the cone whose megasporangium contains one large megaspore, apparently an advanced stage in heterospory; a 1931 paper in which he adds to our knowledge of the pteridosperms in identifying the seeds borne on the foliage of *Alethopteris rubescens*; and a 1950 paper in which he examines the calamitean cones of *Huttonia spicata*—partially preserved compressions that yield significant information through grinding and microexcavation.

The work of another Czech, Jiří Obrhel, is especially significant in view of the present general interest in early land vascular plants. In 1962 he described some very intriguing specimens of *Cooksonia* from the late Silurian of central Bohemia. To the best of my knowledge this is the smallest and simplest land vascular plant that is known with some degree of clarity, and Obrhel's Silurian record is the most ancient. A good review of the other (Lower Devonian) localities in which this genus occurs is given in Høeg's account on the psilophytes in Volume 2 of the *Traité de paléobotanique*. In 1961, Obrhel brought out a good paper on certain Middle Devonian plants from central Bohemia; this account includes descriptions of fertile specimens of *Psilophyton*, *Barrandeina*, and *Protopteridium*.

It has been my opinion that one of the very great achievements in the last quarter of the nineteenth century was Christian Ernst Weiss's two volumes (1876–1884) on some Coal-Age articulates—the arborescent horse-tail rushes. It has puzzled me why his illustrations have not been used more often in textbooks. They are abundant and beautifully executed; he established such genera as *Palaeostachya* and *Cingularia*, and his drawings of magnificent specimens of the infructescences of *Calamostachys* and *Macrostachya* are quite the best I have encountered. It has been my assumption, although I am not quite certain, that Hirmer's well-known restorations of calamite trees are based in considerable part on Weiss's work. However, the massive nature of the rhizomes of the calamite trees, as shown by Hirmer (1927), seems questionable. I think the more recent interpretation of Eggert on this point (1962) may be more exact. Seward did draw rather heavily from Weiss's studies of the articulates in the first volume of his *Fossil Plants*.

Weiss (1833–1890) came from Eilenburg, Germany, and studied at the University of Halle and Frederick Wilhelm University in Berlin. Sterzel (1892) says that he was especially noted for his great competence in mineralogy, having derived inspiration in this direction from his uncle, Samuel Weiss, a noted crystallographer. Christian Weiss was also responsible for numerous maps of the Saar, Thuringia, and Harz regions. Gothan (1950) does not regard Weiss as a great botanist; he says that Weiss was carried away in the matter of establishing new species but that his precise observations and meticulous drawings are of lasting value. I will leave him with my opinion that his memoir on the calamites will long remain a great landmark in paleobotanical literature.

The monumental *Culm Flora* (1875–1877) by Dionys Stur (1827–1893) appeared at almost the same time as Weiss's volumes. It presents much of botanical interest and is important in view of the scarcity of information on Lower Carboniferous plants at that time and the particular interest in that age now. Three of the plates in the 1875 volume illustrate fine series of *Rhodea* and *Rhacopteris* fronds, showing apparent stages in the evolution of the megaphyllous leaf. There are also several plates of *Calymmatotheca* that show the stems and parts of attached fronds. In this context Stur says: "If one takes but a quick glance at the ferns of the Culm period set forth here, and then looks at the ferns of the Carboniferous, the careful observer can scarcely overlook the fact that the ferns of the Culm period were in large part apparently differentiated into smaller divisions, whereas the Carboniferous ferns exhibited larger divisions" (foreword, vol. 1, p. xii; translation).

Stur was concerned with the continuity of several fern families from the Carboniferous to the present. He notes the relatively minor foliar changes, which he illustrates, in several *Calymmatotheca* species which were derived from several successive horizons. On the other hand, he discusses the vast changes that have taken place over a long period of time in the articulate and lycopod groups, noting that *Equisetum* is the sole survivor of the former, and that the diminutive *Isoetes* (quillworts) form a contrast to the great arborescent lycopods of the Carboniferous. Stur seems quite clearly to have been an evolutionist, but whether he subscribed to any particular philosophy I have not been able to determine.

The Culm flora takes its name from the city of Culm in the northwestern part of Bohemia, although most of the plants that Stur described came from Moravia and Silesia. A review of Stur's flora in the *Geological Magazine* for 1884 (probably written by Kidston) points out some of the errors in the work but concludes with the comment: "Although there are some points, in which we cannot agree with Dr. Stur, his communication deserves the careful study of all those interested in Fossil Botany, as it contains an amount of valuable information regarding the fructifications of Carboniferous Fossil Ferns now for the first time brought together in book form" (p. 332).

Stur also did some collecting in the Triassic and produced a comprehensive work on the Lunz flora. He was interested in Fontaine's monograph on the Older Mesozoic flora of Virginia when it appeared and noticed a resemblance between the Virginia plants and those of Lunz. His paleobotanical publications came somewhat late in his career, after he had made notable geological contributions. He was born at Beczko, northern Hungary, and studied at the Polytechnical Institute in Vienna, where his attention was turned especially toward mineralogy and geology. He later studied at the Mining Academy in Schemnitz and was one of the first to be employed in the Royal Austrian Geological Institute when it was established in 1849. He devoted many summers in his early years to mapping work in the alps; in 1865 he brought out the general geological map of the Principality of the Steiermark, and the corresponding text appeared in 1871. In the course of his mountain mapping he developed a strong interest in the alpine flora as well as fossil plants. His more serious study of Carboniferous plants began in the early 1870's and he devoted considerable time to them in the museums of Germany, Belgium, and France. Stur died on October 9, 1893.

A. C. Seward, as I have recounted previously, got his start in paleobotany with Williamson and he then spent a year on the continent, where he had occasion to meet many of the leading people of the time. Among them was Stur, who apparently took a liking to Seward. Stur could speak no English, and in 1888 he asked Seward to accompany him as interpreter to the International Geological Congress in England. In the course of this trip Seward and Stur visited Williamson in Manchester. Walton gives an amusing account of the meeting:

> "Neither of the old gentlemen could speak a word of the other's language" writes Seward, and he had a difficult time as interpreter because he dared not always translate some of Stur's rather uncomplimentary criticisms with complete accuracy for fear of upsetting Williamson too much. Williamson was, somewhat, intolerant of the views of others and Seward tells of an occasion when he was talking to Williamson about the pteridosperms and

> had ventured to say that Kidston's work on impressions did not support some view of Williamson's. Williamson instantly burst out with "Confound Kidston's impressions!" [Walton, 1965, p. 5]

One must be cautious and accurate, in a book of this sort, in using superlatives to present the various leaders and the studies for which they are renowned. Perhaps because there have been a good many keen minds involved in the study of fossil plants over the past three centuries my efforts to comply with this dictum may have fallen short. A particular case in point is the man I present next, Hermann, Count Solms-Laubach (1842–1915). As a purely botanically oriented paleobotanist he was certainly one of the most distinguished, productive, and influential figures that appeared on the scene up to his time. I think it is fair to refer to him as the continental counterpart of the British botanists D. H. Scott (1854–1934) and William Carruthers (1830–1922), whose life spans, it may be noted, bracket his. He brought in a new era in fossil-plant studies in Germany, studies that had a strong botanical-evolutionary emphasis.

Solms was born on December 23, 1842, at Laubach in Upper Hesse, the seat of the ancient family of Solms. His father was Count Otto of Solms-Laubach, and his mother was a princess of Wied; his family connections included Carmen Sylva, the poetess queen of Romania, and Prince Maximilian, known for his botanical explorations in Brazil.

Solms's father did not consider, initially, that Hermann had done well enough in school to proceed to the University, but he later relented. The son was probably influenced by an uncle and an older brother, both of whom were botanists, and at the University of Giessen he was attracted by Rudolf Leuckart's lectures; later, at Berlin, he attended those of Alexander Braun. He also spent a term at Freiburg with Heinrich De Bary and took his degree there in 1865. Following an excursion to southern Portugal, where he interested himself in the distribution of mosses, he returned to work with De Bary, who was then at Halle. He served in the war of 1870, apparently in a hospital unit. Following the war he was appointed to a professorship first at Strasbourg and later at Göttingen. In 1883 he visited Melchior Treub's laboratory in Buitenzorg, Java; this experience in the tropics greatly expanded his botanical knowledge, especially in plant anatomy, and he was able to gather together materials for his own future investigations and for his students'. Solms, like Brongniart but unlike many others before him, seems to have had the best of opportunities available at the time and he used them to the best advantage.

He was a good systematic botanist and contributed monographs for the comprehensive works of Adolf Engler and Augustin de Candolle on such families as the Rafflesiaceae, Hydnoraceae, and Lemnaceae; and he was probably the leading authority of his time on parasitic plants. He had a strong interest in the history of cultivated plants and investigated figs, wheat, tulips,

Hermann Graf zu Solms-Laubach. Courtesy of the Royal Society.

strawberries, and others. According to Jost (1915) he knew of Mendel's work sometime before its rediscovery but he apparently was not attracted to genetics. He carried out important investigations with the algae, working out the reproduction of *Batrachospermum*; he studied the corallines of the Gulf of Naples, and in 1895 the Linnean Society of London brought out his monograph on the Acetabularieae, an important work dealing with recent and fossil forms.

Solms-Laubach's scientific progress followed much the same pattern as Goeppert's, although the former was a botanist from the beginning. Scott had the following to say about Solms's textbook (his first great effort in paleobotany):

> The "Einleitung in die Paleophytologie," published in 1887, and translated into English four years later, under the title "Introduction to Fossil Botany," certainly made a great impression on botanists, and to many was the beginning of their interest in the fossil record. The present writer was among these, and well remembers how the reading of Solms' book, on its first publication, revealed to him, what he had never quite realised before, that fossil plants really matter to a botanist. [1918, p. xxii]

Although most of Solms's works prior to the appearance of this book had been on living plants, it is evident that he had gained a considerable mastery of fossil-plant literature. The introductory chapter deals with coals and peats, the second is a lengthy one on the thallophytes, and then he goes to the conifers and other gymnosperms, followed by the pteridophytes. The plants that are now included in the pteridosperms are discussed in several places; he gave considerable thought to the medullosas, *Lyginopteris*, and *Heterangium* as "intermediates" in their affinities; he clearly recognized that major groups of plants were present in the fossil record about which only fragmentary information was available at the time. The only drawback to the book is the very sparse use of illustrations. In their biographical sketches both Scott and Robinson comment on Solms's difficult manner of writing in contrast to his attractive spoken style. His textbook is not exactly fireside reading, but I have found it highly informative and not as difficult as I expected. His lecturing style was vigorous to the point of being eccentric. Scott says that he paced up and down like a caged lion, and Professor J. Lotsy offered the following: "I think I can state with justice that I have never had a better teacher than him, in the years I spent at Göttingen. Solms had very curious peculiarities in lecturing; it even happened that in his enthusiasm the lamps above his catheder [cathedra?] came down, but he knew how to rivet our attention to such a degree that such things were hardly noticed by us" (quoted in Scott, 1918, p. xxiv).

One of Solms's more important memoirs is his treatise, written with Capellini in 1892, on the Italian specimens of the Bennettiteae. This presents the first evidence for the bisexual character of the bennettitalean "flowers," anticipating the discoveries of Wieland about a decade later.

Also in 1892 he initiated a series of papers on the Lower Carboniferous plants of Falkenberg in Silesia. Of four parts that appeared, the first dealt with *Zygopteris* and *Lepidodendron*, the second with *Protopitys*, the third with *Archaeocalamites*, and the fourth, which came out in 1910, described some of the remarkable polystelic (and still not very well known) fossils including *Voelkelia*, *Cladoxylon*, and *Steloxylon*. His study of *Medullosa leuckarti* in 1897 was a significant addition to the knowledge of the genus, later to be recognized as one of the more important groups of seed ferns.

In 1895 he described some Middle Devonian plants from the Lower Rhine, and although this is not one of his most important studies, it is of interest to note that he recognized the significance of Dawson's *Psilophyton princeps* and was suspicious about the validity of Dawson's restoration. This is considered in more detail in Chapter 10.

A note from Robinson's biography gives one a feeling for Solms's personal charm; it suggests the kind of person that one would like to have known:

> Count Solms never married. His household was long directed by his sister and after her death by two nieces, the Countesses Sophie and Anna of Solms-Rödelheim. . . . At his home, which was one of simplicity and great charm, he and his nieces practised a quiet and delightful hospitality. A notable member of the household was a parrot, an accomplished and gentlemanly bird, that came to the table on the Count's forefinger and there, supplied with napkin and appropriate small dishes, did full justice to a *table d'hôte*, partaking of all courses from soup to desert, though showing particular enthusiasm for the salad. [1925, p. 656]

A near-contemporary of Solms-Laubach's, and, like him, a well-trained and highly competent botanist, was Henry Potonié (1857–1913). His paleobotanical contributions, which poured forth in great abundance between 1890 and 1913, were related largely to Carboniferous plants; he is also noted for his lively interest in philosophy—of a "pure" nature as well as that applied to botanical subjects.

Potonié was born in Berlin of a French family. When he was five he was sent to Paris, where his father had business connections, for his education, but he was there for only a few years, for in 1866, near the end of the German-Austrian war, his father moved the family permanently to Berlin. He anticipated the coming German-French war and regarded Berlin as the safest place to be; Henry became a citizen of Germany and remained there most of his life.

His earlier and basic studies were of a floristic and systematic nature, having been guided by such noted botanists as Eichler and Ascherson. In 1880 he was appointed to an assistantship with Eichler at the Royal Botanic Gardens. Apparently he was not progressing as rapidly as he had hoped, for in 1884 he joined the Geological Institute, serving for a short time as assistant to Christian Weiss. After Weiss's death in 1890 Potonié became Lecturer, and in 1900 received the title of Professor.

Potonié's early work was concerned with living plants, and in this connection he is probably best known for the major role he played in bringing out the *Illustrierten Flora von Nord- und Mitteldeutschland* (Illustrated Flora of North and Middle Germany). He produced a great many original contributions on such topics as the aerenchymatous tissue in the bark of the arborescent lycopods, fern anatomy, the nature of the distinctive aphlebia of certain Carboniferous fernlike foliage, fossil araucarian woods, and amber, just to mention a few; in 1900 he brought out an article reporting the *Glossopteris* flora in German East Africa which was based on the Bornhardt collections.

Potonié also had a fine facility as a compiler of technical information and was equally adept at popularizing scientific discoveries and reports for a wider reading audience. When I was working on my own doctoral dissertation I recall referring to his *Abbildungen und Beschreibungen fossiler Pflanzenreste* (Illustrations and Descriptions of Fossil Plant Remains), which came out in nine parts from 1903 to 1913. He prepared several sections for Engler and Prantl's encyclopedic work on natural plant families, including those on the fossil ferns, the Sphenophyllales, Cycadofilices, and lycopods. His greatest work in this general category is his *Lehrbuch der Pflanzenpaläontologie* (Textbook of Plant Paleontology) of 1897–1899.

He devoted considerable time to the study of the vegetation and peat of the Prussian moors, and knowledge gained in this pursuit was applied to others bearing on the origin of the brown coals and the harder coals of the Carboniferous. The moor lands had an asthetic as well as a scientific interest for Potonié and he was active in efforts to preserve some of these unique areas. His study of, and love for, the moors was an intense one which he engaged in without thought to the physical exertion involved, and Gothan attributes his death, at a rather early age, to malaria which he contracted in these investigations; this led to leukemia, which brought about his death in 1913 after a lingering illness.

There have been rather few father-son sequences in the general area of fossil botany in which both have been highly distinguished. Since Henry Potonié's son Robert is as well known to present-day geologists and botanists as his father was in his own time, I chose to include a few lines concerning him and his work, although the latter falls outside my main theme.

Robert Potonié (1889–1974) was a leading authority on coal chemistry and structure and one of the great pioneers in pre-Pleistocene palynology. He was born in Berlin and received most of his formal education there. In the early 1930's he initiated, with some of his students, a series of studies of the spores and pollen of the Tertiary brown coals. These, as well as some of his more strictly synoptic works, laid much of the foundations of modern palynology. He served on the Editorial Committee of the International Code for Botanical Nomencalture. He was for many years attached to the Prussian State Geological Institute in Berlin and also served as a Professor at the Technical University of Berlin and, after World War II, at the University of Bonn.

It seems natural at this point to bring in Walther Gothan (1879–1954), a very distinguished protégé of Henry Potonié's; but since his studies were mainly oriented toward stratigraphic paleobotany my brief account will hardly do justice to his work as a whole.

Gothan was born on August 26, 1879, the son of a potter in the small Mecklenburg town of Woldegk. He attended the University of Berlin, where he came in contact with Engler and Ascherson as well as Henry Potonié. I met Gothan only once at an International Botanical Congress, but our "literary contact" was much earlier. As a beginning graduate student I attempted to identify some fossil coniferous woods that I had collected in various parts of the country and the focal reference seemed to be Gothan's 1905 work on living and fossil gymnosperm woods (in German). This publication established him as a respected specialist in this area of paleobotany, and as a result Nathorst turned over to him the fossil woods collected by the Swedish Arctic and Antarctic expeditions. In the next few years he produced at least a dozen papers on fossil woods from King Karl's Land, Spitsbergen, Seymour Island, and other localities.

From 1903 on, Gothan was active in the Geological Institute, and when Henry Potonié died in 1913 he took his place in both research and teaching capacities. He was drafted as an infantryman in World War I but was later placed in a position where he served as a geologist for the military services. Gothan's use of fossil plants in detailed Carboniferous stratigraphical studies were among his greatest contributions. He was always anxious to work out a botanical continuity in the floristic sequences, and practical geological results developed from his knowledge of the fossil plants. He wrote: "one practical success in this area might be brought to mind, in that I succeeded by means of vigilance and knowledge of my speciality—despite the influence of opposing geological specialists—in discovering the much-discussed bituminous coal basin of Dobrilugk; at the same time, this is the first time that a German geologist has found a completely new, if not exactly great, occurrence of bituminous coal in Germany, in a place where no one had suspected anything of the sort" (quoted in R. Potonié, 1955, p. xxxii; translation).

Although Gothan's studies were primarily directed to the Carboniferous they were not entirely so confined, and in 1914 he produced the results of a study of a Mesozoic flora from the vicinity of Nuremberg. This is the so-called Rhätflora which he showed to be of Liassic age. Harris has commented as follows on this work: "Gothan (1914) made an exceedingly important contribution when he recognized the existence of two distinct floras in the 'Grenzschichten,' the older of which he regarded as Rhaetic, the younger, which is much more abundantly represented, he regarded as basal Liassic (belonging to the *Planorbis* Zone). I am in complete agreement with Gothan's conclusions" (Harris, 1937, p. 81).

His interest in fossil plants extended well beyond the Carboniferous and in 1921 he revised Henry Potonié's *Lehrbuch der Pflanzenpaläontologie* which went through several editions, the most recent one appearing in 1954. And like his great predecessor Goeppert, he evinced a strong interest in the history of paleobotany; in this context I am grateful to him for his survey "Die Palaobotanik in Deutschland in den letzten 100 Jahren" (1950), which begins with the work of Goeppert, Geinitz, Gutbier, and others in the 1830's. Although I do not agree entirely with the relative emphasis he places on the work of the various German paleobotanists through the nineteenth century, it is an extermely valuable summary. He also worked with the National Conservation League, which was established in 1922, and he played an influential part in initiating the Heerlen Carboniferous Stratigraphical Congresses that began in 1927.

Gothan is characterized by his biographers as being blunt and forthright in his criticism of others—criticism, however, that was well-meant and without malice. He often disagreed with traditional pedagogy and his style of writing has been described by one opponent as "frightful," but to most it seems to have been taken as spontaneous, fresh, and lively. Art and music were important to Gothan; he was a masterful player on the flute and violin, and Thiergart (1955) says that the piccolo accompanied him on his botanical and geological excursions and its sound always kept him in good humor. He was a pipe-smoker, and Maurice Wonnacott relates the following incident that seems worth recording: Shortly after W. N. Edwards completed his studies at Cambridge he spent six months with Gothan, and the two developed a lasting friendship. Later, at the close of the war, Gothan got word through to Edwards that he would very much appreciate it if he could send him a pound of his favorite tobacco. In spite of the restrictions of the times Edwards managed to supply his needs, much to Gothan's pleasure; hoping to avoid any involvement with the authorities he simply wrote to Edwards thanking him "for the opportunity to examine the fine specimens of *Nicotiana diplomatica*!"

My treatment of the fossil algae, or, rather, of those paleobotanists who have been concerned with that branch of our science, is admittedly far less complete than some of the other areas. This is due in part to my own lack of knowledge of the fossil algae, although, like D. H. Scott, I have long thought the living algae to be a most interesting and instructive assemblage of plants. And the places where they are found in abundance—swamps, streams, ponds, and the tidal pools of rocky coasts—add much to their lure. It is a special and very large assemblage of plants and significant studies of the fossils must be based on an extensive knowledge of the living forms. Although I am slighting the group as a whole I cannot leave it without devoting a few lines to Julius Pia.

Julius Pia (1887–1943) was probably the greatest student of the fossil algae of his time; this is recognized in some degree by Hirmer's selection of Pia as the author of that section of his *Handbuch der Paläobotanik* (1927). And the broader aspects of his knowledge are indicated by his introductory chapter in the *Handbuch*.

He was born in Purkersdorf, near Vienna, and after completing his early education he served with an artillery regiment for a year; in the fall of 1906 he entered the University of Vienna. Pia started early with his studies of the fossil algae and never wavered from this course for the remainder of his life. His initial instruction, under Professor V. Uhlig, was concerned with rock-forming limestone algae in the northern Alpine region; this was published in 1912 and was a point of departure for all of his subsequent studies.

In 1913 he was appointed to a permanent position in the Museum of Natural History in Vienna; he served the museum for many years, being interrupted only for four years during the war. He served with an artillery unit in Galicia and Poland in the campaign against Russia, and later on the Italian front; he always took advantage of any breaks to explore and collect fossils in the areas available. Following the war he traveled and collected extensively in various parts of Britain, including Bristol, Gloucester, and Yorkshire, and on the continent, through Croatia, Bosnia, and Dalmatia, where rich collections were gathered. But his biographer notes that "his favorite area of stratigraphic and tectonic-geological research had long since been the dolomites, with their noteworthy richness in fossils, their relatively favorable accessibility and the enchantingly beautiful landscape; the dolomites held him under their spell, like many a previous scientist, and challenged him again and again to the posing and answering of new questions of the Earth's history" (Trauth, 1947, p. 26; translation).

Pia's interests were not confined entirely to the algae; scattered through his long bibliography are studies of Liassic cephalapods, a description of a new species of warthog from southwest Africa, and Miocene whales. His chief interest was, however, with the fossil algae, and two aspects of this seem especially important: his detailed studies of the Dasycladaceae, a structurally complex and distinctive group and one that he demonstrated to be of considerable stratigraphic significance; and his comprehensive accounts such as the *Pflanzen als Gesteinsbilder* (Plants as Rock-builders, 1926) and *Die rezenten Kalksteine* (The Recent Limestones, 1935). Pia also possessed artistic ability and prepared his own rather well known restoration drawings.

It has proven especially difficult for me to gather together information for a summary account on the contributions of Polish paleobotanists. Few countries in the past two centuries have suffered so much from the direct ravages of war and political domination by neighboring states. Many or most of the Polish botanists and geologists in the present century have had their work interrupted and their collections, publications and manuscripts destroyed during the two World Wars. Very few have been able to devote their research activities entirely to fossil botany other than in practical applications to investigations such as Carboniferous stratigraphy where it related to coal studies. I believe that in general their potential was considerably greater than is indicated by the publications they have been able to produce.

The language problem has also entered in—to both the dissemination of the written works of Polish paleobotanists and to my own ability to review it! Moreover, I have not visited Poland, and my only personal contact has been with Maria Reymanowna at international meetings. I am especially indebted to Dr. Krystyna Juchniewicz of the Museum Ziemi in Warsaw for information and translation assistance. I have relied on her aid for my account of the important studies of Hanna Czeczott and Władysław Szafer in Chapter 16. And partly from assistance that she has given me I have selected a few people whose works I have some knowledge of to include here.

Marian Raciborski (1863–1917) was one of Europe's distinguished botanists of his time, and some twenty-five of his nearly two hundred publications dealt with fossil plants. In 1894 he described a rich Jurassic flora under the title *Flora Kopalna Ogniotrwaycg glinek Krakowskich* (Fossil Flora of the Fire Clays of Krakow), and in 1891 there appeared his *Permokarbónska Flora Karniowickiego Wapienia* (Permocarboniferous Flora of the Limestones from Karniowice). Among his other studies are ones on plants in amber and Rhaetic plants of Poland. Raciborski was born in Brzostawa near Opotów. In 1892 he went to Germany, studied in several of the leading universities there, and served as assistant to Professor Goebel. He received the doctorate in Munich in 1894 and when Professor Treub in Buitenzorg was seeking a botanist to work on the pteridophytic flora of Java, Goebel recommended Raciborski. He spent several years there, becoming involved in sugarcane research and for a time directing the experimental station for the cultivation of tobacco in central Java. He left there in 1900, spent some time in Penang and Ceylon, and then returned to Poland to become Professor of Botany at the Agricultural Academy in Dublany-by-Lemberg, and in 1912 he went to Krakow, where he organized a new botanical institue.

In 1939, Tadeusz Bochenski (1901–1958) brought out an important work on the reproductive organs of the arborescent lycopods: *On the Structure of Sigillarian Cones and the Mode of the Association with Their Stems*. This and related studies are cited by Chaloner in the Lycophyta section of the *Traité de paléobotanique*, Volume 2. Bochenski studied geology and paleobotany at the University of Krakow and in 1929 became an assistant at the Academy of Mining and Metallurgy. His work, like that of others, was interrupted during the war years of 1939–1945, after which he organized laboratories for the study of coal deposits at the Institute of Geology in Warsaw. His paleobotanical studies were chiefly in palynology, and stratigraphically oriented.

Bronisław Rydzewski (1884–1945) presents something of a link with the great age of paleobotany in France. He studied at the universities of Warsaw and Lvov and later at Krakow, where he received the doctorate with a thesis on the Carboniferous flora of Brzeszcze (1911). He then spent several years with Professor Zeiller in Paris at the School of Mines, returning to Poland in 1920, where he became Professor of Geology and Paleontology at the University of Vilnius. His publications were on Carboniferous plants and stratigraphically oriented, two of them being comprehensive treatments of the floras of the Dabrowa and Krakow basins.

A great deal of paleobotanical research has been conducted in recent decades in the geographical area that I have been concerned with in this chapter. Some of it is mentioned in other chapters and much of it is too voluminous, too diverse, and too recent for me to attempt a summary. Following my basic plan I have tried to bring the story up to within about a quarter century of the present, and I leave it to others to appraise the work of more recent years. I therefore choose to close my account here with a brief tribute to Richard Kräusel, whose name will certainly stand in the records of paleobotany as one of the greatest of fossil hunters. We have seen that some paleobotanists have made extraordinary contributions within the confines of a single major group of plants—Rudolf Florin and the conifers is the finest example; Seward was great for the tremendous breadth of his coverage and an ability to synthesize information to the benefit of both paleobotanists and the lay reader; there are those who are preeminent in a certain geological age, and here few can be compared with Tom Harris in his work with the plants of the Mesozoic; and some have contributed notably to both evolutionary botany and stratigraphy, such as Kidston. Kräusel impresses me as being unique in still a different way in that he made outstanding contributions in three distinct areas—in studies of fossil woods, angiosperm floras, and early land plants; and, in addition, he produced several comprehensive works that have served paleobotanists as well as the layman.

Richard Kräusel was born in Breslau on August 29, 1890, the son of a theatrical agent. He died on November 25, 1966. He studied botany and paleobotany at the University of Breslau under Professor Ferdinand Pax, who had been one of Goeppert's students, and thus this great line continues from the start of the chapter. In 1920, after serving in the war, he started his long association with the University of Frankfurt and the Senckenberg Museum. In a review of Kräusel's career it is not convenient to separate the three areas of fossil botany with which he was concerned. In 1912 he traveled with Pax to Siebenbürge and the Transylvanian Alps gathering specimens for his dissertation which appeared in 1913 as *Beiträge Zur Kenntnis der Hölzer aus der Schleisen Braunkohl*. More than twenty other studies dealing with coniferous and angiospermous woods followed, some of them very comprehensive ones such as "Die fossilen Koniferenhölzer," which appeared in *Palaeontographica* in 1919, and another large work on the Silesian woods in 1920. The study of fossil woods constituted a main theme throughout his life and led to extended trips to South Africa and South America. In 1964 he wrote a short but interesting account of his travels in Africa—in the back country, where riverbeds served as roads much of the year, where collecting was carried on in sand storms and intense heat, and where gasoline for his truck was hard to come by.

Kräusel made a few mistakes, like the best of men, and to me one source of his greatness is that he left them behind where they belonged and continually forged ahead. One anecdote bearing on this characteristic seems typical. In 1928 Kräusel described a presumed angiosperm wood from a Jurassic horizon, and I asked him, many years later, what his current opinion was about the wood and its age; he replied, with at least a slight touch of humor, that it was a long time ago and he would rather not be reminded of it!

Kräusel collaborated in many of his studies with Hermann Weyland, and among their more important works is a series of three papers entitled *Beiträge zur Kenntnis der Devonflora*, which came out in 1923, 1926, and 1929. These are basic to much that has been done in recent decades and include original descriptions and restoration drawings of *Asteroxylon elberfeldense*,

Richard Kräusel, at the Montreal Botanical Congress, 1959.

Aneurophyton, *Hyenia*, *Calamophyton*, and *Cladoxylon*. These plants were investigated chiefly from compression fossils, but they found some petrified fragments, of which Banks says:

> Kräusel & Weyland were among the few to recognize the vital significance of occasional bits of pyritized vascular tissue on compression specimens. From these pyritic remnants they labored diligently to obtain anatomical structure and their early illustrations of the anatomy of *Aneurophyton* have been modified but little by subsequent workers. What is more important is that subsequent students of the Devonian have extended their work and have learned the anatomy of numerous genera by preparing unpromising bits of pyrite. [1968, pp. 178–79]

It is sad to note that most of these Devonian fossils have been lost. It is my understanding that they were moved to the basement of a castle for safe storage during World War II. Apparently the Allies had reason to believe that the castle was being used for military purposes and they bombed it, so that the fossil-plant specimens were destroyed. A few years ago I examined a series of *Asteroxylon elberfeldense* specimens in the Stocklholm Museum (duplicates from the original collections) and was impressed with the accuracy of Kräusel and Weyland's restoration of the plant.

The *Beiträge* papers followed shortly after the great studies of the Rhynie plants by Kidston and Lang, and the combined work of these four paleobotanists left no doubts about the unique nature of early land vascular plants and the importance of pressing on with such investigations. We have learned much since the time of these four men, and much remains to be discovered, but their efforts marked the start of a great research epoch in plant morphology. Although I have some respect for "pure philosophical speculation" in attempting to work out evolutionary schemes, from this time on it should have been clear to any botanist that the fossil record was more dependable than armchair sketching in arriving at the truth.

Among Kräusel's works that do not fall strictly within the three categories mentioned above I would single out *Die palaobotanischen Untersuchungsmethoden* (Paleobotanical Techniques, 1929) and *Versunkene Floren* (1950), a semipopular book with sixty-four plates that include a host of restoration drawings and are a very valuable teaching aid.

Kräusel made an extended visit to the United States and Canada in 1928. He collected at many places that continue to be productive, including the Devonian horizons in West Virginia and along the Gaspé coast, and in 1941 published an article (in German) on plant remains from the Devonian of North America. He was back again in 1959 for the International Botanical Congress in Montreal. I had an opportunity to become somewhat better acquainted with him at meetings and field trips in India; he was working at Lucknow during the year I spent at Poona as a Fulbright lecturer.

Kräusel was an extraordinarily productive worker, as evidenced by his many writings. My overall impression of him is that he was a man with a rather stern exterior but most kind and helpful when one became better acquainted with him. He had little patience with those who lacked industry and competence. I greatly enjoyed our several meetings in India.

It is hardly possible to separate the two names of Richard Kräusel and Hermann Weyland; rarely have two men worked together so productively over such a long period of time. I have found that opinions vary somewhat as to the relative equality of their contributions to the many joint publications. It is my feeling that Kräusel was a more knowledgeable botanist, as might be surmised from the general careers that the two followed. But Weyland was a highly competent man; he contributed much to the field work and was responsible for many of the fine drawings in their works. My only personal meeting with him was an afternoon spent in his laboratory at Cologne in company with Suzanne Leclercq. We spent some hours going over certain of his Devonian plant collections in conjunction with a research project that Miss Leclercq and I were engaged in. This was conducted in a slightly confused mixture of French, German, and English, but it was a very pleasant experience and I was impressed with Weyland's modesty and his knowledge of fossil plants.

Weyland was born on March 25, 1888, in St. Ingbert, on the Bavarian side of the Saar, and was given an informal introduction to science by his grandfather, including field trips to mines where he first became acquainted with fossil plants. This interest was further developed by field excursions with Professor Friedrich Kinkelin, who brought him into contact with the Senckenberg Natural History Society. However, his formal studies were focused in the general area of pharmacy and botanical physiology, and his doctoral dissertation at Jena under E. Stahl in 1912 was on the nutritional physiology of mycotrophic plants. He also studied under Wihelm Pfeffer and had planned a career in botany, but the exigencies of the times after World War I made it necessary for him to seek other employment, and from 1934 to 1952 he served as a chemist in the pharmaceutical and physiological departments of the paint factory of Bayer and Co. in Elberfeld (later I. G. Farben). Paleobotany with him was a sideline but certainly a very important one.

Gothan gives him credit for the discovery of the Devonian plants of Elberfeld, in 1922, which were found near his laboratory. It was fortunate that the chairman of the Board of Directors of I. G. Farben, Professor Duisberg, took an interest in Weyland's paleobotanical investigations and allowed him the necessary time for field work. This assistance is acknowledged in the name of one of Kräusel and Weyland's Devonian plants, *Duisbergia*. Duisberg also contributed toward publication costs.

Weyland later devoted more and more time to botany and paleobotany; in 1931 he was named Honorary Professor at the University of Cologne and served as provisional Director of the Botanical Institute.

14

Paleobotany in Western Europe: Some Contributions from France, Belgium, and Holland

As in the previous chapter, my selection of persons and their achievements here has been guided by my own interests and knowledge. If there is not a central unity in the chapter as a whole there are at least several subunits in which we find some continuity of important paleobotanical action. Again, the number of people involved in the age and area considered is large and I have selected a few that seem especially important to the main streams of progress. My starting point is in the mid-years of Adolphe Brongniart's life, with four men who were quite closely associated with him, who were his students in a broad sense, and the successors in France to his great heritage: Bernard Renault, Cyrille Grand'Eury, René Zeiller, and the Marquis Gaston de Saporta. Of these four I am intentionally emphasizing the life and work of Renault; I feel that his studies typify the botanical-evolutionary approach that is my central theme.

The Belgian trio that I bring into the story are of a more recent vintage; their work has focused chiefly on Devonian horizons, and I have known them personally. Their work stands as another significant pulse in the progress of fossil botany. The people concerned are Armand Renier, Suzanne Leclercq, and François Stockmans. And I will introduce Willem J. Jongmans to represent work that has gone on in Holland; although he was strongly oriented toward the use of fossil plants in stratigraphic studies, his achievements and influence were broad. There have been many others, a few of whom I mention briefly, not for any lack of regard for their efforts but because I feel less competent to judge their contributions critically.

When I was a graduate student I was fortunate enough to have been able to purchase a set of Bernard Renault's two-volume work, *Bassin Houiller et*

Permien d'Autun et d'Épinac. I have long treasured it partly because it originally came from Sir Albert Seward's library and because it is one of the great works on petrified plants of the Paleozoic, written by one of the most illustrious and least appreciated botanists of his time. Renault accomplished in France largely what Williamson did for Britain; I do not imply that there was overlapping of results but rather a complimentary series of studies which together marked a tremendous advance in our knowledge of petrified Paleozoic plants. It is unfortunate that during much of their careers they were in dispute over the significance of the cambium in plant classification. It was a dispute in which Williamson was proven correct although Renault was the better botanist.

Bernard Renault (1836–1904) was born at Autun, received the Bachelor of Science and Bachelor of Letters degrees at Autun College in 1854, and soon after joined the Institut Brenot at Dijon, where he became a professor of chemistry and physics. He received a doctorate in 1867 at Paris for his thesis on electrolysis and then for a time taught at Cluny. In 1870 he worked with the Committee of National Defense helping to organize the defense of various points of the Saône-et-Loire in the 1870 war with Prussia.

He developed an interest in the Permo-Carboniferous silicified fossils that were found in the vicinity of Autun and at St. Étienne, and his studies of the plants contained in them are among our great classics. It is thus of interest to include a few anecdotes on the collecting activities of Renault and his collaborators. Paul Bertrand, in his inaugural address of 1947 at the Paris Museum, says that at Autun every spring following the plowing, and every fall after the harvest, the collectors gathered the silicified specimens in the fields. The best ones were selected for study, and a time-honored procedure in geology and paleontology was applied: for lack of an immediate supply of water to wash the specimens, they were licked with the tongue. The local folks, as usual, found such a procedure amusing and called the collectors "les lichus d'piarres." The Abbé Lacatte, one of Renault's helpers, was especially noted for the great length of his tongue ("Entre tout l'abbé Lacatte était célèbre pour les dimensions de sa langue, qui s'allongeait démesurément hors de sa bouche"; Bertrand, 1947, p. 120).

Several others assisted in the collection of these fine petrifactions, including the Abbé Landriot, M. Faivre, and Renault's friend and biographer Auguste Roche (1827–1905), who also prepared a large number of the ground thin sections on which Renault's studies were based. As might be expected, in view of Brongniart's eminence at the time, Renault communicated his results to him. The latter's duties included official inspection tours at the Normal School at Cluny where Renault was in charge of teaching chemistry. The welcome opportunity to see and talk about Renault's fossil plants gave Brongniart some consolation in his round of inspection duties, and in 1872 he invited Renault to come to Paris. For four years he held a position as "préparateur" in the Museum of Natural History there, and in 1876 he was appointed "Assistant Naturalist", an appointment attached to the Chair of Botany, Organography, and Physiology. This was the only post that he held

Bernard Renault. Reproduced from the *Journal of the Royal Microscopical Society*, 1906.

for the remainder of his life, and Scott says: "Few men so distinguished have received such miserably inadequate recognition from their official chiefs, an injustice which his learned countrymen cannot speak of without indignation" (1906, p. 132).

The pitiful conditions under which most of Renault's great studies were conducted are best recorded in D. H. Scott's first-hand account:

> Renault had no laboratory at first; after a time, however, he induced the architect of the museum to put up for him, under the head of "repairs," two little glazed wooden boxes, one on each side of the portico of the botanical and geological department. Chevreul, the venerable director of the museum, was indignant when he first saw this inartistic excrescence on the architecture of his building. . . . One was the workshop where Renault cut his fossil sections, [the other] was the laboratory proper, where his microscopic work went on, and where he received his scientific visitors. "Renault's cage," as the laboratory was jocosely called, constituted to the end of his life his only official quarters. It looks exactly like a porter's lodge, and the savant whom it sheltered was often addressed by strangers as the concierge, a comic situation which he thoroughly enjoyed. Visitors to the laboratory, as I know to my cost, were sometimes the victims of the same mistake, and were not always equally capable of dealing with the position. [Scott, 1906, pp. 132–33]

It was here, with a foot-powered saw and grinding lap of his own design, that Renault prepared some seven thousand thin sections; considering the extremely hard nature of the silicified material that he worked with the task was a much greater one than that of his English colleagues of the same period who were working largely with carbonate coal-ball petrifactions.

Renault's own published works are numerous and include many short papers as well as several large volumes which are among the great landmarks on the anatomy of Paleozoic plants. In addition, he collaborated in some great ventures with other leading French paleobotanists. I can do little more than list the works and I will then select a few specific lines of investigation for which Renault is well remembered and which continue in full vigor to the present day.

In 1878 he brought out his *Recherches sur la structure et les affinitiés botaniques des végétaux silicifies recueillis aux environs d'Autun et de St.-Étienne*, a volume of 211 pages and thirty plates describing articulate cones, *Sphenophyllum*, the vegetative and fertile organs of *Zygopteris* and *Botryopteris*, and a few others. In 1879 his thesis *Structure comparée de quelques tiges de la flore Carbonifère* was published at Clichy. This includes eight plates and deals with the anatomy of *Lepidodendron*, *Sigillaria*, *Cycadoxylon*, *Poroxylon*, and *Cordaites*.

Through the offices of the physiologist Paul Bert, Renault gave a course of lectures during 1879–1883 on vegetable palaeontology which Scott says was "undoubtedly the most important course of lectures ever given on this subject." The lectures were based largely on the results of his own investigations and appeared in five volumes as Renault's *Cours de botanique fossile*. A considerably smaller volume was published in 1888 under the title *Les Plantes fossiles*.

At the outset I mentioned his work on the fossil plants of the Permian coal beds of Autun and Epinac; the text volume of 570 pages came out in 1896 and the atlas of sixty-two plates appeared in 1893. By any standard of judgment this is one of the great botanical works of the nineteenth century. It brought together much of the information that Renault had published earlier in numerous shorter papers.

Bernard Renault laid much of the foundation of our knowledge of the diverse and intriguing complex that we refer to as the coenopterid ferns. In 1875 he established the genus *Botryopteris*, a group of small, probably epiphytic, plants with simple stem anatomy and sporangia organized in huge spherical aggregates. Although much remains to be learned about *Botryopteris*, many people since Renault have contributed to the story, and it is better known than many living genera. In 1939, William Darrah described the sporangial aggregates from specimens of Iowa coal balls. In 1950, Sergius Mamay and I made a comparable study of the stem and frond anatomy, the availability of the peel method at that time being advantageous. In 1957, William Murdy and I made a study of some especially well preserved fructifications; in 1962, Henry Holden described a *Botryopteris antiqua* from the Lower Carboniferous of Scotland which afforded additional information on the morphology of the stems and leaves; in 1954, Theodore Delevoryas and Jeanne Morgan contributed to the leaf morphology; K. R. Surange added to this in 1952; Natasha Snigirevskaya added more information in 1961 and 1962 based on Russian coal balls. Tom L. Phillips has a great fund of information on both American and European species, and the first part of his detailed study came out in 1970. This is by no means a complete listing of the results to date.

In 1906, D. H. Scott gave us a succinct and very readable account of Renault's life and works in his presidential address to the Royal Microscopical Society. He thought well of Renault, and this in itself is high praise. In 1906 the state of paleobotanical thinking about any fossils that were at all "fernlike" was in a state of flux because of the establishment of the pteridosperm group in 1904. Scott puts Renault's work into this context as follows:

> Apart from the Botryopterideae, a group which some would separate from the Ferns proper, the last refuge of the true Ferns has hitherto been found in the genus *Pecopteris*. This stronghold, it is true, was rudely shaken when, in April of last year, M. Grand'Eury discovered fronds of *Pecopteris Pluckeneti* laden with seeds! The species, however, is an aberrant one, and cannot decide the fate of the genus as a whole. A large number

of species of the form-genus *Pecopteris* are known to have borne fructifications of the type commonly recognised as Marattiaceous. The sporangia are somewhat massive, without a definite annulus, and, in the more characteristic cases, are united together, like the carpels of a multilocular ovary, to form compound fructifications known as synangia. While our knowledge of the external characters of such fructifications is chiefly due to Stur, Zeiller and Grand'Eury, it is to Renault, more than anyone else, that we owe an acquaintance with their internal organisation, which he described in *Scolecopteris polymorpha*, *Ptychocarpus unitus*, *Pecopteris geriensis*, *P. oreopteridia*, *P. exigua*, *Sturiella intermedia*, and others (Renault, 1883 and 1896). The last case mentioned is particularly interesting, as here the sporangia, though grouped in definite synangia, possess a kind of annulus, thus showing a remarkable combination of characters. Renault's observations undoubtedly tended strongly to confirm the idea of the Marattiaceous affinities of the fossils in question. Whether this view, once, as it seemed, so unassailable, can hold its ground, or whether we shall have to admit that all these fructifications, like *Crossotheca Höninghausi*, were but the pollen-bearing organs of Pteridosperms, cannot yet be decided. In either event, the ultimate decision will rest, in no small degree, on the accurate data afforded by Renault's researches. [Scott, 1906, p. 137]

In Chapter 4 I mentioned Brongniart's posthumous *Researches on Fossil Silicified Seeds*, which came out in 1881. Renault worked closely with Brongniart and he almost certainly had much to do with the production of this great monograph. He brought out several works on petrified seeds himself, and since I have not had access to all of them I will again take a few pertinent comments from Scott's address of 1906:

In the course of his work with Brongniart on the wonderfully preserved silicified seeds of Autun and St. Étienne, Renault discovered the constant presence of a pollen chamber—i.e. a definite excavation in the tip of the nucleus [apparently a misprint for nucellus], adapted for the reception of the pollen-grains. Guided by his observations on Palaeozoic seeds, he was able to detect the same organs in those of their nearest living allies, the cycads, in ignorance of the fact that this discovery had already been made by Griffith, thirty, or perhaps even forty, years before. . . .

In examining a seed named *Aetheotesta*, probably belonging to a Pteridosperm, Renault, in 1887, observed the minute structure of the multicellular pollen-grains in the pollen-chamber. He found that both in the outer wall of the grain, and in the septa between its cells, there were perforations. He suggests that the latter may represent the traces of insertion of multiple pollen-tubes, or—and this is the important point—"that they

> served for the passage of mobile bodies analagous to antherzoids" (Renault, 1887, p. 156). . . .
>
> At the close of the paper he adds: "We do not regard as impossible the existence in the past of pollen-grains, which, instead of effecting fertilization by means of a tube, discharged into the pollen-chamber of the appropriate seeds antherozoids capable of performing this function. . . . "
>
> Thus the study of the phenomena presented by Palaeozoic fructifications led Renault to anticipate, by some ten years, the great discovery made in Tokyo by Ikeno and Hirase, that in the lower seed-plants the cryptogamic or animal mode of fertilization by mobile male cells still persists. . . . The case affords a striking example, both of the acumen of Renault himself, and of the value of palaeobotanical evidence in its bearing even on the more minute points of morphology. [Pp. 140–41]

Renault's considerable interest in fossil microorganisms was brought together in *Sur quelques microorganisms des combustibles*, a rather large volume which came out in 1900. The emphasis in this work is on microorganisms found in peats, lignites, and coals of various kinds, and it includes studies of bacteria, infusoria, fungi, and other thallophytes. Ward referred to it as "a superb work on a very difficult, but at the same time very important subject." It reveals Renault's great breadth of knowledge, and in view of the rather primitive state of microbiology at the time, it strikes me as quite remarkable. He also concerned himself with the problematical "boghead" coals, but I will defer a discussion of them to a later page in this chapter.

A considerable part of Renault's work was intertwined with the researches of C. E. Bertrand, Brongniart, Zeiller, and Grand'Eury; it was a great period of distinguished men and important advances in paleobotany. Renault worked with this group in the field as well as in the preparation of publications. Scott records an interesting incident which makes a good introduction to Grand'Eury:

> One day, when Renault was accompanied by his friend Grand'Eury, another illustrious palaeobotanist, who happily is still with us in full vigor, a curious incident occurred. Grand'Eury, as they approached a good locality, asked his friend, "What do you want me to find?" "A *Sigillaria* with its bark on," replied Renault. In a minute or two Grand'Eury picked up a block, and without examining it said, "There's the *Sigillaria* you asked for." Chance had favored him. Renault cleaned the specimen according to custom; it revealed itself as a magnificent *Sigillaria spinulosa*, which formed the subject of one of his best known memoirs, written in collaboration with Grand'Eury (Renault, 1874). It was ten years before another piece was found, and no more has been met with since then. [Scott, 1906, p. 131]

Paul Bertrand has said that while Brongniart founded paleobotanical stratigraphy it was Grand'Eury who applied it masterfully to the Carboniferous horizons. It is my impression that Grand'Eury's knowledge of Carboniferous plants and their distribution was considerably greater than his published record indicates. This combined with the fact that my emphasis here is not with stratigraphic achievements and literature probably results in slighting the overall importance of his contributions.

François-Cyrille Grand'Eury (1838–1917) was born at Houdreville (Meurthe-et-Moselle); he attended the Ecole Loritz in Nancy and then the School of Mines at St.-Etienne. For several years he worked as an engineer at the mines of Roche-le-Molière and then for reasons of health he took a post at the School of Mines at St.-Etienne. For much of his time there he served as Professor of Trigonometry.

He is probably best known for his *Flore Carbonifère du Départment de la Loire*, which came out in 1877. It reveals his special capacity for assembling the isolated parts of fossil plants and putting them together to make whole individuals. His restoration of a cordaite tree is one of the best known and has been reproduced many times; the same holds for his corresponding drawing showing the manner in which the "inflorescences" are borne. There is perhaps a disadvantage in especially fine restorations in that they may discourage, or imply a lack of necessity for, comparable attempts by other paleobotanists. The cordaite group is a large one and important from the standpoint of its dominance at certain Upper Carboniferous horizons. I recall just a few years ago finding an abundance of great slabs of shale at an open-pit mine in New Brunswick, Canada, that were covered with fine specimens of a large-leafed *Cordaites*, and it was easy to visualize the great trees, as Grand'Eury did, growing in an ancient landscape in which the vegetation was very different from that of the present. It is probable that different species and genera in the cordaite group displayed different growth habits, and it is refreshing to encounter Arthur Cridland's concept (expounded in 1964) of one as a small tree similar to the modern mangrove (*Rhizophora*). His drawing shows the plant supported by its stilt roots growing in what may have been a subtropical saline-influenced swamp.

Grand'Eury's restorations were based on careful observation of the various parts, which he then "put together" only as the evidence seemed to warrant. One of his biographers says that he "was convinced of the necessity of using special generic names for the various organs of a plant. He originated the terms *Cordaianthus* for the inflorescence of the Cordaites, *Cordaicladus* for the axis, *Cordaifloyos* for the bark, and *Cordaixylon* for certain woods. His reconstructions of *Lepidodendra*, of Cordaites, and of other plants from isolated remains of leaves, trunks, and seeds were scientific masterpieces that are still discussed in works on paleobotany" (Stockmans, 1972, p. 498).

I mentioned in Chapter 7 something of the important part that Grand'Eury played in the development of the "pteridosperm concept." Over a period of many years leading up to the 1904 discoveries of Oliver and of Kidston, Grand'Eury had observed the constant association of certain seed and leaf types. There can be little doubt that, like several others of his time, he was reasonably sure that such evidence indicated a distinct group of seed plants, and he helped to settle the matter with his report in 1905 of seeds attached to *Pecopteris* foliage that he found at St.-Etienne.

It is not possible to sharply define and assign appropriate credit to the various works of Brongniart, Renault, Zeiller, Grand'Eury, and others with whom they were associated in their various studies. And some of these had great practical as well as scientific results. In his biography of Zeiller, Douville (1917) says that in his earlier years Zeiller, working with Grand'Eury, helped solve a stratigraphic problem of the coal seams in the Grand'Combe and Ste.-Barbe areas. Because of their combined and extensive knowledge of the floras they encouraged the mine operators to continue their boring at a particular point, with the result that two or more productive seams were discovered. Armand Renier once told me of a similar experience that he had and it is related on another page.

René Zeiller (1847–1915) was one of the more important contributors to paleobotanical progress during the last quarter of the nineteenth century. He produced several fine floristic works on Carboniferous plants, his studies of structurally preserved fossils were extensive, his interests were wide-ranging, and his overall knowledge is clearly evinced in his excellent textbook *Eléments de paléobotanique*. He was born at Nancy and educated at the Polytechnic School and the School of Mines, where he became interested in fossil plants and established himself as an authority at an early age. His father died when he was fourteen, but he seems to have been fortunate in the associations which helped to direct the course of his development and which contributed to his broad interests. His maternal grandfather, Charles-François Guibal, a judge who cultivated several sciences, as well as poetry, influenced him; and he followed the excursions of the botanist Godron, often accompanied by his grandfather and the future artist Galle. Like others of the time, Zeiller was involved in the turmoil of the 1870 war, during which he worked to establish underground communications between Paris and the forts in the south of France.

As to the nature of his official duties, in 1871 he was assigned to supervise work on the Orleans railway; he returned to Paris in 1874 with the same administration for the next ten years. In 1882 he transferred to the Service de Topographie Souterraine des Bassin Houillers de France (the Mine Service) and rose through the ranks to become President in 1911.

Zeiller's earliest work was in geology with a special bent toward metallurgy; in 1873 he published a memoir on the eruptive rocks and metalliferous veins of the Chemnitz district. His first publication in fossil botany was a review of Schimper's *Traité*, which is discussed on a later page.

He was concerned with the practical use of fossil plants in stratigraphic work; E. W. Berry says:

> Professor Zeiller was one of the first to demonstrate the precision with which fossil plants can be used in stratigraphic geology and in the numerous large memoirs on the Carboniferous and Permian floras of the coal basins of Grand'Combe (1884), Valenciennes (1888), Blanzey and Creusot (1906), as well as his work on the fossil plants, which form part 2 of Vol. 4 of "Explication de la carte géologique de la France" (1979) he displayed a philosophic interpretation that had never been equalled. [Berry, 1916, pp. 201–202]

Zeiller had a special interest in the ferns of the Carboniferous, and in this capacity Scott refers to him as "perhaps the first authority." His description of the Valenciennes flora, which deals with a particularly important coal field of France, includes a fine discussion of the group and is well illustrated. He also handled most of the ferns in the great volumes on the Autun and Epinac flora; the first part, by Zeiller, came out in 1890 and the second, by Renault, appeared a few years later, as noted above. The first part of Renault's text is a continuation of the work on the ferns.

Zeiller gave an elaborate treatment of *Psaronius* in a joint work with Renault on the flora of Commentry (1888–1890), which is beautifully illustrated with an atlas of large plates. I have used some of their illustrations in my teaching and regard it as one of the outstanding classics in paleobotany, combining, as it does, the efforts of these two great men. His 1892 work on the flora of Brive included a description of the fructification of *Zygopteris*, extending a knowledge of the plant which had been established previously by Renault. In 1911 he described the cone *Lepidostrobus brownii* (Unger) Schimper in the *Memoires of the Academy of Sciences of Paris*. The sixty-seven-page account is accompanied by fourteen splendid plates and is an outstanding item in the literature on the lycopods. He was the first to describe the pteridosperm male fructifications *Crossotheca* and *Potoniea*, and I believe he merits the credit for first demonstrating the connection of the cone

Bowmanites dawsoni to a stem of *Sphenophyllum*. Zeiller also contributed to the matter of settling the significance of the cambium in the classification of the arborescent lycopods. Although at first following Brongniart's philosophy that the cambium was strictly the mark of a gymnosperm, he was, I believe, the first in the French group to recognize the validity of Williamson's work, and, indeed, Zeiller supported Williamson's viewpoint in his demonstration that the cone of *Sigillaria* was that of a vascular cryptogam. This is discussed more fully in Chapter 5.

Zeiller's textbook *Éléments de paléobotanique* (1900) is one of the best ever written and still merits a place in the useful section of a paleobotanist's bookshelf. It starts with a very brief historical statement, proceeds to preservation and nomenclature, and continues with a well-balanced treatment of the various plant groups. Its general organization is still a good model for such a text.

Scott has summarized Zeiller's overall philosophy rather well:

> As regards the broader questions involved in palaeontological studies, Zeiller was a thorough evolutionist, but not a Darwinian, for he was a believer in abrupt changes (mutations or rather saltations) leading at once not merely from species to species but even from group to group. Though, in the opinion of the writer, he went too far in this direction, Zeiller undoubtedly did good service to Science in dwelling on the immense difficulties and complexities which still stand in the way of a continuous phylogeny of fossil forms. [1916, p. 77]

I have the impression that Zeiller was a man whom one would have valued highly as a friend. Scott adds, "Personally, Zeiller had a great charm of manner, his dignity and courtesy winning the hearts of all his colleagues, many of whom have the pleasantest recollections of his warm kindness and hospitality."

The last of the four great disciples of Brongniart, Louis Charles Joseph Gaston, Marquis de Saporta (1823–1895), was also a man of broad interests and great productivity in paleobotany. He was born at St.-Zacharie, Var, in the south of France, of a family of Spanish origins and was educated in the Jesuit college at Fribourg in Switzerland. In his boyhood he was in communication with the entomologist Boyer de Fonscolombe, but he devoted himself mostly to literature and history. He later developed an interest in the living plants of Provence, and two incidents in his life, when he was about thirty, seem to have been largely responsible for guiding him into the varied and numerous researches that he undertook in fossil botany. In the course of his botanical exploration of the countryside he found fossils in the gypsum quarries of Aix and took them to Brongniart, who encouraged him to continue his studies in this direction. And before he was thirty his young wife died; thus he had an even greater incentive to intensive study as a diversion in his grief.

The estates owned by his family seem to have been sufficient to allow him to conduct his studies free from financial burdens, and Williamson (1896) notes the following relative to his scholarly interests: "A Science Academy having been formed at Aix, where Saporta had a princely home, I need scarcely add that he soon became the head, not only of it but of other institutions in the town. The Aix Academy is not only a scientific one; it exists also for the encouragement of literature, agriculture, and so forth" (1896, p. 212).

After some ten years of work in the field, herbarium, and laboratory the first part of his *Etudes sur la végétation du Sud-Est de la France à l'époque tertiaire* (Studies of the Tertiary Vegetation of the Southeast of France) appeared in 1862. This resulted from extensive collections from many localities including Aix, Gargas, Castellane, St.-Zacharie, Marseilles, and Manosque. And while he was still deeply involved with the Tertiary plants he initiated a very extensive investigation of the Jurassic floras of France. These appeared in the *Paléontologie française* in forty-seven fascicles from 1872 to 1891. They are usually found today bound in four volumes with a total of more than two thousand pages and three hundred plates. Zeiller (1895) states that Saporta was a skillful draftsman and all of his plates were prepared from his own drawings.

Saporta kept in touch with Ettingshausen and Heer, and when the latter died in 1883 the Mesozoic collections from the Geological Survey of Portugal were sent to Saporta. He brought out a preliminary report on them in 1888, and in 1894, the year before he died, the Survey published his *Flore fossile du Portugal*, which is probably his most important single work. This dealt with some of the oldest records of flowering plants and at the time was apparently regarded as an important publication relative to the problem of the origin of the angiosperms. Some additions and revisions were made in 1948 by Carlos Teixeira.

Saporta had a strong and successful flair for popular writing and in 1879 he brought out *Le Monde des plantes avant l'apparition de l'homme* (The Plant World before the Appearance of Man). From 1881 to 1885 Saporta brought out, with A. F. Marion, a three-volume work, *L'Evolution de règne végétale*. And his continued interest in living as well as fossil plants is evinced by a small volume of 1888: *Origine paléontologique des arbres cul-*

tivés ou utilisés par l'homme (The Paleontological Origin of Trees Cultivated or Used by Man).

It is my impression that our colleagues of the past were more concerned with the matter of bringing the results of their researches to the layman than is the case now. Many of the great people—Seward, Stopes, Knowlton, and Berry—wrote books that were well suited to the general reader. Such books should be written by the people who are actively involved in research. At the expense of being immodest I would note that I believe my own book *Ancient Plants* is one of my most useful accomplishments and after thirty-one years it is still on the publisher's sales list—in part at least because others have not come along to replace it! Shortly after *Ancient Plants* appeared I received a letter from Oakes Ames, the distinguished Harvard orchidologist. Previously I had spent a sabbatical period working in the quiet atmosphere of the Botanical Museum at Harvard; Professor Ames's kindness added to the pleasure of my visit, and I have treasured his letter which I include here because of the delightful way in which he expresses the need that I mention:

August 15, 1947

Dear Dr. Andrews:

I wish to thank you for "Ancient Plants"; not only because it is a timely book, but because it proves that you are still willing to be human. —It is so easy to dodge behind the drab pillars of science to be dull. After all we who are carrying one of the burdens of civilization should take time out now and again to understand that what we are doing is to educate not to discourage. Well, I am still pleased to see the present through the past; but if I were a paleobotanist I should long for the night to come when in my dreams I might see what now I am supposed to imagine.

It was good of you to remember me!

With all good wishes
Oakes Ames

I have puzzled over just where to bring in a few words about W. P. Schimper, whether the previous chapter or the present one is more appropriate. The shifting political boundaries of nineteenth-century Europe enter into the picture. Wilhelm Philipp (Guillaume Philippe) Schimper (1808–1880) is unusual in the annals of paleobotany; his original contributions are minimal, the only comprehensive one being a monograph with A. Mougeot on the Triassic flora of the Vosges (*Monographie des plantes fossiles du Grès Bigarré de la*

Chaîne des Vosges), a large work with forty-two plates. However, his great *Traité de paléontologie végétale*, which was published between 1869 and 1874, is one of the monumental works in fossil botany. It consists of three text volumes with a total of 2,600 pages and an atlas of 110 magnificent quarto plates. I confess that I have never gone through the text carefully, but the atlas is something that will always be a pleasure to thumb through. It is largely of historical interest now and probably best appraised by a contemporary of Schimper; Lester Ward said in 1885:

> The "Traité" is unquestionably the most important contribution yet made to the science. Although necessarily to a large degree a compilation of the work of others, still it is by no means wanting in originality, and contains a great amount of new matter. Its chief merit, however, is in its conception and plan as a complete manual of systematic paleobotany. The classification is highly scientific and rational, and the discussion of abstruse points in defense of it is acute and cogent. [Ward, 1885, p. 375]

Schimper was a fine botanist of the time, best known as a bryologist. He was born at Dossenheim, near Saverne, and at an early age showed a strong interest in natural history, making collections of flowers and butterflies—and painting them. He was influenced and probably encouraged in this direction by a cousin, Karl Schimper, who was a noted naturalist, and a little later by Alexander Braun. His father was a Protestant pastor who gave his children their first lessons in Latin and Greek, and it was apparently due to his strong feelings in the matter that Wilhelm studied at Bouxwiller and Strasbourg with the intent of following in his father's footsteps. He received a bachelor's degree in theology in 1833 and took his own pastorate.

He found rather soon, however, that this was not the life he wanted to follow and, probably with some opposition from his father, he left the religious life. We have seen that he was not the first to make this change but I think that Schimper's biographer, Charles Grad, words it a little better than most when he says that "theology never attracted him. The observation and study of plants used to captivate his attention much more than the commentaries of the Bible, and he used to prefer the interpretation of the grand poem of nature with its delightful harmonies to the chanting of the psalms" (1880, p. 354; translation).

With the aid of the bryologist Philipp Bruch and the geologist Voltz, Schimper obtained a curatorial position at the Natural History Museum in Strasbourg. This was a low-paying post and it was not until many years later, in 1862, that he received the Chair of Geology and Mineralogy at the University of Strasbourg. He developed what seems best described as a passionate interest in the mosses, and for two decades he traveled extensively over Europe making large collections of mosses, and also fossil plants. Like several of our other friends he suffered an encounter with shipwreck; it is reported that one of his finest collections, made in Scandinavia, was lost on the north coast of Norway, but apparently Schimper was not with the collections at the time. He visited with Leo Lesquereux several times before the latter left his home for the United States. He also conducted some geological studies with Voltz, and participated in the glaciological investigations of Agassiz. His publications on the mosses were numerous, probably the best-known being the *Brylogia Europaea*, which was written with Bruch and Theodor Gümbel; this was a six-volume work with 654 plates, most of the drawings being prepared by Schimper.

Some of the vast collections that he gathered together in the museum in Strasbourg were destroyed by the bombardment of the city in the Franco-Prussian war, and many of the faculty were scattered when Alsace became German territory. But Schimper remained there although he was offered a post at the Botanical Garden in Paris. Richards's summary statement (1975) seems a just one: "His work covered a vast field, but he was a supremely competent observer and describer rather than an originator of new ideas." It is also of interest to note that he was the father of the plant geographer A. F. W. Schimper.

The Bertrands, a father-and-son team, follow in sequence the disciples of Brongniart that I have presented above. In his dual biographical sketch, T. S. Mahabale says: "Charles-Eugène Bertrand, and his son Paul Bertrand, form a unique pair in the history of Botany in France. They practically ruled the subject of plant Anatomy, Morphology and Paleobotany in France for quite a long time by their ceaseless work and vast influence over a wide circle of disciples and admirers" (1954, p. 445).

Charles-Eugène (1851–1917) was born in Paris in the year that Hofmeister's great work *The Higher Cryptogamia* was published. He was educated at the Sorbonne and in 1878 took the Chair of Botany at Lille, where he remained for the rest of his career. Among his students who also made significant contributions to paleobotany were Maurice Hovelacque and Octave Lignier. His most important studies were with the anatomy of living plants and included studies of the Gnetales, Coniferales, and Lycopodiales, but he also added to our understanding of the anatomy of some of the coenopterid ferns such as *Zygopteris*, *Botryopteris*, *Anachoropteris*, and *Tubicaulis*. He collaborated with Renault in a study of *Poroxylon*, produced a contribution on the female fructification of the cordaites, and prepared a fine study of the anatomy of *Lepidodendron harcourtii*.

Charles Bertrand was strongly attached to Lille and his work there, for he remained in the city after the German invasion in 1914, continuing his studies and teaching at his home at one time when there was bombing in the vicinity of the University.

Paul Bertrand (1879–1944) was born at Loos-les-Lille and studied with such men as Jules Gosselet, Charles Barrois, and his father. He became a laboratory assistant at the Coal Museum in Lille in 1906 and received an appointment at the University, where he remained until he took a professorship in 1938 at the Natural History Museum in Paris. In 1937 he wrote a booklet entitled "Notice sur les travaux scientifiques de Paul Bertrand," and a supplement came out in 1943. It is essentially a "scientific autobiography," is nicely illustrated, and includes summary statements on all of his major publications; it is a most helpful piece of literature for anyone preparing a historical account such as the present one.

He states that at the outset of his career he started to study the living cycads, but he was outdone by the work of Henry Matte, and then acting on the advice of his father he took to a study of the ancient ferns.

My own personal contact with Paul Bertrand was brief but pleasant. As a graduate student I spent a day with him in the summer of 1938. In the morning we took a short train journey to a colliery where we spent an hour or so—a rather short time in which to accomplish very much, but it was raining and he did not seem to be enjoying himself especially; the afternoon spent in his laboratory was more memorable. He regaled me with some of his theories of plant relationships, which, in their briefest form, are illustrated by diagrams given on pages 34 and 35 of his "Notice." The essence of his viewpoint is that the major groups of vascular plants have originated directly from a primitive vascular plant stock which he labels "Rhyniales." This is worded very clearly in the "Notice": "All the large groups of vascular plants are independent of one another. Each group has accomplished its evolution (that is, has reached its highest level of differentiation) with its own energy, in its own way. None is derived from the others" (1937, p. 26, trans.).

In our afternoon conversation I especially remember his emphasis on the angiosperms, their apparent sudden origin in Cretaceous times being explained by this direct route from a much more primitive stock. With many of the groups involved it seems to me that more recent work tends to support his general idea, but in the case of the angiosperms we are still pretty much in the dark as to just how they evolved from the primitive stock; there must have been a considerable assemblage of intermediate forms. More of his theoretical views are given in a small book *Les Végétaux vasculaires*, which came out in 1947.

Paul Bertrand's bibliography also includes a large number of solid factual

Paul Bertrand. From a photo received from Madame Bertrand.
Studio Phocion, Lille.

contributions, the total number being some 150. His doctoral thesis (*Etudes sur la fronde des Zygoptéridées*) appeared in 1909, with 300 pages and an atlas of sixteen plates. He brought out many shorter accounts on the coenopterid ferns and laid much of the foundations of our knowledge of this heterogeneous and problematical assemblage. He produced about ten papers on the Cladoxylales, which culminated in a comprehensive account in *Palaeontographica* in 1935.

The Bertrands, as well as Bernard Renault and others, were interested in the so-called boghead coals, and although coal studies in general fall outside my central theme, this type of fossil material is so distinctive and so many paleobotanists have been interested in it that I will include a few brief comments.

Charles Bertrand and Renault in 1892 described Permian coals referred to as "bogheads" that they considered to have been formed from an accumulation of algae belonging to the Protococcaceae. This conclusion was confirmed by similar studies made by Henry Potonié with certain north German coals. Paul Bertrand devoted some time to a study of the boghead coals of Autun. E. C. Jeffrey stoutly opposed Renault's interpretation of the supposed algae (which Jeffrey considered to be spores) in the boghead coals, and I remember hearing R. E. Torrey support Jeffrey's views with equal vigor! In 1924, in his otherwise interesting and useful memoir *The Origin and Organization of Coal*, Jeffrey wrote: "It is quite inconceivable on rational grounds that organisms so delicate as Algae should have maintained their integrity, when the structure of wood under the same conditions is entirely obliterated" (p. 23). People such as Renault and Paul Bertrand had a broader knowledge of fossil plants than Jeffrey did and I think were aware of the fact that fossilization processes do not necessarily conform to man's idea of what is "rational." The Russian paleobotanists also had an interest in the boghead coals, and in his history of paleobotany in the U.S.S.R. Kryshtofovich says that the work of Mikhail Zalessky "brilliantly proved the incorrectness of the American scientists Thiessen and Jeffrey, who saw in those formations spores, which he correctly considered to be algae" (p. 38; translated). It is my understanding that, although spores do occur in the boghead coals, more recent studies have upheld the earlier conclusions of the French paleobotanists, although the exact classification of the algae involved may be questioned.

Two additional men made important contributions: Elie Antoine Octave Lignier (1855–1916) and Alfred (Abbé) Carpentier (1878–1952). Octave Lignier served as a teaching assistant at Lille under Charles Bertrand. He received a doctorate in natural science from the University of Paris and was appointed to a professorship at Caen in 1889.

Lignier brought out a series of rather comprehensive articles under the general title *Végétaux fossiles de Normandie*. These included studies of a *Bennettites*, a Liassic flora of Ste-Honorine-la-Guillaume, a *Cycadeoidea*, and some remains described as *Propalmophyllum liasinum* from the lower Jurassic. The last describes fragments of a presumed palm leaf which show the base of the lamina and part of the petiole. Since few, if any, generally accepted fossils of flowering plants have been found below the Cretaceous they have been looked upon with doubt. In his *Plant Life through the Ages* Seward says: "Some French fossils described by Prof. Lignier as *Propalmophyllum* bear a striking resemblance to pieces of a palm leaf, though they cannot be accepted as proof of the existence of the palms in a Jurassic flora" (p. 366). Roland Brown, in his account of *Sanmiguelia*, felt more certain; he said, "I, myself, see no reason for doubting Lignier's identification of his leaves as palms."

In 1912, Lignier wrote a detailed article on *Stauropteris oldhamia*, which he regarded as quite a primitive plant; it would be most helpful and interesting to have more information on this genus in view of the striking nature of the heterosporous species *S. burntislandica*.

Lignier's work was chiefly with living plants and especially concerned the reproductive morphology of certain families of flowering plants. But from a perusal of his eighty-five or so titles, and such of his works as I have read, I think Lignier's importance is chiefly as a theorist; whether right or wrong in his concepts he caused many of his colleagues to pause and ponder a bit. In his brief biographical sketch Stockmans says: "what attracted the attention of the world's leading paleobotanists were Lignier's theories on evolution and proposed classification. Among his theories were the cauloid concept, involving an undifferentiated fundamental element with dichotomous ramification, which underlies the theory of the present telome, and the meriphyte concept, which attributes such great importance to the fibrovascular system of the leaf in phanerogams" (1973, p. 354).

Alfred Carpentier was born at Avesnes and in 1897 entered the University of Lille, where he studied in the faculties of science and theology. He then took charge of the science instruction at the College of Notre Dame at Valenciennes, but in 1905 returned to Lille to take the place of Abbe Boulay, also a paleobotanist of some note. In 1920 Carpentier was appointed Professor of Botany.

During both World Wars, Carpentier found it necessary to leave Lille for some time. In 1917 he went to the Catholic University in Angers, where he taught botany, and he carried on paleobotanical explorations to the north and south of that city. He was evidently a man with a considerable breadth of knowledge. He had a particular interest in mosses and lichens and published several articles on them.

Carpentier wrote a considerable number of short papers on pteridosperm seeds and male pteridosperm or fernlike fructifications, including studies of seeds associated with *Linopteris* foliage in 1912; some interesting compression specimens of *Zeilleria, Discopteris*, and *Renaultia*, written with G. Depape in 1914; seeds associated with *Neuropteris* and *Alethopteris*, in 1925; and a study of *Crossotheca* specimens containing spores, in 1934—to mention only a very few.

I am greatly indebted to several Belgian paleobotanists for aid in my early days, as well as much later when I was able to initiate research with some of the very fascinating plants of the Belgian Devonian. In the summer of 1938, following my year with Hamshaw Thomas at Cambridge, I spent a month in Belgium with a small grant from the Belgian-American Educational Foundation. My time was spent partly in the Natural History Museum in Brussels and partly on numerous field excursions to coal mines with Professors Armand Renier and Suzanne Leclercq. We traveled about, visiting and collecting fine plant materials which I used throughout my teaching career. At the time I am afraid that I did not fully appreciate the generosity of the Chief of the Mine Service of Belgium in taking so many of his days to entertain and educate an American graduate student. It was a lasting pleasure to have been able to make the acquaintance of Suzanne Leclercq, and this ultimately led to a visit of several months with her in 1958 when we collaborated in a study of one of her Middle Devonian plants.

Armand Renier was born at Verviers on June 26, 1876. He wrote several short papers of botanical interest, including one on the articulate cone *Calamostachys ludwigi*, but his paleobotanical studies are chiefly of a stratigraphic nature. His knowledge of the Carboniferous compression floras in Belgium was extensive and he put it to good use. During the course of our excursions in the summer of 1938 he told us of an experience he had had that bore this out very well. The coal seams in Belgium in places are twisted and distorted, and stratigraphic relationships are not always easy to determine. At a particular mine the operators were in doubt as to whether it would be profitable to sink their shaft any deeper. But Renier was able to correlate the plant assemblage present at the lowest level with another horizon in a different part of Belgium, where coal was known to be present below it. Thus, acting on his advice to go deeper, the operators extracted another quarter million tons of coal.

The collecting that summer was quite good, and as there were several of us involved it occasionally became quite spirited and somewhat competitive. One morning near the top of a wagonload of shale that had been brought up for us Miss Leclercq found a fine specimen of a *Calamostachys* infructescence—one with several cones attached—and I must admit that I had a slight feeling of jealousy. But she said "Keep digging," and after an hour or two, when we were near the bottom of the load, I was fortunate enough to find the counterpart to her specimen. It has long been a prized specimen and most useful in teaching.

I believe that Armand Renier was the first to recognize coal balls in Belgium, and Suzanne Leclercq initiated a study of the petrifactions in 1925 with a comprehensive account *Les Coal Balls de la Couche Bouxharmont*. Several more of her studies followed during the next fifteen years, a short one on the fructification known as *Sphenophyllum fertile* being especially important. Scott had studied this some years earlier, using ground sections, and he gives a restoration drawing on page 100 of the third edition of his *Studies in Fossil Botany*. It is incorrect as shown there—one of Scott's few mistakes I should think, and an understandable one in view of the minute size and complex nature of the cone. Using the peel method, Miss Leclercq demonstrated that each sporangial complex consisted of about sixteen slender pedicels, each terminated by a pair of sporangia, and she gave a restoration figure in a short paper in 1936. The extraordinary care and precision evidenced in this study are typical of all of her work, with the result that her publications are not numerous but are of the highest quality.

My real introduction to the study of Devonian plants came in the summer and early autumn of 1958 when I had the opportunity to collaborate with Suzanne Leclercq at Liège in a study of one of her Middle Devonian plants, a species of *Calamophyton*. Harlan Banks had been there for several months, working with her on their classic study of *Pseudosporochnus*, and we "overlapped" by a few weeks. It was a very pleasant experience, exciting and enlightening to me, although I do not especially recommend Liège as a summer resort. My family toured the countryside from Holland to Switzerland, occasionally returning to Liège on the weekends to take me to a castle or two that I wanted to see in eastern Belgium.

I was introduced to the "degaging" technique—working under a low-power binocular microscope with a small hammer, needles, and a syringe to blow away the dust. Some plant fossils, such as dicot leaves, are revealed in their entirety when one first splits open a piece of shale or sandstone. But with most of the Lower to Middle Devonian, and even many of the Upper Devonian plants, one is dealing with fossils that are preserved in three dimensions. In a brief tribute to Suzanne Leclercq's research, Harlan Banks has written: "She early recognized the obvious but frequently neglected truth that part and counterpart constitute a plant—*not* one of these alone. This led her to the equally obvious, but again usually neglected, truth that the three-dimensional character of a living plant lies buried in the matrix" (Banks, 1972, p. 1).

Suzanne Leclercq, at the Berryville, Illinois, coal-ball locality, 1950.

I was not allowed to touch any of the precious Middle Devonian plants until I had served a short apprenticeship with some Carboniferous compressions of lesser importance. I have always enjoyed hand-work along with the head-work, and paleobotany requires both in the laboratory as well as in the field. For my choice the degaging technique is the most fun of all; it requires extreme care and patience and every new plant affords a "microadventure"—

working slowly down into the rock matrix to expose morphological structures never seen before.

We worked very closely together on the *Calamophyton* specimens. Miss Leclercq was reluctant to let me get ahead of her with the investigation, and I was occasionally transferred to some other plant when her schedule took her away from the work for a day. She was not the first, of course, to take a hammer and microchisel in hand to excavate a fossil from the rock matrix, but she demonstrated better than most others the great potential of this simple method. It is especially useful with fossils that are not sufficiently well preserved to allow one to dissolve away the rock matrix, although it does present difficulties if the rock is excessively hard. It is a method that must be experienced to fully appreciate—and probably with Devonian plants! Following my stay in Liège in 1958 I spent a week in Stockholm, where I lived with the Florins. I discussed our work on *Calamophyton* with Rudolf Florin, but he was not very sympathetic, especially with the lack of mechanical devices. Even with the great precision that Florin brought to paleobotany I do not believe that he understood the unique nature of some of the Devonian plants and the extreme care that had to be employed in excavating them. In my own laboratory we have used electric engraving machines with some of the very hard rock from northern Maine, but this would never have been allowed in Liège!

Shortly before Harlan Banks left we were taken to the ancient quarry near Goé, east of Liège, where Miss Leclercq's plants were found. Although the sediments there are much twisted they contain an amazing wealth of plant material which will one day add much more to our knowledge of the Middle Devonian flora.

Among Suzanne Leclercq's earlier studies is one on the Middle Devonian *Hyenia*, on which additional information has been provided more recently by Hans-Joachim Schweitzer. Among the fossils from the Goé quarry, specimens are occasionally found with fragments of petrified axes, and in 1968 Miss Leclercq and K. M. Lele added significantly to our knowledge of *Pseudosporochnus* with a study of the stem anatomy. Her 1951 study of the Upper Devonian *Rhacophyton zygopteroides* is a fine piece of work and basic to more recent investigations. Following exploratory work by Jim Schopf, Tom Phillips and I were fortunate in finding some especially well preserved material of *Rhacophyton* in West Virginia, and this was described in the Linnean Society's special volume that was presented to Tom Harris at the time of his retirement in 1968. The matrix in which the West Virginia plants are preserved is similar to that of Goé and nicely suited to the degaging method.

Miss Leclercq once told me that she took up the study of the Middle Devonian plants somewhat reluctantly at the urging of Armand Renier. She had intended to go on with studies of coal-ball plants, and although much remains to be done with the latter, there are few deposits in the world as richly rewarding as the Goé quarry. I am quite sure that she did not regret the shift in her research program.

During the course of my period of study with Miss Leclercq in Liège in 1958 I spent a week going through some of the fossil-plant collections in the Brussels Natural History Museum. The collections are abundant, superb, and readily accessible to the visitor, all of which is due in large part to the efforts of François Stockmans, and his cordial hospitality always made it a pleasure to work there.

Stockmans is the author of many studies dealing with the Paleozoic plants of Belgium, of which I will select two as representative. He has in a sense bracketed the Middle Devonian work of Miss Leclercq with comprehensive studies of the Lower and Upper Devonian. The early Devonian localities are found for the most part in a line running through central Belgium west from Liège through Namur to Mons. The plants include *Taeniocrada, Sciadophyton, Psilophyton, Drepanophycus*, and *Sporogonites exuberans*.

I was especially attracted by a large collection (some sixteen drawers) of *Sporogonites* specimens. I went through them rather quickly one day, and on the train back to Liège that evening I began to wonder why, in so many of the specimens, the slender axes, each of which bore a terminal sporangium, always had a parallel alignment. The next day I went through the collection very carefully and found several specimens in which the axis appeared to arise from a carbonaceous compression. Dr. Stockmans kindly allowed me to investigate the matter and the result was a restoration drawing that now appears on the cover of Volume 2 of the *Traité de paléobotanique*; the plant is referred to in the volume under both "Bryophyta Problematica" and the family Sporogonitaceae of the psilophytes. I have assumed that the authors involved were as puzzled as I was, and still am, about the nature and affinities of the plant; and I hope that someone else will one day go through the collection to check my concept of the plant.

In 1948, Stockmans brought out a fine memoir on the Upper Devonian which includes descriptions of some very interesting plants. Some of these, such as *Archaeopteris, Rhacophyton*, and *Aneurophyton*, are rather well known; others are less well known and exciting, such as *Barinophyton*; and there are primitive seed (?) organs (*Moresnetia* and *Condrusia*) included in the collections. Stockmans' interests are strongly stratigraphic and I suspect that much more remains to be learned, in a morphological vein, about these plants. He has given us a tantalizing lead.

The most important representative of fossil botany in Holland is Willem Josephus Jongmans (1878–1957)—a distinctive personality and a man who contributed much to several aspects of paleobotany. He was born at Leyden and entered the University there in 1898 to study pharmacy, but his interests soon developed in a more purely botanical direction and were continued at

W. J. Jongmans, 1957. Courtesy of H. W. J. van Amerom, Geologisch Bureau voor het Mijngebied.

Munich, where he received the doctorate in 1906. During his late student days he collaborated with Professor J. W. G. Goethart in the preparation of some 450 distribution maps of Dutch plants. In 1906 he was appointed curator of the public herbarium in Leyden. In the same year he made the acquaintance of W. A. J. M. Van Waterschoot van der Gracht, who was in charge of mineral prospecting, and this meeting started Jongmans on his long career as a Carboniferous paleobotanical stratigrapher. In 1919 he left the herbarium post for service with the Government Geological Service, which led to a position as head of the Heerlen office.

Jongmans' travels were extensive. He visited coal basins in West Africa, North America, England, Belgium, France, Austria, Russia, Spain, and Turkey, and as his reputation developed he received collections from many places, including East Asia, China, New Guinea, Brazil, and South America.

In 1908 he went to Edinburgh, where he spent some time with Robert Kidston, as I have noted previously.

Jongmans was responsible for some unique publications. One that I think ranks very high in the annals of fossil botany is his small book *Het wisselend aspect van het bos in de oudere geologische formaties* (The Changing Aspect of the Forest in the Older Geological Formations). This appeared in 1949 as part of *Hout in alle Tijden* (Wood over the Ages), a grand collection of many of the classic landscape restorations from paleobotanical literature, individual plant restorations, and photographs of fossil trees and stumps in place. It deserved to be translated into other languages, but the illustrations are extremely informative and useful.

In 1913 he brought out the first part of the plant section of his *Fossilium catalogus*. This is the most comprehensive encyclopedic work that exists in paleobotany, and many people have contributed to the preparation of the various parts dealing with particular plant groups. Part 30 came out in 1957, the year Jongmans died, and after that it was edited by S. J. Dijkstra. The most recent number that has come to my attention is Part 87, *Ginkgophyta et Coniferae* (1975). Jongmans is also especially well known for the part he played in organizing the Heerlen Carboniferous Stratigraphical Congresses, the publication of which has been referred to as "The Heerlen Bible of the Carboniferous."

That Jongmans retained a vigorous interest in living plants is well borne out by the following incident, related to me by Dr. H. W. J. van Amerom; it took place sometime in 1943 or 1944 when Jongmans, then in his sixties, and several colleagues were searching in a small pond for specimens of *Isoetes* (the quillwort):

> It was a cold day and it was not pleasant to stand in the water and not find the plants. Finally Dijkstra found it, because as a botanist he was familiar with the plant. Jongmans was glad and honor was saved! and he could put on his clothes. At that moment Dr. Diemont, the forester of the State Forest Administration, passed by, saying "Who the ——— is trampling upon our *Isoetes*!" Then he saw it was Prof. Jongmans. Several years later he married one of his daughters. [Personal letter]

I have been informed that Jongmans developed rather strong likes and dislikes for people at the first meeting, and that he lived pretty much unto himself. Walther Gothan seems to have been one of the few that he worked closely with and van Amerom says: "This co-operation lasted over the years and could stand as an example. When both gentlcmen did not know how to determine a fossil, it was Gothan who came to a solution by exclaiming 'Neue Sorte' [new species or 'art']. This was not according to the character of Jongmans, who made indeed only a few new forms and arts" (personal letter).

His biographer Van Rummelen (1958) says that Jongmans did not concern himself very much with the affairs of society, for which he could not spare the time from his many activities. And although from 1932 on he held a professorship in paleobotany at the University of Groningen, his interest in students seems to have been minimal. His lectures are said to have been presented with little sympathy for the beginner, and attendance soon dropped to the vanishing point. I inquired of Dr. van Amerom about Jongmans' hobbies or interests outside of his scientific life, and the answer I received was that "his hobby was paleobotany."

15

Fossil Botany in the Soviet Union

THE tremendous area encompassed by the Soviet Union, combined with the great explorations and scientific developments in recent decades, has produced a very extensive paleobotanical literature, and I will not even apologize for the inadequacies of my treatment. At best I can only mention a few of the leaders and refer the reader to some literature sources that may contribute to a more detailed understanding of paleobotanical developments there. Rather than attempt to give any semblance of a "history," I will present a few personal impressions of present activities and summary notations taken largely from historical accounts by Afrikan Kryshtofovich and several others. Unless otherwise indicated the quotations in this chapter are translations of passages in Kryshtofovich's *History of Paleobotany in the U.S.S.R.* (1956); quotations from other Russian works have also been translated. Page numbers refer to the Russian texts, but titles are given only in English.*

In the autumn of 1958 I spent ten days in Leningrad and Moscow visiting paleobotanists with whom I had been in correspondence for some years. Ten days is a short time but I think I saw some representative samples of what was going on and I examined some of their fine collections. In Leningrad I spent several days at the Komarov Botanical Institute. Pavel Dorofeev showed me his collections of Quarternary and Tertiary seeds and fruits, and I shall have more to say of his work, as well as that of his predecessor, P. A. Nikitin, in Chapter 16. Valentina Samylina showed me the interesting and diverse collections of Cretaceous plants from areas close to the Arctic circle. Antonina Turutanova had large collections of Jurassic plants, some of which looked familiar while others were quite new to me. Since I was at that time

* I wish to make a special acknowledgment here of my indebtedness to Frederick T. Pope, Jr., of Alexandria, Virginia, for his aid over the past few years with the Russian literature. His translations have been invaluable and his interest in my project has been most encouraging.

working primarily with petrified Upper Carboniferous plants I was especially interested in Natasha Snigirevskaya's work with coal-ball plants from the Donets, to which I will return on a later page.

The Komarov Botanical Institute and its associated Garden is a large and impressive organization, and for its 250th anniversary several of the paleobotanists prepared a summary account of the fossil-plant collections (published in the Soviet *Botanical Journal*, volume 50, for 1965). Their holdings, which I believe are the largest in the Soviet Union, included some thirteen hundred collections at that time. Some forty-one collections were deposited in the museum of the Botanical Garden during the nineteenth century, about twice that number were added during the first seventeen years of the twentieth, and following the Revolution in 1917 the number has increased at a very much faster rate. The authors give most of the credit for the accelerated work of the first decade and a half of this century to I. V. Palibin. Under his leadership a Paleobotanical Section was established in the Department of Systematics and Geography of Higher Plants. During the years of World War II progress was, of course, greatly slowed, and several of the more important fossil botanists, including K. K. Shaparenko and A. V. Karmolenko, died in the defense of Leningrad. The severity of those times may be realized to some extent by the fact that Leningrad was under siege for nine hundred days, and although the German army never succeeded in entering the city, there was vast destruction. During my visit I took an afternoon off from paleobotanical pursuits to visit the Tsar's Summer Palace, located on the shores of the Baltic, about a forty-minute drive away. At the outskirts of the city we passed through a no man's land, a five-kilometer band of desolation and destruction that existed between the Russian and German lines.

Hundreds of collections have been added at the Komarov since about 1950 from various geographical areas and geological horizons of the Soviet Union. To mention only a representative few: a collection of strobili of the Williamsoniaceae; rich Lower Cretaceous collections from the Lena and Kolyma river basins and some of the Arctic islands, gathered by V. A. Samylina, I. N. Sveshnikovaya, and L. Y. Budantsev; a collection of some two thousand cuticle preparations prepared largely from Mesozoic plants and a comparable collection prepared from living gymnosperms and flowering plants; and large collections of fossil woods and coal balls.

After considering several ways of presenting this short account I thought it most informative to deal first with several of the key people who have developed Soviet paleobotany in the present century and then go back in time to the pre-Soviet period of the nineteenth century. Among the numerous contributors in the first half of the present century three seem to stand out: Maria

F. Neuburg, Mikhail D. Zalessky, and Afrikan N. Kryshtofovich. Of these, the last is one of the very great people in the three-hundred-year history of paleobotany, and I will accordingly devote what I hope is appropriate attention to his many and varied contributions to botany, geology, and paleontology.

Afrikan Kryshtofovich (1885–1953) was born at Krishtopovka, near Padlograd, his father being a bank teller. In 1903 he entered the Natural Sciences section of Novorossisk University in Odessa, and completed his studies there in 1908. His botanical studies and publications began early; as a student he collected plants in the vicinity of Krishtopovka and in the Crimea and wrote two works before completing his university courses: one on the spring vegetation of Krishtopovka in 1906 and another on the vegetation of Laspa and the Baidar valley in 1908. At the suggestion of G. I. Tanfil'ev and V. D. Laskarev he began a study of Jurassic and Tertiary plants of the Ukraine in 1907. It is, however, something of a paradox that Professor Laskarev informed Kryshtofovich that he knew nothing about fossil plants and that he would be "on his own." Thus, like many others who have made great names for themselves in fossil botany, Kryshtofovich gained much of his knowledge by his own efforts.

His extensive travels brought him into contact with many of the leading people of the time and enabled him to acquire a vast first-hand knowledge of fossil plants in the field. One of his biographers says: "In the period before the first World War the iron curtain was in one of its temporary states of disrepair, and the young Kryshtofovich took the opportunity, between 1909 and 1915, of visiting Egypt, Japan, Austria, France, England, Germany, Italy and Sweden" (Edwards, 1956, p. 127). In Egypt, Japan, and Europe he had an opportunity to study many of the classic collections and make the acquaintance of Henry Potonié, Zeiller, Seward, Arber, and E. Iokoyama. In 1915 he spent some time in Stockholm where he met Nathorst, Florin, Halle, and Ernst Antevs; and in 1921–22 he was in the Philippines studying the flora of Luzon, thus gaining a considerable knowledge of living tropical vegetation. He described a fossil shark's tooth from Mindoro Island and collected foraminifera which were turned over to Professor K. Yabe.

In 1912 he was appointed assistant geologist to the Geological Committee, thus initiating a paleobotanical program in that organization. This led to various expeditions to Central Asia, to a teaching post in Vladivostok in 1922–1924, and in 1924 he began work at the Botanical Garden in Leningrad. That institution evolved into the Komarov Institute, where he served until 1950.

Kryshtofovich's studies were directed chiefly toward Tertiary and Cretaceous floras. Among his total of some 350 works are a considerable number of general studies of both national and international importance. He produced a series of geological descriptions of the various regions of the U.S.S.R., which resulted in *A Geological Review of Countries of the Far East* in 1932; and in 1941 he brought out a *Catalog of Fossil Plants of the U.S.S.R.*, which

Afrikan Kryshtofovich. Courtesy of the Komarov Botanical Institute.

included a listing of all known fossil plants with a detailed bibliography of the works of the paleobotanists involved. He was an accomplished linguist and compiled an Anglo-Russian geological dictionary, but I do not know whether this was ever published. He translated Potonié's book on the origin of coal, a work on the geology of China by Lee-siu-huan, and, in collaboration with V. M. Kryshtofovich, Seward's *Plant Life through the Ages*, to which comprehensive additions were made. As a result of his teaching at Odessa, Leningrad, and other universities he brought out his *Course of Paleobotany* in 1933, which went through several editions, later taking the title *Paleobotany*, and is a most detailed and comprehensive textbook. His students were numerous and include A. I. Turutanova, M. I. Brik, V. A. Prinada, and V. A. Vakhrameev, who have contributed much to our knowledge of Mesozoic floras; and P. I. Dorofeev and K. K. Shaparenko, who have worked with various aspects of Tertiary floras.

Aside from his own numerous contributions to botany and geology Kryshtofovich is probably equally important for his general influence on the development of these natural sciences in the Soviet Union. Budantsev notes in his "One Hundred Years of the Investigations of Fossil Floras of the Arctic":

> "Under the direct influence of A. N. Kryshtofovich were undertaken, beginning in 1959, perennial investigations of fossil Arctic floras. As a result, an enormous phytopaleontological material was gathered and partially developed, including several thousand imprints of leaves, the remains of woods, coniferous cones, and so forth, from deposits of the Lower Carboniferous, Lower and Upper Cretaceous, and the Paleogene of Spitsbergen, Franz Josef Land, New Siberian Islands, delta of the Lena River, etc. . . . In particular, on Franz Josef Land a rich Aptian-Albian [?] flora was discovered, large-leaved cycadophytes and conifers. Ideas about the composition of Late Cretaceous floras of the New Siberian Islands, first studied by I. F. Schmalhausen, were significantly augmented; these floras turned out to be incomparably richer and more variegated than one could guess. On Spitsbergen at the time of two expeditions in 1959 and 1967 there were found several new deposits with the richest remains of Tertiary plants. This provided the opportunity to undertake a complete revision of former determinations of extinct species of plants that existed in the Paleogene on Spitsbergen. [1969, pp. 488–489]

Of his own numerous and varied contributions, Kryshtofovich relates the following in his *History*: "But no matter how important these finds were, perhaps the most remarkable event for the history of Russian paleobotany, was the discovery of a Cretaceous flora with angiosperm plants, formerly completely unknown in our country. If such plants had been found earlier, then they were usually considered as Tertiary (O. Heer), so great was the skepticism concerning the presence of Upper Cretaceous floras in Russia" (1956, p. 42).

W. N. Edwards concludes his short biographical sketch of Kryshtofovich with a well-earned tribute: "He inspired numerous pupils and assistants, and in his friendly geniality as well as in range of interests, approach to palaeobotanical problems, and prolific output, he may be considered as the Russian counterpart of our A. C. Seward" (1956, p. 128). It may be fair to go even a little farther than this; both Seward and Kryshtofovich were incredible compilers and synthesizers of paleobotanical information and literature, but I believe that Kryshtofovich was somewhat greater as an original investigator.

Immediately after visiting the Komarov I went to Moscow with the primary objective of spending some time with Maria Neuburg in her laboratory at the University. Although Moscow has much of interest for the visitor in the Kremlin, St. Basil's, the subway system, the theaters, and so on, it seemed to me much like any other big city, without the distinctive charm of Leningrad. This impression was tempered by my visit with Mrs. Neuburg; although a woman with a very stern appearance she proved to be extraordinarily kind and I learned a good deal in the two or three days that I spent with her. She was one of the important contributors to fossil botany and it is a pleasure to present a brief summary of her many achievements.

Maria E. Neuburg (1894–1962) was born of a family exiled in Krasnoyarsk. Her childhood and youth were difficult times; she lost her father, mother, and older brother when she was fourteen and had to make her own way after that time. Her family seems to have had little enthusiasm for the Tsarist regime. Her older brother took part in a revolution in 1905 and was exiled to Turukhansk, and during her student years Maria was active in the social-democratic movement. In spite of her difficulties she managed to obtain a good education through upper-level women's courses at Tomsk, and she later taught at the University there.

While still a student, in 1916, she took part in an expedition to northwestern Mongolia, conducting botanical and geological studies in the region of Khan-Keko, and gathering paleobotanical materials which were used in her doctoral dissertation.

From 1921 to 1933 she served with the Geological Committee in Leningrad. She engaged in several expeditions and developed an extraordinarily fine understanding of the paleobotany and stratigraphy of the Carboniferous Kuznetsk Basin. This program, which lasted from approximately 1921 to 1933, established her reputation as a leading stratigraphical paleobotanist. She demonstrated that the Kuznetsk included a great series of deposits, from the Carboniferous to the Jurassic, and her work resulted in a monograph, *Upper Paleozoic Flora of the Kuznetsk Basin*, published in 1948, of which Meyen says: "The significance of this monograph is difficult to overestimate. In it under one point of view are decisively adduced all the materials on the Upper Paleozoic flora of the basin, both gathered by the author herself and obtained from other sources. In contrast with the often hurried and careless descriptions that were scattered through the works of many authors of that time, this study was masterfully carried out, with consideration for all necessary rules of monographic description" (1963, p. 152).

Mrs. Neuburg came into sharp conflict with the previous views of Zeiller, and into even stronger conflict with Zalessky, who did not recognize this great stratigraphic diversity. Feelings apparently ran rather high about this matter, and Meyen refers to "violent criticism on the part of M. D. Zalessky which sometimes appeared in the press as unscientific attacks."

Zalessky established a good many new species and genera on the basis of fossil-plant material of questionable value, and I do not believe that there can be any doubt about Maria Neuburg's superiority as a knowledgeable and very precise worker. And from my own visit with her, although she was most cordial, she impressed me as one who would vigorously defend her own views regardless of who her opponent might be.

In 1944 she began an investigation of the Upper Paleozoic flora of the Pechora coal basin. The sections on the ginkgophytes and lycopods were completed; those covering the horsetails, cordaites, and other seed-plant groups remained in manuscript. Kryshtofovich summarizes her work in these great stratigraphical studies rather nicely: "she appeared in the study of the Kusbass and Angarida as a brave innovator, who experienced many confrontations with representatives of traditional points of view, brilliantly proved her own correctness, and thus charted the right paths for further works in the study of the Carboniferous basins of Siberia and Kazakhstan" (quoted in Meyen, 1963, p. 153).

In a vein quite different from the larger floristic works is her splendid 1960 memoir on the mosses from the Lower and Upper Permian sediment of the Kuznetsk, Tunguska, and Pechora basins. This is a unique piece of work, one of the twentieth-century classics in paleobotany, and I believe by far the most comprehensive study to date of fossil mosses, at least from the older geologic periods. It is based on specimens taken from drill cores, is a work of 104 pages and seventy-eight plates, and describes fourteen species, including twelve that were new, and nine new genera. Eleven species are placed in the Bryales or are regarded as closely related to that order; of special interest are three species that are classified in the Protosphagnales and are either primitive sphagnums (peat mosses) or immediate progenitors of the Sphagnales. A condensed English version was published in the *Journal of the Palaeontological Society of India* in 1958.

During the course of my visit with Mrs. Neuburg she showed me the specimen which forms the basis of her short paper on the curious and unique seed plant that she has called *Vojnovskya paradoxa*. I included in my textbook

Studies in Paleobotany a figure of the plant which is taken from her original article of 1955. *Vojnovskya* is quite unlike any other seed plant and the erection of a new order, the Vojnovskyales, seems thoroughly justified. It is to be hoped that additional specimens may be found in the future that will yield more information about this distinctive order.

Mikhail Dimitrievich Zalessky (1877–1946) was the oldest of the trio that I am considering here. I never met Zalessky, but I have had occasion to examine a considerable number of his works, some casually and some more carefully. I am not sure that I am correct in this appraisal but he impresses me as having played a part in Russian-Soviet paleobotany that was somewhat similar to that of E. W. Berry in this country. He was impressed by and concerned with the overwhelming volume of work to be done in a period when paleobotanical investigators were rather few, and encouragement and financial support from the powers-that-be were minimal. He described, perhaps hurriedly, many new species and genera of fossil plants that are of questionable value, but he worked to establish a basis for those who followed him to build upon.

Zalessky was born in the city of Orel in central Russia. He studied at the gymnasium in his native city and then taught natural sciences at St. Petersburg University, where he also graduated in 1900. He also taught for a short time at the High School of Mining in what is now Dniepropetrovsk and it was there that he initiated his study of Carboniferous floras. From 1918 on he lived much of the time in Orel, where he served as a staff member of the Geological Committee. Like many of his countrymen his life was not entirely an easy one. His own splendid library and much of his collections were lost in the burning of Orel. Kryshtofovich says in his biography: "In 1943 he was taken by the retreating German army to Berlin, where he suffered all the horrors of siege, hunger, and hiding in cellars during the bombardments of the last days. Broken in health, he returned to Moscow in 1945, where he underwent a serious surgical operation, afterward passing some months in the best sanatorium in the country. He was longing to go back to his studies and early in December 1946 returned to Leningrad, where he died" (1949, pp. xlii–xliii).

In a somewhat brighter vein, the following is taken from a letter of January 2, 1946, that I received from Professor F. W. Oliver:

> In the spring of that year [1910?] I had arranged a small international palaeobotanical excursion to visit the principal English localities: coal measures in Lancashire; Mesozoics on the Yorkshire coast and Whitby, and tertiaries in the Cromer beds. I remember Bertrand *fils* and Zalessky being of the party. Zalessky, from interior Russia, had never seen the sea,

so to speak, and when we reached the Cycadeoidia localities on the sea shore he went mad and collected all the seaweeds he could discover—I recall him loaded with Fucaceae—with trailing laminaries, Araria, etc. He announced his intention of forsaking fossils for seaweeds; however he reverted on returning to Russia!

M. D. Zalessky. Courtesy of the Komarov Botanical Institute.

From 1901 to 1910 Zalessky produced a series of works on the Donets Basin which included two monographs on *Sigillaria* and *Lepidophyllum*. He also published numerous works of a morphological and anatomical nature. However, his major area of research was centered on a general elucidation of the Paleozoic floras of the Donets and Kuznetsk basins. He was also interested in the structure and origin of coals.

Kryshtofovich relates the following about Zalessky's work:

In spite of the divergence in points of view with other paleobotanists on the age of several Paleozoic floras, the services of Zalessky relative to

> their study are invaluable not only from a systematic, but also from an anatomical standpoint. Working until 1918 in St. Petersburg, Zalessky afterward moved all his work to his own home town of Orel, where, having an excellent library and set of instruments; [and] . . . going out annually for field work until the last years, when he was already in old age. One cannot overlook the fact that in his work Zalessky remained almost isolated, and although his works lay at the base of all further works of paleobotanists who studied our Paleozoic flora, he had no students, except his colleague of many years, E. F. Chirkova.
>
> Zalessky's great achievement was that he made Russian paleobotany known to the outside world, and, conversely, introduced to his fellow-countrymen the discoveries of foreign paleobotany. [P. 59]

According to Kryshtofovich's *History*, there are several pre-nineteenth-century references to fossil plants in the area now encompassed by the U.S.S.R. The rather faltering development of scientific knowledge throughout the world at that time seems to have followed a similar pattern in Russia, but I think we have heard less about it because of the language barrier and the rather sparse amount of communication between Russia and the rest of Europe. I will simply cite what seem to be a few representative examples.

The distinguished Russian scientist Michael V. Lomonosov (1711–1765) "repudiated any mysticism and Biblical mythology for the explanation of the nature of organic fossil remains in the earth's crust." In his work *On the Earth's Layers* of 1763 he dealt with the processes involved in the formation of hard coal, peat, and amber. He described peat as being composed of some kind of "underground moss"—and seems to have had a pretty good understanding of its origin. He regarded hard coal as having been formed under the influence of pressure and heat from peat. He understood amber to have originated from the resin of trees rather than being of inorganic origin, a theory many geologists of the time subscribed to. He noted the inclusion in it of insects and plant fragments.

I. I. Lepekhin (1740–1802), a botanist by profession, wrote in his "Daily Notes" of 1771 of a fossil tree near Simbirsk on the Volga, and of fossil trees in a mine in the Orenburg province.

In the course of a six-year trip in Siberia (1768–1774), P. S. Pallas (1741–1811) reported a petrified tree near Krasnoyarsk, and he is credited with having the honor of "the first factual determination of fossil flora—leaf imprints on the river Talakoka (now Talovka), which he determined to be alder leaves."

With the start of the nineteenth century and the initiation of what is often called the "scientific period," several names of importance enter into the development of Russian paleobotany. Some of the literature from the western European countries was reaching Russia, although how widely it was distributed I have not determined. For example, A. N. Karpinski, in 1829, translated a part of Brongniart's *Prodrome*.

Kryshtofovich refers to Y. G. Zembnitskii (1784–1851) as the pioneer in Russian paleobotany. In 1825 Zembnitskii published *A General Review of Fossils*, and in 1830 *A Review of Fossil Plants*. He recognized that many of the fossils are quite different from modern plants and that different geologic horizons contain different assemblages of plants. He also noted that, "according to the distinction of fossil species, naturalists conjecture about past changes in the globe, about the differences of its composition at that time, when other plants were growing, and with greater or lesser accuracy draw conclusions about the temperature, about the extent of continents and waters, about the characteristics of the soil and the atmosphere" (p. 16).

Zembnitskii was a professor of botany at St. Petersburg University as well as a teacher of the Corps of Mining Engineers. His contributions seem to have been chiefly those of a compiler of information, while several others who followed shortly after him initiated original investigations. Among these were G. I. Fisher de Waldheim (1771–1853), K. E. Merklin (1821–1904), S. S. Kutorga (1805–1861), and Eduard Eichwald (1795–1876).

Fisher held the Chair of Natural History at Moscow University and, although primarily concerned with entomology and fossil animals, he described a series of fossil plants from the environs of Moscow.

Kutorga described a Permian flora of the Priural'e in 1838. He was a zoologist, geologist, and paleontologist and Kryshtofovich says that "he remained a creationist to the end of his days. This did not prevent him, however, not long before his death, from acquainting his students with Darwin's theory, which had just been published."

Eduard Eichwald seems to be the most important figure of this period in the development of paleobotany in Russia:

> He was the first scientist who made a great contribution to Russian paleobotany and who actually laid its foundations. He described a great number of plants from the Carboniferous and the Permian systems of the Ural region and the Donets Basin, as well as the fossil flora of Moscow's environs, Mangyshlak, and the Aleutian Islands. Eichwald deserves credit for the discovery and description of the Jurassic flora of Kamenka near the city of Izyum, one of the classical locations. In his four-volume work *Paleontology of Russia* (*Lethaea Rossica*), which came out in 1860–1868, Eichwald reviewed all the fossil flora of Russia, accompanying his descriptions with excellent drawings.

Eichwald was born in Litau, Latvia, his father being a private tutor to the family of a Baron of Courland and later a lecturer in modern languages and natural history. Eduard began his university studies at Tartu in 1814 but soon transferred to Berlin University, where he studied medicine and natural science. He traveled and studied in Switzerland, France, and England and returned to Russia in 1819, receiving an M.D. degree at the University of Vilnius. He practiced as a physician for two years and then spent several years as a professor at Tartu, Kazan, and Vilnius lecturing on botany, mineralogy, anatomy, and obstetrics.

He was clearly a man of broad education and interests and well-traveled. In addition to the journeys of his student days, he later visited Scandinavia, Italy, Algeria, and Persia, where he made extensive natural-history collections. His primary fields of interest were geology and paleontology; he became the leading figure in paleontology in Russia, and as his reputation grew his own collections were swelled by others sent to him by various people.

As to his basic philosophy, his biographer Tikhomirov (1971) says:

> Eichwald several times changed his ideas on the development of the organic world. Originally, in the 1820's, his views were transformist: he thought that all types and classes of animals originated from a primal protoplasm. Later, under the influence of proponents of the catastrophist theory, Eichwald wrote of the periodic destruction of every living thing and the subsequent appearance of a completely new fauna and flora as the result of an act of divine creation, this new life being more highly organized than the one that had previously existed. Following the publication of Darwin's *Origin of Species*, however, Eichwald renounced his catastrophist and creativist concepts and became an adherent of evolutionary theory. [P. 309]

K. E. Merklin (1821–1904) was one of the founders of paleobotanical research in the Komarov Institute. In 1855 he started work on a collection of fossil woods gathered from various parts of Russia; the results were published under the title *Paleodendrologicon rossicum* and the collections are preserved in the Komarov.

A considerable portion of nineteenth-century Russian paleontological studies were conducted by outsiders. In some cases foreign geologists were brought in to conduct explorations and in other cases collections made in the country were sent to paleobotanists in other European countries. It is my assumption that this policy followed the usual pattern in "undeveloped" countries, and, whether or not it was justified, the Russian government seemed to favor the great names of western Europe. Men such as Humboldt and Murchison made expeditions into Russia, and the Russian scientists themselves were not always able to participate in them. Fossil plants from Russia reached Brongniart, Geinitz, Stur, and other eminent workers. The collections were not always precisely and carefully labeled, and, as might be expected, this led to numerous mistakes as to locality and geologic horizon. Oswald Heer received a significant number of collections, which he rather freely assigned to his favorite Miocene period! Some of these later proved to be of earlier Tertiary or Cretaceous age, and thus a certain amount of error and confusion entered into geological concepts of the time. However, in his *Catalogue* of 1941 Kryshtofovich does pay tribute to Heer:

> Heer's description of the Jurassic flora of Irkutsk province and of the Amur territory, based on collections by F. Shmidt and P. Glen, and his descriptions of the floras of Simonova, Bureya, Poset, and Khanka, taken by him as Tertiary, not only lay at the foundation of all further works on the paleobotany of Asia, but also served as the cornerstone of all paleobotany, in spite of a series of mistakes and omissions, due in part to the insufficient accuracy of collectors, in part to mistakes in classification which led to an incorrect determination of age (Tertiary instead of Cretaceous on Sakhalin and in Bureya and Simonova). [P. 11]

In what Kryshtofovich refers to as the third phase in the development of paleobotany in Russia, a systematic geological survey began to be carried out, first within the limits of European Russia. It was organized by the Geological Committee, founded in 1882. Soon after, in conjunction with the construction of the Siberian railway, geological survey parties began to work in Siberia. The most outstanding figure in paleobotany in this period is Johannes T. Schmalhausen (1849–1894). Although his life was not long, he accomplished much.

Schmalhausen recognized and described (1879) a Jurassic flora from the Kuznetsk, Tunguska, and Pechora basins that had previously been regarded as Upper Paleozoic in age, and he made an important contribution in his study of the anatomy of the Cretaceous fern *Protopteris*. Of particular importance are his contributions to the knowledge of the Tertiary floras in the European parts of the country. Prior to this time the available information was chiefly limited to isolated fossil remains from Usha near Kamyshim, from Molotych in Kursk province, and some limited data from the work of Professor A. S. Rogovich on the Ukranian Paleogene.

He also described a Lower Carboniferous flora of the Urals and a Permian flora of the Ural region (1887). His last work (1894) was concerned with an Upper Devonian flora from the region of the River Karakuba in the Donets Basin, and included descriptions of two species of *Archaeopteris*.

Schmalhausen was born in St. Petersburg where his father was an assistant in the Royal Library. He entered the University there and as a student won a gold medal in 1870 for a study of the flowering stage of certain grasses. His education was continued in Zurich, where he studied with Oswald Heer, and at Strasbourg in De Bary's laboratory, where he presumably advanced his knowledge of plant anatomy. In 1876 he returned to Petersburg where he was appointed "elder curator" of the herbarium in the Royal Botanic Garden; in 1877 he went to the University of Kiev as Professor of Botany and Director of the Botanical Garden. He seems to fit into our story as an important "transition" leader between the days of the old and the new regimes. He died at an early age in 1894, leaving no direct protégés. His only student, insofar as I have been able to determine, was N. V. Grigor'ev, who began a study of the Carboniferous and Jurassic floras of the Donets Basin but was drowned in 1899 in the course of his field studies.

Aside from Zalessky, the most important immediate successor to Schmalhausen was Ivan Vladimirovich Palibin (1872–1949). I think his contributions are best summarized in the following lines from Kryshtofovich's *History*:

> Palibin, like Zalessky, had no teachers in the field of his specialty, although he encountered the paleobotanical work already left by Merklin. His studies on Tertiary floras of Russia began in conjunction with the works of the Geological Committee, with which he maintained a close connection all his life, and of which he was an employee for two years (1931–1932). The object of his first work was the Tertiary flora of the Kursk region (1901), in which, continuing the work of Schmalhausen, he attempted to arrive at a decision on the question of the age of white sandstones of southern Russia. Having given attention to Paleocene flora of Mt. Usha near Kamyshin and to Paleogene floras of southeastern Russia (1905), I. V. Palibin transferred to the study of Tertiary floras of the Asian part of the country and, later, of the Caucasus, in which region he has the honor of being the first investigator of fossil floras.
>
> By this time fifty years had passed since O. Heer had described the first seven species of Tertiary plants from the Priural'e. In two works (1904 and 1906) Palibin significantly enriched the number of forms known then. One after another he published works on Tertiary flora of the Commander Islands, of Sikhote-Alin, and of Manchuria, which appeared a long time after the publication in 1878 of the first reports about the flora of the Far East.

After the work of Palibin, Kryshtofovich, and Maria Neuburg, the numbers of paleobotanists and the diversity of their achievements in the Soviet Union expand far beyond the scope of my account, as stated at the beginning of this chapter. However, certain contributions and the workers concerned are mentioned in other chapters.

Natasha S. Snigirevskaya, at the Botanical Congress, Leningrad, 1975. Photo by James M. Schopf.

I come back to the starting point of my own brief travels in Russia, and claim an author's privilege to conclude with a few lines about a Soviet paleobotanist who has been most helpful to me, whose work I respect, and who has generously served many in her own and other countries—Natasha Snigirevskaya. She was born at Ufa in Bashkiria on August 9, 1932, and studied at Leningrad State University from 1950 to 1955. The following year she took a position at the Komarov Botanical Institute; although Zalessky, some twenty-five years earlier, had studied coal-ball plants, I think she should be regarded as the real pioneer in this phase of Soviet paleobotany.

She produced a fine article on the lycopod leaves much like the work that was done some years ago by Roy Graham in this country. That was followed by a demonstration of gametophytes in several species of *Achlamydocarpon*. In 1961 and 1962 she brought out two papers on *Botryopteris* and then several on *Sphenophyllum* cones. She has a good understanding of European and American coal-ball floras and has also written articles on techniques and popular articles on coal-ball plants.

When I visited the Komarov in 1958 I received a very cordial reception from many of the botanists there, but I am especially indebted to Natasha Snigirevskaya for making my visit both scientifically rewarding and pleasant. She is a good linguist, a very intelligent woman, and most generous with her time; if it were not for the latter and problems with her health in recent years her publication record would be much longer. I know that she spent a great deal of time on the preparations for the 1975 International Botanical Congress in Leningrad. Many of us are indebted to her for her hospitality. Quite recently Professor Tom L. Phillips spent some time with her at the Komarov making comparative studies of the Soviet and American coal-ball floras; it is to be hoped that this collaboration will continue.

Beyond this brief sketch of certain people and lines of investigation in the U.S.S.R. I can only refer the reader to several summary accounts for more detailed historical information.

To the best of my knowledge the most comprehensive source is Kryshtofovich's 92-page *History of Paleobotany in the U.S.S.R.*, which came out in 1956. Among the shorter accounts that I have found especially informative are: "The Progress and Development of Paleobotany in the U.S.S.R. for 25 Years" written by Kryshtofovich in 1943; "Results of the Investigations of Paleozoic Floras of the U.S.S.R. during 50 years (1917–1967)", by S. V. Meyen; and "One Hundred Years of the Investigations of Fossil Floras in the Arctic," by L. Y. Budantsev (1969). All of these were published in Russian, but I have deposited English translations in the Paleobotanical Library of the U.S. Geological Survey.

16

Of Seeds and Fruits and Flowers

Although I have tried to develop some semblance of a logical sequence, or sequences, in preparing this account, there are some aspects of paleobotanical history that do not fit perfectly into the various patterns that I have considered possible or useful. The subject material presented here could have been spread through several other chapters, but I think its novel features warrant making a separate section of it. Fossil seeds and fruits are distinctive in their manner of preservation, special techniques have been developed to study them, and those who begin by examining them casually tend to become specialists in this branch of the science. Like most of the other chapters this one is a sampling—and I hope a representative one—from the many studies that have been conducted over the past two centuries. I am concerned especially with the work that has gone on in southern England because it was pioneering and of excellent quality; but I am aware of many omissions and only regret that time and space do not permit a more thorough survey.

I have chosen to violate my time rule by bringing in some very recent studies of fossil flowers because they present an important development in fossil botany, one that may have a very significant bearing on the problem of angiosperm origins, and because, after all, the seeds and fruits came from flowers!

The focal point of much that follows is the renowned volume, *The London Clay Flora*, by Eleanor Reid and Marjorie Chandler (1933). This book includes much historical information on the Tertiary floras of England, especially that of the London Clay; I have drawn freely from it and am also indebted to Miss Chandler for additional information through correspondence and visits to her lovely home in the village of Powerstock.

The London Clay flora is known from several hundred species of pyritized seeds and fruits found in Eocene deposits along the shores of the Island of Sheppey on the southeast coast of England. Among the conspicuous elements of the fossil assemblage are the large fruits of the *Nipa* palm; there are many others, a considerable portion of which suggest a relationship with the forests of Indo-Malaya and the Malay Islands. Some of the fossils have been difficult or impossible to identify, and as in any investigation of such magnitude there are undoubtedly mistakes in Reid and Chandler's account. But in large part I believe it contains accurate identifications which indicate a much warmer climate for that part of England in early Tertiary times. Their work initiated or accelerated a general interest in seed and fruit floras of continental Europe which continues to the present. Thus the history of the London Clay fossils is of special significance.

Aside from an anonymous work that appeared in 1709, the earliest study of consequence seems to be that of James Parsons (1757). I mentioned this in Chapter 2 in connection with Parsons's speculation as to the time of year of the Flood, which was based on his notion of the state of maturity of some of the fossil fruits.

Some pyritic fossils tend to disintegrate in a few months or years, and the present method of preventing this is to keep them in jars of glycerin. Parsons encountered the problem: "I thought it necessary to make drawings of them while in a sound state, in order for engraving, if the [Royal] Society shall think fit" (1757, p. 396).

The next comprehensive study of the fossils is found in a manuscript of 1810 by Francis Crow; it is preserved in the British Museum and describes over seven hundred specimens. It is entitled "A Catalogue of Rare Fossil Fruits from Sheppey Island, etc., in the Collection of Francis Crow of Faversham, 1810, Being the Result of upwards of Twenty Years Collecting." Reid and Chandler quote Crow as saying: "I hope I may be acquitted of the charge of vanity in assigning credit for more experience in the investigations of Fossil Fruits than any Man and for an Invention my researches have led to which has facilitated the business of ascertaining the species more than any circumstance whatever. I mean that of opening them and displaying their internal structure. Notwithstanding which I have found great difficulty in ascertaining varieties from the Twelve following causes. . . . " These "causes" include such problems as the immaturity of some fruits and the overmaturity of others. Apparently Parsons did not make a comparable observation, or if so it did not inhibit him in determining that the great Flood took place in the autumn.

Crow concludes: "The specimens here Delineated are selected from various productions of both animals and vegetables originally of a Country and Climate very different from our own, from the appearance of both animals and vegetables one would be inclined to think that they once belonged to a Tropical, or high southern Latitude" (quoted in Reid and Chandler, 1933, p.

6). He also offers an amusing apology for his accompanying drawings ("The person that figured them never was learnt to draw"), and as a result of their quality Reid and Chandler add that it has been possible to identify only a very few specimens.

In 1840, James Scott Bowerbank (1797–1877) brought out his classic book *A History of the Fossil Fruits & Seeds of the London Clay*. It includes seventeen plates of the fossils and, according to a "prospectus" on the last page, was intended as a five-volume work, the result of examining some twenty-five thousand specimens selected from an original collection of one hundred and twenty thousand. The engravings were executed by James de Carl Sowerby. But the single volume, or part, of 1840 is the only one that ever appeared. Many of Bowerbank's species are, according to Reid and Chandler, based on rather minor distinctions, but he was responsible for stirring up a good deal of serious interest in the fossils.

Bowerbank seems to have been the epitome of a man who, although having had but little formal education, used his talents and financial resources to the best advantage. He was born in London and at the age of fifteen entered his father's distillery, where he worked until he was about fifty. This was a lucrative business which allowed him considerable leisure during his working years and freedom from money problems when he retired. He participated in the activity of many scientific societies, founded new ones, gave lectures, and engaged in research in diverse areas. During the early 1820's he delivered courses of public lectures on botany, and later on human osteology. Entomology was among his earlier interests, and he made several contributions in this direction; he also conducted studies of sponges. He was one of the founders of the Royal Microscopical Society and served as President. He was noted for his cordiality and encouragement of others in the study of science, and, according to one of his biographers, "Indeed, so many men came at this time to his house in order to examine their specimens by his instrument [microscope], that he was compelled to fix one night a week for their reception, and thus originated the celebrated '*Monday Night Meetings*,' where so many eminent men used to assemble and always received a kindly greeting" (Tyler, 1878, p. 29).

About 1836 "The London Clay Club" was formed by Bowerbank, in company with John Morris, Searles Wood, Alfred White, Nathaniel Wetherell, James Sowerby, and Frederick Edwards, with "the purpose of illustrating the Eocene Mollusca." It strikes me that this enthusiasm for fossils at a particular horizon must have been uncommon in the annals of London clubs! The aim expanded, however, to include other fossils such as the seeds and fruits, and in turn gave birth to the Palaeontographical Society—one of the great British organizations long devoted to the publication of important paleontological monographs.

J. Starkie Gardner (1844–1930) enters on the scene as an investigator of English Tertiary floras shortly after Bowerbank's time, and among his several publications is a joint one with Ettingshausen, the "British Eocene Flora," published between 1879 and 1882. His identifications have not stood the test of time but he seems to have been an extraordinary collector. A popular account that he wrote in 1884, *Fossil Plants*, makes rather good reading and, as I interpret it, carries the very significant message "collect while the collecting is possible." He deplores the lack of interest on the part of both professionals and the general public in preserving fossil remains and comments on the loss of them in places such as quarries as well as through natural causes along the rapidly eroding seacoasts. Even with the popular interest in collecting fossils today this message still holds true to a degree.

Some years ago several of my English colleagues kindly took me on an excursion to Sheppey, and although it was a most worthwhile experience, an afternoon of probing through the mud and fog and rain yielded only a few fragmentary specimens. Gardner's impressions of the collecting there are worth recording:

> From the vast quantities of fruits which have been collected—and I have myself several thousand—it might be anticipated by geologists visiting Sheppey, that one has only to take a bag to the right spot to fill it with choice specimens; but, if this is the belief, disappointment will ensue, for a day's collecting will probably yield but a few indifferent specimens of the commoner forms, unless the search is very diligent. The only plan likely to succeed in a brief excursion is to visit the cottages on the cliff and inquire for fossils, which are always picked up by the copperas and cementstone collectors, and in this way a number may be bought during a day's visit for little money. Returning by the beach, if the tide serves, the collector should select spots where finer particles of pyrites have drifted, light his pipe, and sprawl, for I know no other term so apt, when many rare seeds, etc., will probably reward his search. [1884, p. 306]

Collecting along the cliffs was not without its tense moments, for Gardner also says: "In one instance a considerable excavation that was the result of two months' constant work, was filled by a landslip with only a few seconds' warning, and so rapidly that I and a companion, although having time to get clear of the course of the great blocks that tore by, were buried up to our armpits in sand, and were only dug out in time to avoid a still greater fall that occurred a few seconds later. My tools are still buried underneath" (p. 310).

By profession Gardner was a metal worker; he designed and made the wrought iron gates and screens at Holyrood Palace, and he also made and erected the Victoria Gate in Hyde Park. He was a busy man with many interests, and his writings include works on English enamels and armour.

Clement Reid. From the *Journal of Botany*, 1917.

Many more recent naturalists have played greater or lesser parts in exploring the Tertiary rocks of southern England for fossil plants, and among those who have been especially relevant to my own story are Clement Reid, his wife Eleanor Mary Reid, and Marjorie E. J. Chandler. They worked together as a very productive unit, and their lives as well as their scientific publications are of great interest.

Clement Reid was born in 1853, the son of a London goldsmith and of a niece of Michael Faraday. In his early years Clement was deaf as a result of scarlet fever, and being unable to play in a normal way with other children he spent much of his time in solitary rambles, developing a great love of nature that was encouraged by his mother. His formal education was minimal, and, for economic reasons, it was necessary for him to take a job in a publisher's office when he was fourteen. He disliked the work, and after seven years, upon learning of a vacancy in the Geological Survey, he prepared himself to the best of his ability and received the appointment.

Reid was always a vigorous field geologist and a pioneer in his explorations of plant deposits and in the development of techniques for extracting new information from them. In his early days with the Geological Survey he spent considerable time surveying on the Norfolk coast, and this work resulted directly in *The Geology of the Country around Cromer*, published in 1882, and laid the foundations for much of his later research. I think he merits the credit for being the first to fully appreciate the botanical value of late Tertiary and Quaternary fossil plants—remains of a type not always as immediately attractive as those found in some older horizons.

One of Clement Reid's best known works is *The Origin of the British Flora*, written in 1899. It reveals his interest in Ice Age and modern floras, and Hamshaw Thomas credits it with initiating many investigations by later workers in different countries. A few lines from this work of Reid's reveal the profound knowledge acquired by a man who was essentially a self-made botanist and geologist:

> this life spent principally in field, and moor, and forest has forced me to observe how each changing season is marked by corresponding adaptations in the animals and plants, such as enable the species to preserve themselves, to multiply, and to spread; or, if adaptation fails at any point, through some climatic irregularity, how sweeping and rapid may be the extermination of all except some few accidentally favoured individuals. While collecting seeds and fruits for comparison with the fossils I was compelled particularly to observe their many adaptations for dispersal, and also their times of ripening, and the abundance or scarcity of ripe seeds. [1899, p. 2]

In 1897, Reid married Eleanor Mary Wynne-Edwards. She was born at Denbigh, Wales, in 1860, attended Westfield College, and took her B.Sc. degree while teaching mathematics at Cheltenham. It is not evident that she had any special interest in botany or geology before her marriage, but quite clearly she entered wholeheartedly into her husband's work. Of their years together James Groves says: "it was with that lady's assistance and co-operation that most of [Clement's] subsequent work at fossil plants was accomplished. Immense quantities of 'matrix,' sometimes actually amounting to hundred-weights, were dealt with by their united efforts, being washed and treated by various methods, and subjected to such careful examination as to insure that minute, often almost microscopic, organisms, should not escape notice" (Groves, 1917, p. 147).

When the Reids started their studies of the Tertiary seeds and fruits they recognized the lack of an adequate collection of comparable modern material. Actually, Clement had gathered considerable information and specimens during his earlier studies of Pleistocene and Recent deposits, but a much more comprehensive collection was now needed. Marjorie Chandler, who enters the story a little later, had the following to say of this developing collection in a letter to me of September 10, 1973: "Many great Herbaria and Botanical gardens as well as seedsmen and private individuals and collectors contributed. This unique collection was given . . . on [my] retirement in 1967, to the Palaeobotanical Department of the British Museum (Natural History) where it was hoped it would help Palaeobotanists who came from all over the world to study the fine collections of Tertiary fossils housed in that Department."

Clement Reid's work in the Tertiary was largely confined to the Pliocene, and as a result of extensive field work he brought out a paper in 1890 on the Pliocene deposits of Britain. He visited Tegelen in Holland in 1905 and two years later produced *The Fossil Flora of Tegelen sur Meuse*, a joint work with his wife. They later investigated rich plant deposits in Limburg, Holland, and an adjacent locality over the German border. This resulted, in 1915, in a large quarto monograph, *The Pliocene Floras of the Dutch Prussian Border*; they reported 189 species, many of which were regarded as having their closest living representatives in the Himalayas, China, and Japan. Eleanor Reid's biographer says that this study "demonstrated the fundamental basis of the Reids' work: that the critical study of fruits and seeds preserved as fossils not only gave reliable generic and specific identifications, but illustrated the composition of past floras more completely than leaf impressions, which normally represent only the deciduous elements of the vegetation" (Edwards, 1954, p. cxl).

Among the Reids' other collaborative studies was a report on the seeds and fruits from the lignites of Bovey Tracey, in Devon which came out in the *Philosophical Transactions* in 1910. The fossils were preserved in clays renowned for pottery uses. The flora includes remains of magnolia, grape, and tupelo (*Nyssa*), as well as members of the borage and rose families and a few conifers. They regarded it as being of Upper Oligocene age and they note: "In another way the Bovey flora is of interest, for it shows the gradual dying

Eleanor Reid and her cat at Pinewood, 1921.
Courtesy of Marjorie Chandler.

out of the tropical or warm-temperate plants and the incoming of a few northern genera, probably washed down from the surrounding uplands of Dartmoor" (p. 162).

The Bovey Tracey plants had been studied previously by Oswald Heer and a Devonshire geologist, William Pengelly, both of whom published reports in the *Transactions* in 1863. Pengelly is an interesting character in his own right. He was born at East Looe in Cornwall on January 12, 1812, and received his only formal education in that village as a child, for he went to sea with his father when he was twelve. After some ten years of this life of hardships, including a shipwreck in which he nearly lost his life, he returned to the land, educated himself, and established what seems to have been quite a successful school at Torquay. One of his biographers notes that "he was one of the first to introduce the use of the chalk and black-board in imparting instruction" (1894, p. 238). He made numerous original contributions to several areas of geology and paleontology and was especially well known as a popular lecturer. He died at Torquay in 1894.

Clement Reid retired in 1913 to the village of Milford-on-Sea in South Hampshire, where he built a house that he christened "One Acre." It was located within a few minutes' walk of the famous Hordle Cliffs where the Lower Headon beds are exposed. His continued interest in living plants as well as fossils was evident in his garden, which included many plants generically related to the fossils. Unfortunately, he lived only three years to enjoy this new home. Mrs. Reid sold it shortly after his death and bought a smaller house known as "Pinewood," where she lived, later joined by Marjorie Chandler, for the remainder of her life. It was a center for research and the continued development of the comparative collection of seeds and fruits, aided by Kew Gardens, the British Museum, and numerous other institutions and individuals.

The most detailed biographical sketch of Clement Reid that I have found was written by James Groves, a close friend of Reid's. Groves says, in reference to their friendship:

> It came about in this way: in 1913, my attention had been drawn to the curious remains of Characeae found in the Middle Purbeck Beds of Dorset, and I was attracted to make a study of the early history of the group. Years before, Reid had sent us Chara-fruits from the Cromer Forest Bed for examination, but we had not pursued the matter further. He and I had always been on very friendly terms, and I naturally turned to him for assistance in obtaining specimens and information. He entered into the matter with his usual zest. We borrowed all the specimens we could of the Middle Purbeck cherts, and in the spring of 1914 paid a visit to Durlston Bay to collect more. [1917, p. 149]

The charophytes, or stoneworts, include two living genera, *Chara* and *Ni-*

tella. They are curious aquatic plants, usually classified with the algae but thoroughly distinctive with their jointed stems, very complex male organs, and a large egg cell that is surrounded by several spiral filaments. The female organ is quite often found as a fossil.

James Groves was born in London in 1858, and a few years after his father died in 1869, James entered the employment of the Army and Navy Stores, where he remained until he retired in 1918. James and his brother Henry had a strong interest in natural history from their early years. They became acquainted with B. Daydon Jackson and many other leading botanists of the day. Although they were concerned with several aspects of the British flora, they gradually devoted much of their time to the charophytes and published a number of significant works on that group. When Henry died in 1912, James carried on; while most of his work was with the living charas he collaborated with Clement Reid in a study entitled *Charophyta of the Lower Headon Beds of Hordle Cliffs* in 1921.

Following Clement Reid's somewhat untimely death his wife Eleanor was determined to continue the work with the Tertiary fruits and seeds. She brought out several works under her own name and one or two in collaboration with the French paleobotanist Pierre Marty. These were concerned chiefly with the Pliocene floras such as the cromerian of Norfolk, the teglian of Limburg in the Netherlands, and the Pont-de-Gail of Cantal, France. In 1920 she gathered this information together in *A Comparative Review of Pliocene Floras*, which deals with their history and distribution, as well as climatic and geographic relationships.

But the task of accomplishing even a portion of her objectives was a large one and assistance was required; she therefore wrote to Professor Seward at Cambridge, asking him to suggest an appropriately trained student. The student that he sent did not like the work and decided not to stay. Marjorie Chandler was also at Pinewood at the time (1920), using the Reid collections in connection with a research project that she had started as a student at Cambridge. Mrs. Reid offered her the job; Miss Chandler accepted, and she remained with Mrs. Reid until the latter's death in 1953. This was one of the great partnerships in the entire history of botany, resulting in several monumental contributions by the two. I have had the very good furtune to visit with Miss Chandler at her home in Powerstock, Dorset, and she has supplied me with considerable information on her personal and scientific life. This is a unique and memorable chapter in the annals of fossil botany and, with her kind permission, I would like to record some of it, extracted from her letters:

> *Early Days.* As one of a family of 6 children we had little for luxuries so made our own amusements, walking and cycling miles to collect wild

flowers and, as far as the limited resources of our locality permitted, fossils. But the Lias only was available.

The Canal Bank was a grand habitat for water plants and we had the

Marjorie E. J. Chandler at her home in Powerstock, Dorset, 1975.

added entertainment of watching the horse-drawn gay barges and the rough-looking families on board.

My next sister and I were among the first pupils of a new fee-paying school offering secondary education. A percentage of free places was available for able children from Primary Schools, so we mixed with all and sundry. But High School fees were prohibitive. Nevertheless our small staff was mainly Oxford and Cambridge Graduates. To one Teacher I owe much of my scientific interest and so developed a fixed determination to study Science at the University. This teacher gave up her free periods to help me as far as the limited facilities of the school permitted. But to gain a scholarship it was necessary to sit for it in English Language and Literature which the school was better equipped to teach. The scholarship paid all fees (£ 90 per annum: board, tuition and bench fees) and I went to Newnham College, Cambridge, in 1915, to a University bereft of most of its students by the 1st World War.

At Cambridge. Life at Newnham College was lived in those days at Boarding School level. We were protected and bound by rules which would not be tolerated to-day. We could never go out after 7 p.m. (dinner time) except with special permission and accompanied by a chaperone. Thus when I went to play in a quartet a few hundred yards away I had each time to have the Tutor's approval.

Women were not members of the University till after my Cambridge days. But we were tolerated at University lectures and were required to sit in the front row and to behave correctly lest we damage the cause of Women's Higher Education. I took Ist Class Honours in the Science Tripos (Pt. one then the equivalent of a degree). Part II was not set in my time as I was the only student. There were no Ph.D. degrees in those days.

Professor Marr, Woodwardian Professor of Geology at the Sedwick Museum, was by this time nearly blind and tore hectically between the desk and the blackboard drawing undecipherable sections and diagrams. But I liked him and it was really he who started me on the Palaeobotanical Road because of his interest at the end of his career in the Cambridge Quaternary Gravels. So when a 4th year student discovered sub-arctic plant beds at Barnwell I was switched in my 5th year to Milford-on-Sea to identify these plants by comparing them with Mrs. Reid's Reference collection of Recent fruits and seeds. Incidentally Mrs. Marr was not in favour of Higher Education for women but was generous to me because I talked about cooking.

At Milford-On-Sea. Having joined Mrs. Reid temporarily in 1920 I stayed with her till she died in 1953 and we worked together until 1933 when, worn out with the arduous study of the London Clay, she gave up and retired into private life while I soldiered on alone and acted as a daughter to her, cooking, washing and ultimately nursing. I had the run of her workroom and library (in an attic at the top of the house, icy in winter and scorching in summer). Our equipment would be despised by modern workers, for it was home-made including the camera (gas pipes and boards and a tape measure and a real professional bellows and lenses). We used winkle saucers for picking over our washed residues which after sifting were examined under water and anything showing form and pattern was removed for study (by a paint-brush). [I don't believe that present-day paleobotanists would "despise" the old camera, but they would marvel at the photos it was induced to yield. Miss Chandler showed it to me a few years ago, tucked away in a corner of her kitchen; I had to get onto a stepladder to look down into it and operate the chain that regulated the bellows. Having used some primitive photographic equipment myself in the earlier years I could appreciate the patience that went into the preparation of Miss Chandler's and Mrs. Reid's many fine photos.]

One of the interesting incidents of life at Milford was a visit from time to time by the palaeobotanist Dukinfield Scott, a wealthy man living in North Hampshire. He appeared with his wife to call without warning causing a flutter in our little dovecot. I used to be despatched to the village to provide the necessary fare for a simple luncheon. We indulged in friendly chat and professional talk followed by a walk to the cliffs to glance at the Hordle plant beds and the classic Alum Bay section across the Solent.

One real friendship which grew out of this work was formed with Pierre Marty who lived in Cantal. This because Mrs Reid had been identifying Pliocene fruits and seeds he had collected. He entertained us in his lovely old but neglected Chateau and took us round to see the scenery and productive sections of the volcanic area in Cantal and the Auvergne. We lost sight of him when France was overrun in World War II and learnt after the war that he had died. Another contemporary worker was Professor Laurent of Marseilles who had studied the leaf impressions from the same area. Unfortunately the leaves gave a different picture from the fruits and this caused considerable friction between him and Mrs. Reid for at that time no satisfactory explanation of the discrepancy had been thought out.

Miss Chandler goes on to mention correspondence with other notable paleobotanists including Nikitin, Dorofeev, Kryshtofovich, Kirchheimer, and Kräusel, who enter into my story on other pages. In reference to the start of her collaboration with Mrs. Reid, Miss Chandler writes: "about 1925 Dr. Bather of the British Museum (no doubt inspired by his palaeobotanist W. N. Edwards) invited [Mrs. Reid and me] to catalogue fossil fruits and seeds in the National Collection. This was an annual appointment only, financed by a Treasury grant but it lasted for about 40 years."

The first result of their work together was *The Bembridge Flora*, which came out in 1926. It was based on several collections that had long been stored away in the Museum; the largest was one that had been gathered together by James Edwin Ely A'Court Smith (1814–1900), who had served with the East India Company and the Merchant Service and in 1859 retired to the Isle of Wight. He then devoted some twenty years to collecting plant and animal fossils on the northwest coast in Gurnard Bay. He sold a part of his collection, which included plants, to the British Museum in 1877. Later the main collection was offered to the Museum but was rejected because the price was considered excessive. When Smith died in 1900 the collection was acquired for a few shillings by a Mr. Hooley, and upon his death in 1925 the Museum obtained the collection, containing several hundred plant specimens, from his widow.

The Oligocene Bembridge flora is considered to be a warm temperate one; Reid and Chandler say:

> The number of recognizable species represented by fruits and seeds is much greater than the number represented by determinable leaves. Those occurring in the greatest numbers are of two kinds.
>
> (1). Fruits and seeds of land plants especially adapted for wind dispersal, such as *Engelhardtia, Hooleya* (an extinct genus of the walnut family), *Abelia*, *Clematis*, and various Apocynaceae.
>
> (2). Fruits and seeds of water plants, such as *Sparganium*, *Brasenia*, *Stratiotes*, *Potamogeton*, *Limnocarpus* (an extinct genus of Potamogetonaceae), and *Ranunculus*, which is probably an aquatic or semi-aquatic species. [1933, p. 6]

In 1933 Mrs. Reid and Miss Chandler brought out their great work, *The London Clay Flora*, a monumental volume accompanied by thirty-three plates illustrating the fossil seeds and fruits. This was based on the Museum collections as well as many years of their own collecting. Of 314 species that came under study 234 have been identified, the affinities of some remaining doubtful. Of the one hundred genera of flowering plants that are described only twenty-eight are extant; thus it is family relationships that are especially critical. Many of the families are either entirely or almost entirely tropical. The ninety-page introduction to this great work includes a valuable summary of many aspects of plant distribution and palaeoclimatology with particular reference to seeds and fruits.

After Mrs. Reid's retirement in 1933 and her death in 1953, Miss Chandler continued to work alone. Much time was devoted to making new collections, and the greater part of her tremendous efforts is recorded in four monographs dealing with the Lower Tertiary floras of southern England, published by the British Museum: a 1961 supplement to the original *London Clay Flora*, with 354 pages and thirty-four plates; a study of the flora of the Pipe-clay series of Dorset in 1962; the study of the flora of the Bournemouth Beds in 1963; and a final number in the series in 1964, which contains a general summary and survey of this great achievement.

Miss Chandler's interests were not confined to seeds and fruits, for among her other works are one on some well-preserved fertile fronds of *Anemia* and *Lygodium* ferns and another in which she described a *Tempskya* specimen from Sheppey.

Miss Chandler gives much credit for the work done in these later years to Dr. K. I. M. Chesters, who came to her to use the Reference Collection in connection with a study that she made of fossil fruits from Africa. Dr. Chesters aided her in many ways, including assisting Maurice Wonnacott in editing Miss Chandler's publications.

The floras of the fossil beds of southern England have long been famous—and not only among professional botanists and geologists. It is not often that paleobotany is favored in poetry, but John Scafe in his *King Coal's Levee or Geological Etiquette* gave the seeds and fruits a few lines back in 1819:

> But could PYRITES, of aspiring soul,
> To shells alone her ardent views controul?
> No; *Botany* her thoughts alternate claim'd,
> And Sheppey's Isle again was justly fam'd:
> There did the plants reward her guardian care
> With fruit and seeds that other climates bear.
> Much in exotics the Queen's passion lay,
> But whence, and how they came, none ever knew but CLAY.
> Earth's widest realms have been explor'd in vain,
> No prototypes of most on earth remain.

Studies of Tertiary seeds and fruits by continental paleobotanists date back to the early part of the nineteenth century, but they are scattered and, until rather recent decades, did not reach the state of development that I have described for southern England. However, when concentrated efforts got under way, in about the 1920's, progress was rapid. A very considerable literature has now accumulated, and again I must be selective. In so doing I have chosen a few exemplary people to represent the many who have now contributed to an elucidation of Tertiary floras in this particular way—a way that seems to be giving us a much more sound and precise picture than had been presented previously.

The so-called brown coals or lignites of central Europe contain some of the largest known accumulations of fossil-plant materials in any geologic period. In some places the beds are several hundred feet in thickness. Probably the outstanding student of the seeds and fruits of these European brown coals is Franz Kirchheimer. His written works are numerous and of excellent quality, and I can only select a very few to illustrate the great scope of his contributions. He was born on July 1, 1911, in Müllheim, Baden, and attended school in Berlin, Frankfurt am Main, and Giessen; at the University of Giessen he studied biology, botany, mineralogy, and zoology, receiving the doctorate in 1933, but as early as 1930 he had been awarded a prize for his studies of the brown coal. In 1933 he was discharged from university service by the national regime of the time as being "politically unreliable," and his life for the next decade was not an easy one. For two years, starting in 1943, he served as director of a private biological laboratory at Bodman, near the Lake of Constance, where he engaged in research on penicillin. Although my own

Franz Kirchheimer. Courtesy of Richard Eyde.

paleobotanical studies have been mostly with the Paleozoic I feel some kinship with Kirchheimer on this point as I also devoted some time to penicillin research during the war years. In 1945 he received a professorship at the University of Giessen and in 1951 at the University of Freiburg. The following year he was also appointed Director of the State Geological Office of

Baden-Württemberg. He has held several other posts, received many honors, and in spite of his difficulties with the Nazi regime and a great deal of administrative work he is the author of more than 240 articles and books, including several on numismatics. In 1975 he retired from his State Geological position and presently holds a professorship at the University of Heidelberg.

Among Kirchheimer's earlier comprehensive works on the Tertiary brown-coal fossils is an informative article (1936), "Beiträge zur Kenntnis der Tertiärflora: Fruchte und Samen aus dem deutschen Tertiär" (Contributions to the Knowledge of the Tertiary Flora: Fruits and Seeds of the German Tertiary). This is concerned with the lignitic fruits and seeds of a brown-coal deposit at Salzhausen in Upper Hesse. It is a locality that has been known for a long time and is of historical as well as botanical interest. The brown coal was discovered there in 1812 and was first used to heat the evaporating pans in the salt works, and a few years later was sold for household fuel. Some of the fossils were turned over to both Brongniart and Sternberg for study and the latter's *Flora der Vorwelt* includes some descriptions. Several noted paleobotanists in later years entered into the picture. Kirchheimer notes that Alexander Braun (1805–1877) first became acquainted with the fossils as a student, and during 1850–51, when he held the Chair of Botany at Giessen, he devoted his research primarily to the fruits and seeds—beginning with specimens extracted from the coal used for heating in his own home! Braun wrote: "I put no piece of brown coal in the oven without looking at it, and I find something almost daily; I also receive contributions from all sides. It was a special pleasure to discover the existence of the grape-vine in the brown coal" (quoted in Kirchheimer, 1936, p. 75, translation). Although Braun lectured on the fossils he apparently never brought his studies to the point of publicaton.

Unger, Ettingshausen, Goeppert, Heer, Conwentz, and Hermann Engelhardt are among others who have been interested in the brown-coal fossils. The German philosopher Goethe owned a small collection, probably presented to him by Schlotheim and Sternberg.

The following from Kirchheimer points to the reason for this general interest in deposits such as the one at Salzhausen:

> This well-known "carpolith or fruit-coal" was occasionally up to 1.5 m thick. In some places it disappeared, or reached a thickness of only a few centimeters. . . . The quantity of fruits and seeds contained in the Salzhausen brown-coal surpasses the imagination. Thus a cross-section sample taken from various pieces of the "carpolith-coal" in an area of only 125 cm^3 contained the following: 87 stone cores of *Symplocos gregaria*, 12 stone cores of *S. jugata*, 9 *Brasenia* seeds, 6 *Vitis* seeds, 3 *Magnolia* seeds

and one seed of *Stratiotes*. The investigation of only nut-sized pieces of ordinary humus-coal will frequently yield several small seeds, also. [1936, pp. 77, 78, translation]

(Different parts of the coal bed as a whole have been named in accordance with the dominant type of plant remains; thus terms such as "carpolith" or "fruit" coal, as translated here). The abundance of seeds in this deposit is well shown by some of Kirchheimer's plate illustrations—great aggregates of walnut fruits (*Juglans ventricosa*), *Prunus langsdorfi, Sapindoidea margaritifera*, and several others.

After many years of research devoted to the brown-coal fossils, and scores of publications, Kirchheimer brought out his classic summary *Die Laubgewächse der Braunkohlenzeit* (Foliage Plants of the Brown-Coal Period). Although Kirchheimer had intended it to appear in 1948, problems arising from the war delayed its publication until 1957. In addition to the description of the fossil seeds and fruits, this book contains much that is of general interest. Professor Władysław Szafer, the distinguished Polish botanist, says that, although he did not agree with it in all details, "This does not by any means belittle the value of that synthetical work, which is undoubtedly the most precious modern publication on the Tertiary flora of Central Europe" (1961, p. 165).

Kirchheimer is generous in his praise of the work of Mrs. Reid and Miss Chandler and sharp in his criticism of the general neglect of recognition of the Tertiary floras among previous writers, especially those in the United States. This is a significant point, for some of the better textbooks touch only lightly on angiosperm floras. I think there is some justification for this omission: it is difficult to filter out the more significant aspects of research on Tertiary floras, the numbers of species involved is enormous, and the literature is large and encumbered with much that is of questionable value.

The investigations of fossil seeds and fruits have formed an important part of the efforts of Soviet paleobotanists in recent decades. The outstanding pioneer seems to be Petr A. Nikitin (1890–1950), and I will introduce his work with a passage from the foreword of one of P. I. Dorofeev's studies. It should be noted that Dorofeev coined the term "paleocarpology" to refer to the study of fossil seeds and fruits.

P. A. Nikitin's large monograph *Pliocene and Quaternary Flora of Voronezh Province*, published in 1957, is not only a summary of the study of the Voronezh Pliocene and Pleistocene; it is the first in our country and a beautiful introduction to the paleocarpological study of Tertiary and Quar-

> ternary deposits, a reference book for every paleocarpologist. Moreover, this book dispelled the myth of a rich Tertiary flora, supposedly preserved in the Russian Plain up to the Pliocene, and wrote a new chapter into the history of our flora, having to a great degree determined the further fate of the Pleistocene flora." [1966, p. 3; translated]

Nikitin was born in Glazok, near Michurinsk. His family moved to Moscow in 1894, and, as his parents died soon afterward, he was brought up and educated by philanthropic societies. In 1909 he graduated as a technician-mechanic from a technical institution in Moscow, but he gradually transferred his interests and work into botanical science and later studied at the Voronezh Agricultural Institute. He also taught for some years at the Institute and at Voronezh University. In 1934 he was invited to Tomsk University, where he was in charge of the department of systematics of higher plants. His interests became directed more toward the study of fossil plants, and in 1939 he went to work in the geological office at Novosibirsk, where he organized a paleocarpological laboratory, and continued to work there until the end of his life.

I think it is of historical interest to insert here a few lines from a letter that I recently received from Miss Chandler: "You may not know that P. A. Nikitin was a devoted pupil and admirer of Eleanor Reid and I think it was really she who started him, helping him with determinations and about the characters used in determination. . . . His letters to her always began 'My dear Teacher.' He gave her many packets of seeds which, bearing the initials P.A.N., eventually went into her collection at the British Museum."

Dorofeev gives much credit to Nikitin for his refinements of techniques, particularly for extracting the smaller elements such as minute seeds and megaspores. I think it is also significant to note that Dorofeev, like Kirchheimer, gives considerable credit to Clement and Eleanor Reid for the foundations they laid for this new branch of fossil botany.

In using a greatly increased source of evidence Nikitin was able to elucidate more precisely the nature of late Tertiary floras and correct some previous mistakes in identification that had been based on an examination only of the larger fossil remains. He also demonstrated errors due to the examination of "mixed" floras—that is, deposits in which fossils from an older, eroded horizon had become mixed with a more recent deposit. Owing to his administrative duties much of his work was unpublished at the time of his death, and a considerable portion of his studies were with Pleistocene deposits which lie somewhat outside the scope of my account.

P. I. Dorofeev, a longtime staff member of the great Komarov Botanical Institute in Leningrad, has continued and developed in a very productive manner the work that was initiated by Nikitin. His studies began to appear in the early 1950's, and among his more important works is *The Pliocene Flora of the Matanov Garden on the River Don* (1966).

The Pliocene "Matanov Garden" flora (the name is one given to the place by local inhabitants) was collected in considerable abundance by Dorofeev in 1948–49 from beds exposed along the river Don near the Cossack village of Nagavakaya. It is of special interest in that it is an herbaceous flora, contrasting sharply with the usual arborescent, forest floras that contribute so large a portion of our knowledge of the landscapes of the past. "This flora," according to Dorofeev, "is interesting in that it reproduces an already treeless vegetation, greatly similar to the contemporary steppe, while the greater portion of the Pliocene flora is characterized by a forest flora, basically a taiga flora." The following plants are representative of this distinctive assemblage: the water ferns *Salvinia* and *Azolla*; the bur reed (*Sparganium*); several species of pondweed (*Potamogeton*, as well as *Najas*, another member of the same family); two members of the Alismaceae, the arrowhead *(Sagittaria)* and *Alisma*, the water plantain itself; several sedges, including *Cyperus*, the bulrush (*Scirpus*), and the ubiquitous *Carex*. Among the herbaceous species that suggest a somewhat drier habitat are *Rumex* (dock), *Ranunculus* (buttercup), and species of *Rubus* and *Potentilla* in the rose family. Although shrubs and trees form a distinct minority, willow, birch, and grape are represented.

Significant progress has been made by the Polish botanists in the area of paleobotany that Dorofeev calls "paleocarpology"; two of particular importance are Hanna Czeczott and Władysław Szafer.

Hanna Czeczott's contributions range rather widely both botanically and geographically. In summarizing a part of her work I wish to emphasize the courage that she has brought to her studies, which extended over many years. For much of the following biographical information I am grateful to Krystyna Juchniewicz, one of Mrs. Czeczott's colleagues at the Warsaw Museum.

Hanna Peretiatkowicz Czeczott was born on January 3, 1888, in Leningrad, or St. Petersburg as it was known at the time, one of eight children of a Polish family. She completed high school there in 1905, then went with her family to Warsaw, but in 1910 she was back in St. Petersburg, married to Henry Czeczott, a mining engineer. She attended natural science courses for women over a period of some years, frequently interrupted by travels with her husband as well as by World War I and the Russian Revolution. The travels took them to Canada, the United States, various parts of Europe, and into Mongolia. Some of her later publications reflect her many keen observations along the way.

They returned to Poland in 1922, and she remained in Krakow until her husband's death in 1928, when she moved to Warsaw. During 1939–1945 her work was again interrupted, and a large part of her rather vast collections as well as her library was lost when the German army destroyed Warsaw. It has not been my good fortune ever to have met Mrs. Czeczott but several of my colleagues who have done so have regaled me with accounts of her charm and extraordinary vigor in her researches through a very long life. Krystyna Juchniewicz writes: "Hanna Czeczott is endowed with a very strong personality. Her tremendous drive and energy along with her passion for research, bent everyone and everything to her will. The word 'impossible' never existed in her vocabulary, especially where the execution of a scientific project was concerned. Strict and exacting as she was towards her co-workers, Prof. Czeczott was equally as demanding of herself. Although not physically strong, she did not tolerate weakness either in herself or in others" (personal letter).

Among her more important publications is her contribution to a collaborative work, *The Fossil Flora of Turów near Bogatynia*, which came out in several parts starting in 1959. Turów is a brown-coal deposit, and in the first part she described the seeds, fruits, and leaves of the dicotyledons, with Alina Skirgiello. She had a strong interest in the amber plants and in 1961 brought out a comprehensive study, *The Flora of the Baltic Amber and Its Age*. This was accompanied by a short but useful English summary which includes a historical sketch of amber studies and a critical discussion of the geographical distribution and climatic relationships indicated by the fossil plants and animals.

During Hanna Czeczott's period in Krakow she collaborated with the noted Polish botanist Władysław Szafer. I do not believe that Professor Szafer should be regarded primarily as a paleobotanist, but he produced one comprehensive work, the *Miocene Flora from Stare Gliwice in Upper Silesia*, which impresses me as a real classic in Tertiary paleobotany. The work includes Polish, Russian, and English texts, but whether one reads any of those languages it is a pleasant and rewarding experience to thumb through the numerous fine plates that reveal this remarkable flora. Some 184 species have been studied, including fungi, numerous mosses, and a very great array of conifers (*Abies*, *Taxodium*, *Pinus*, *Cunninghamia*, *Podocarpus*, and *Juniperus*, among others). The remains of flowering plants include *Fagus* (beech), *Castanea* (chestnut), *Quercus* (oak), *Juglans* (walnut), *Ulmus* (elm), *Liquidambar* (sweet gum), *Nymphaea* (water lily), *Liriodendron* (tulip tree), *Magnolia* and a host of others. The work is based very largely on fossil seeds and fruits. Of particular interest is Szafer's discussion of the paleoclimatology, evidence being given for a series, probably four, of warm-cool cycles during the Miocene in that area.

Szafer was born in Sosnowiec in 1886, studied at the University of Vienna

Hanna Czeczott with her assistants at a brown coal mine in southwest Poland, 1958. Photo courtesy of Krystyna Juchniewicz.

with R. Wettstein, and was particularly influenced by Professor Marian Raciborski. After his studies in Vienna and Munich he moved to Krakow, where he was appointed Professor at the Jagiellonian University and Director of the Botanical Institute. He was a very versatile scientist and contributed to several fields, including palynology, plant geography, and especially many aspects of nature conservation. Like so many others in his country he lived through much of the worst of two world wars, and in the second one he helped to organize and operate an "Underground University." Harassed by the occupying forces, jailed, evicted from his home, and refused permission to work in the Botanical Garden, he nevertheless continued to teach and carry on his numerous researches.

Studies of fossil seeds and fruits have not been a conspicuous element of American paleobotany, but we have one locality that is unique in several ways, that has attracted the attention of numerous botanists and geologists, and that now promises to yield important information from Tertiary fossils of this type: the Brandon lignite of Vermont.

In 1848 a deposit of lignite was discovered in the course of exploration for iron ore in Brandon township. The deposit as a whole is distinctive in that it is Tertiary—the only one containing fossil plants north of New Jersey—and is unique as an isolated sedimentary deposit enclosed by the surrounding ancient granites and metamorphic rocks. The lignite is rich in fossil seeds, fruits, wood, and pollen, only a portion of which have been studied critically as yet.

Numerous people were concerned with the deposit over the ensuing century in what I think is fair to call a desultory fashion. In 1850 specimens were sent by a Professor Shedd of the University of Vermont to Edward Hitchcock, who was President of Amherst College and State Geologist for Massachusetts. He visited the locality and issued a brief report in 1853. Hitchcock, in turn, sent specimens to Leo Lesquereux, who stated that an accurate analysis was not possible but he assigned names to twenty-three species of the seeds and fruits. Little went on during the remainder of the nineteenth century, but the deposit was "rediscovered" in 1901–1902 during clay-mining operations, and a national coal strike in 1902 added interest to the possible use of the lignite for fuel. In the same year Frank Knowlton, who had been born in Brandon, described a few fruits and fragments of coniferous and dicotyledonous woods. In 1904, Professor George H. Perkins sent E. C. Jeffrey and M. A. Chrysler some wood specimens from the lignite, one of which they identified as *Laurinoxylon brandonianum*. Perkins himself was an interesting man; he was for many years a professor of botany, zoology, and geology at the University of Vermont. His services to the University were numerous and his biographer, Herman Fairchild, says that he continued with his classes in anthropology at his residence to within three months of his death at the age of eighty-nine.

The Brandon deposit came to the attention of E. W. Berry, and in 1919 he added his bit to the accumulated literature. Several more decades then elapsed, until 1947, when Elso Barghoorn at Harvard University reopened the deposit and initiated the latest phase of studies, which is beginning to yield significant information. In a short preliminary report on the flora Barghoorn and William Spackman in 1949 cited the presence of the following plants: *Nyssa* (tupelo), *Bumelia* (buckthorn), *Persea, Vitis* (grape), *Quercus* (the live-oak group), and *Ilex* (holly), all except the last being represented by seeds and fruits. All who have studied the fossils at all seriously agree that they represent a warm-temperate to possibly subtropical climate—certainly appreciably warmer than the present climate of north-central Vermont.

The first comprehensive report on the Brandon plants was Alfred Traverse's 1955 study of the fossil pollen. His account revealed ten or twelve genera that had not been recognized previously from the larger fossils. These include *Glyptostrobus*, a conifer found at the present time in the Kiangsi and Kwangtung provinces of China; *Engelhardtia* and *Pterocarya*, both members of the walnut family, the former found in subtropical climes while the latter is more of a temperate genus; *Liquidambar*, the sweet gum, a member of the witch-hazel family that gets as far north as Connecticut now; and *Alangium*, which is related to the tupelo.

The Brandon flora is a large one, and in 1963 Richard Eyde started what promises to be a series of systematic studies dealing with the seeds and fruits genus by genus. With Alexandra Bartlett and Barghoorn in 1969 he described a species of *Alangium*, a genus that presently ranges through parts of Asia, Africa, and certain Pacific Islands. And in 1963 Eyde and Barghoorn described five species of *Nyssa* and summarized previous studies of the fossil remains (leaves, pollen, wood, and fruits) of the genus. I think that *Nyssa sylvatica*, or tupelo, is one of the more attractive trees along the lake shores in central New Hampshire, where I live, and it is especially striking in autumn. I am not alone in my love for the plant; George Emerson, a century ago, in his *Trees and Shrubs of Massachusetts* said: "The brilliant color of the green of the leaves, and the rich scarlet and crimson to which they turn in autumn, at which season some of the trees are covered with the bright blue fruit, make it always a beautiful object" (p. 356). *Nyssa* figures quite conspicuously in Tertiary plant literature; it was widespread although not all of the references to it have stood the test of time.

Alfred Traverse and Elso S. Barghoorn at the Brandon, Vermont, lignite "dig," August 1948.

In 1976, Bruce Tiffney and Barghoorn added another chapter to the Brandon story with descriptions of several species of *Vitis* (grape) based on the fossil seeds.

Although the Brandon deposit is unique in terms of New England geology, there are very extensive beds of brown coals in the western states and it is my feeling that they present a little-exploited source of Tertiary fossil plants. I recall visiting such a deposit, the Wyodak Mine, a few miles east of Gillette, Wyoming, in the early 1940's. The coal was about ninety feet thick at one point. It was but little consolidated and the branches and trunks of trees were very clearly defined. My knowledge at the time was not adequate to appreciate whatever potentialities there might have been, and I have often wondered what might have been found with a careful search.

It has been my intention in the general course of this entire account to serve as a historian rather than a prophet. It is precarious to speculate on the future likes and dislikes of mankind as well as to guess what great paleontological discoveries of the future may direct the development of our science. But I do not believe that I am casting discretion aside in concluding this chapter with a short report on some very recent discoveries that may be opening a new and significant channel of information to a better understanding of the earlier flowering plants.

Reports of fossil flowers, including those preserved in amber, are scattered rather sparsely through the literature. Although interesting, they have contributed rather little to a better understanding of the origin of the flowering plants. But some recent studies, taking advantage of keen observations and improved techniques, seem to be the beginning of a great step forward. Especially well preserved flowers have been found in the fine-grained Eocene clays of western Tennessee. In 1975, William Crepet, David Dilcher, and F. W. Potter described staminate catkins under the name *Eokachyra aeolius*, which are compared with modern plants found in the walnut family (Juglandaceae). The individual flowers have a distinctive three-lobed bract, well-preserved perianth parts showing stomates and scales, and pollen-filled anthers. In an article of 1977, Crepet and Dilcher described an inflorescence of a plant assigned to the legume family (subfamily Mimosoideae). The flowers have a shallow disc-shaped calyx, a corolla of four lobes, ten stamens, and a slightly extended, hairy, rounded stigma. The anthers are quite well preserved and contain pollen grains held together into permanent tetrads—an exceptional feature, although one that is not confined to this family. The Tennessee clays contain a great wealth of fossil plants, and with a better understanding of what to look for and what to do with the flowers when they are found, much may be expected from this area in the future.

In 1976, James Basinger described some small petrified flowers under the name *Paleorosa similkameenensis* from the Middle Eocene of British Columbia. The cellular structure is quite well preserved and the plant is thought to be a primitive member of the rose family. And in 1977, Bruce Tiffney reported a flower from a lignite band in the famous Upper Cretaceous cliffs of Gay Head on the island of Martha's Vineyard, Massachusetts. Portions of the calyx, corolla, a gynaecium with a five-lobed stigma, and parts of five stamens with some pollen are preserved.

These are remarkable discoveries by younger paleobotanists who I trust will not be inhibited by those critics of the past who thought of fossil plants as merely "the waifs and strays of a by-gone Flora."

Epilogue

On the dust jacket of a book that was given to me for Christmas—*Night Train at Wiscasset Station*, by Lew Dietz, a nostalgic and beautifully illustrated book about the changing face of the state of Maine—I read the phrase, " . . . without a backward glance change may lose all human value." It is the "human value" I wish to emphasize in this epilogue, which Webster defines as "a concluding section that rounds out the design of a literary work."

Quite frankly, this was not included as an integral part of my manuscript when I sent it off to the publisher. I felt that I had told my story pretty much as I planned when I started to put together all of the information that had accumulated in the form of photos, letters, reprints, and books that line the walls of my study. But there was a suggestion that I might do a little better about bringing it all "up to date." I respected the suggestion and, although I cannot review very much of the recent literature in detail, it did incite me to prepare a few words that may make for a smoother conclusion.

Why write such a story as this? I tried to explain my purpose at the outset, but now, at the end, it may be in order to clarify a few points. I have long enjoyed readings books of a "reminiscing" nature—histories, biographies, and travelogues by naturalists of the past—and I suspect that the present account is partly the result of a long-standing urge to do something in a similar vein. But insofar as I may have succeeded, the credit is due to the people whose lives and works I have looked into. It is my impression that there are not quite as many books of this kind as there used to be, although I hope that I am wrong. Science and even natural history have become very diverse, and it is reasonable that a publisher should be concerned about who will read an account of this kind. The number of people who are interested in, and actively studying, the life of the past has increased greatly, and younger people who have serious thoughts about devoting their lives to botanical or geological pursuits will be better scientists if they have some knowledge of the struggles and adventures of their predecessors. To understand Charles Darwin and the philosophic era that he created, I believe it is much more important to read his account of the voyage of H.M.S. *Beagle* than anything else that he wrote, and it is indeed grand reading. To appreciate and enjoy the excitement of the early days of American vertebrate paleontology books like Henry Fairfield Osborn's *Cope: Master Naturalist* make it difficult to resist the urge to follow the old trails again, and open up new ones. And to know what is perhaps the finest in literary natural history, for which there is no substitute that I am aware of, there is W. H. Hudson's *Far Away and Long Ago* or *Idle Days in Patagonia*.

It seems to me that there is no great overproduction today in really well written scientific literature, and I refer to works of a popular nature as well as those that fill the technical journals. I think this lack is due in part to the fact that many scientists are not concerned with the readability of their work, and if they are good writers they do not take sufficient time with their students to teach them something of the art. There are several unfortunate results of this general deficiency; editors and reviewers are unjustly overburdened, much of the literature makes dull reading, and communications are neither attractive nor effective. The ability to write well is not just a thing that one inherits, although admittedly some have a greater genetic facility than others. It is hard work, but the final result can be worth it. People like D. H. Scott, Marie Stopes, and Albert Seward were great in part because they were effective and entertaining writers. I have long urged my students to read the works of D. H. Scott and Somerset Maugham as an aid to improving their own writing. Not many have taken me seriously as far as Maugham is concerned, but I most certainly meant it.

I have no illusions that the present book is a literary masterpiece but I do hope that it will prove interesting—even entertaining. I have tried to peer into the lives and works of many of our colleagues of the past, and a few of the present, and open a window here and there for others to look through. In so doing I have introduced a certain amount of what might be regarded as gossip. As long as the gossip is of a kindly nature I believe its inclusion is justifiable and significant.

But I also hope that the subject material will be of lasting value, for it has been assembled with very serious intentions. I recently received from a colleague a letter that states my own goal better than I have been able to express it, and he has kindly allowed me to quote from it. He says: "When I first began to work with Devonian materials, especially lycopods, I began to fear that I was to spend the remainder of my life correcting Dawson's errors or misinterpretations. With the accumulation of a lot more years and a very little more humility, and maturity, I find that I see him as a man of his times laboring as I do under a load of current beliefs, wisdom and preconceptions. He was, however, a man of many parts and I have come to respect him."

It is thus an old and well-worn story: A little knowledge about a subject can be erroneous and misleading. Many people have simply ridiculed the curious ideas of the naturalists of the seventeenth and eighteenth centuries, and the mistakes of men and women of a more recent era may attract more attention than their positive contributions. It is quite necessary to have some understanding of the intellectual environment of the times, and of the character of the people involved, to evaluate their work fairly. There have been fossil botanists of the past who were superficial and inconsequential, but in most cases my respect for the people whose lives I have investigated at all carefully has increased, and in a few instances I have been greatly chagrined to find out how very wrong I had been in judging individuals on the basis of scattered bits of information. I therefore hope that what I have brought together in all of the preceding pages will help to reveal the fossil hunters of the past three centuries in their best light.

In a few final words I will try to deal more directly with the matter of the coverage of my story. Temporally as well as geographically I have confined my efforts to facets of paleobotanical history of which I had some knowledge at the start or that I could present in a reasonable period of further investigation. Considerable work has gone on in Japan, and I am aware of Toshimasa Tanai's book *Palaeobotany in Japan* and the "Catalogue of Type-specimens of Fossils in Japan" published by the Palaeontological Society of Japan in 1961. South America also has been neglected, although Sergio Archangelsky, among others, has sent me numerous papers, including his "Paleobotany and Palynology in South America: A Historical Review" (1968). And in that area it was sad to learn of the death of Carlos A. Menendez in 1976. Dr. Menendez was on the staff of the Museo Argentino de Ciencias Naturales in Buenos Aires for thirty years and Head of the Paleobotanical Division for part of that period. I believe that the last work he sent me was a fine memoir on the Bennettitales based on fossils from the Ticó Flora of Argentina.

I retired from my university position in 1975 and have devoted myself almost entirely to the present task for four years. Even if I were to devote another four years to it I suspect that I would still be far from having brought it "up to date." I can only add that I am most grateful for the many colleagues who have continued to send me their publications and I will mention just a sampling which indicate the many paths along which paleobotany is now progressing.

I think we need more general works such as William Chaloner and Margaret Collinson's "Illustrated Key to the Commoner British Upper Carboniferous Plant Compression Fossils" (1975). I cannot avoid an emphasis on recent studies of early land vegetation where many important discoveries are being made. Dianne Edwards in Wales has added much in this direction with her descriptions of *Gosslingia*, *Cooksonia*, and other genera of very simple land plants. And Rudolf Daber in 1960 described a remarkable one, *Eogaspesiea gracilis*, from the Gaspé of Canada. It seems to me that these, and other studies that I have mentioned previously, indicate quite clearly that land vascular plants evolved in the late Silurian, and I am sure that much more will be known about them in the next decade.

I have received numerous informative reports on Carboniferous plants in recent years from W. and R. Remy, as well as Manfred Berthel and Ludwig Rüffle. And one of the most impressive memoirs on the coenopterid ferns that I know of is John Holmes's work on *Psalixochlaena* which came from the paleobotanical laboratory at Montpellier in 1977. I have been especially impressed with the excellent contributions, so well illustrated, that John Townrow has brought out on the Peltaspermaceae and Corystospermaceae; I know that my old mentor Hamshaw Thomas would have been greatly intrigued with these developments.

The fossil algae have been dealt with only lightly, partly because I have little familiarity with the literature and partly because they seem to constitute a world of their own in paleontology. Like D. H. Scott, I regard the living algae as a fascinating group from which one can learn much about living things. There are so many different kinds, and among them the stoneworts (*Chara*) have always interested me. I would like to have reviewed the fine work of Louis Grambast on the fossil charophytes; his death in 1976 was a very great loss to our science. And so it goes.

It is my hope that this will be taken for what it is, fragmentary though it may be, and if there is significance and a further interest in accounts of this kind perhaps others will continue where I must leave off.

IT IS A WHOLESOME THING, AND A HELPFUL, TO REMEMBER THE GOOD WORK MEN HAVE DONE, AND TO THANK GOD FOR IT AND FOR THEM.

—Henry Jackson of Trinity College, quoted in A. C. Seward's biographical sketch of John Ray, 1937

Sources of Information

Articles and books that are chiefly biographical.

ARBER, E. A. NEWELL.

Arber, Agnes. 1918. Geol. Mag. Dec 6. 5:426–431.

Scott, D. H. 1918. Proceed. Linnean Soc. London. Session 131:39–48.

ARNOLD, CHESTER A.

Jones, Kenneth L., and Charles B. Beck. 1977. Memorial Service, Mattaei Botanical Gardens, Ann Arbor, Michigan. 7 pp.

ARTIS, EDMUND T.

Anon. 1849. Quart. Journ. Geol. Soc. London. 5:xxii–xxiii.

BALFOUR, JOHN HUTTON.

Bettany, G. T. 1908. Dict. Nat. Biogr. Brit. 1:976.

BEAN FAMILY.

McMillan, Nora F., and E. F. Greenwood. 1972. The Beans of Scarborough: a family of naturalists. Journ. Soc. Biblio. Nat. Hist. 6:152–161.

BELL, WALTER.

Anon. 1944. The Institute Journ. 23:93–94.

McLaren, D. J. 1969. Proceed. Roy. Soc. Canada, 4th ser. 7:45–48.

BERRY, EDWARD WILBER.

Cloos, Ernst. 1974. Nat. Acad. Sci. Biogr. Mem. 45:57–95.

Stephenson, Lloyd W. 1946a. The Johns Hopkins Alumni Mag. 34:109–113.

——1946b. Geol. Soc. Amer. Proceed. for 1945. Pp. 193–214.

BERTRAND, PAUL.

Mahabale, T. S. 1954. Two French savants: Charles Eugène Bertrand, the botanist, and Paul Bertrand, the paleobotanist. Bull. Mus. Paris, 2 ser. 26:444–453.

BINNEY, EDWARD W.

Anon. 1881. In: Centenary of science in Manchester. Mem. and Proceed. Manchester Lit. and Phil. Soc. 10:447–464.

Binney, James. 1912. The centenary of a 19th century geologist. Taunton, England. 58 pp.

Hunt, Robert. 1908. Dict. Nat. Biogr. Brit. 2:518–519.

BOWERBANK, JAMES S.

Anon. 1877. Geol. Mag., n.s. 4:191–192.

Hunt, Robert. 1908. Dict. Nat. Biogr. Brit. 2:961–962.

Tyler, Charles. 1878. Proceed. Roy. Microscop. Soc. 1:28–30.

BRAUN, ALEXANDER C. H.

Geison, Gerald L. 1970. Dict. Sci. Biogr. 2:425–427.

Ziegenspeck, Hermann. 1953. Neue deut. Biogr. 2:528.

BRITTS, JOHN H.

Andrews, Henry N. 1947. Ann. Missouri Bot. Gard. 34:115–117.

BRONGNIART, ADOLPHE.

de Launay, Louis. 1940. Les Brongniart, une grande famille de savants. Paris: G. Rapilly.

de Saporta, Le Comte G. 1876. Bull. Soc. Géol. France, 4:373–406.

BROWN, ROLAND W.

Mamay, Sergius H. 1963. Bull. Geol. Soc. Amer. 74:79–83.

BUCKLAND, WILLIAM.

Mrs. Gordon. 1894. The life and correspondence of William Buckland. London: John Murray.

Hunt, Robert. 1900. Dict. Nat. Biogr. Brit. 3:206–208.

BUNBURY, CHARLES J. F.

Hooker, J. D. 1890. Proceed. Roy. Soc. London. 56:xiii–xiv.

CARPENTIER, ALFRED.

Delépine, G. 1953. Bull. Soc. Géol. France, ser. 6, 3:427–440.

Depape, George, 1953. Rev. Gén. Bot. 60:545–572.

CARRUTHERS, WILLIAM.

Britten, James. 1922. Journ. Bot. Brit. and Foreign. 60:249–256.

Rendle, A. B. 1925. Proceed. Roy. Soc. London. 97B:vi–viii.

Smith, William G. 1922. Trans. Bot. Soc. Edinburgh. 28:118–121.

Woodward, Henry. 1912. Geol. Mag. n.s. Dec. 4. 9:193–199.

CASPARY, ROBERT.

Anon. 1888. Ann. Bot. 1:387–395.

Pfitzer, E. 1888. Ber. deut. bot. Gesell. 6:xxvii–xxxi.

CHANEY, RALPH W.

Axelrod, Daniel I. 1971. Amer. Phil. Soc. Yearbook, 1971. Pp. 115–120.

Gray, Jane, and D. I. Axelrod. 1971. Memorial to Ralph Works Chaney 1890–1971. Geol. Soc. Amer. 9 pp.

CHARLTON, LIONEL.

Cooper, Thompson. 1908. Dict. Nat. Biogr. Brit. 4:128.

CIST, JACOB.

Mitman, Carl W. 1930. Dict. Amer. Biogr. 4:109–110.

CONWENTZ, HUGO.

Klose, H. 1922. Naturschutz. 3:180–186.

Vogel, Stefan. 1956. Neue deut. Biogr. 3:347.

COOKSON, ISABEL C.

Baker, George. 1973. Rev. Palaeobot. and Palynol. 16:133–135.

CORDA, AUGUST JOSEPH.

Weitenweber, Wilhelm R. 1852. Abhand. könig. böhmischen Gesell. Wissen. Special paper. 39 pp.

COTTA, CARL BERNARD.

Krenkel, Erich. 1956. Neue deut. Biogr. 3:381.

CRICHTON, ALEXANDER.
Boase. G. C. 1908. Dict. Nat. Biogr. Brit. 5:85–86.
CROFT, WILLIAM N.
Edwards, W. N. 1953. Nature, Sept. 19. P. 524.
——1954. Proceed. Geol. Soc. London. No. 1515. Pp. cxxxii–cxxxiii.
DA COSTA, EMANUEL MENDES.
Goodwin, Gordon. 1908. Dict. Nat. Biogr. Brit. 4:1196–1197.
DAWSON, J. WILLIAM.
Adams, Frank D. 1899. Bull. Geol. Soc. Amer. 11:550–580.
Ami, Henry M. 1900. Amer. Geol. 26:1–48.
EDWARDS, WILFRED N.
White, E. I. 1957. Proceed. Geol. Soc. London. No. 1554. Pp. 142–144.
Wonnacott, F. M. 1957. Journ. Soc. Biblio. Nat. Hist. 3:231–237.
EICHWALD, EDUARD.
Tikhomirov, V. V. 1971. Dict. Sci. Biogr. 4:307–309.
FAUJAS DE SAINT-FOND, B.
Geikie, Archibald. 1907. Memoir of the author, in: A journey through England and Scotland to the Hebrides in 1784, by B. Faujas de Saint-Fond. 1:xv–xxxii. Glasgow: Hugh Hopkins.
FLORIN, RUDOLF.
Harris, T. M. 1968. Phytomorphology. 18:174–178.
Lundblad, Britta. 1966. Taxon. 15:85–93.
FONTAINE, WILLIAM H.
Watson, Thomas L. 1914. Bull. Geol. Soc. Amer. 25:6–12.
GEINITZ, HANS BRUNO.
Laube, Gustav C. 1900. Lotos. 48:11–14.
GOEPPERT, HEINRICH ROBERT.
Conwentz, H. 1885. Naturforschenden Gesell. Danzig. 6:1–22.
Tobein, Heinz. 1972. Dict. Sci. Biogr. 5:440–442.
GOLDENBERG, CARL F.
Dechen, H. von. 1881. Verh. mit. hist. Ver. preuss. Rheinl. u. Westfal. (Bonn). 30:58–64.
GOLDRING, WINIFRED.
Fisher, Donald W. 1971. Memorial to Winifred Goldring, 1888–1971. Geol. Soc. Amer. 7 pp.
GORDON, WILLIAM T.
Rowland, E. O. 1976. Reminiscences of Prof. William Thomas Gordon, a typescript of a recording, deposited in the U.S. Geol. Survey Paleobot. Library, Washington, D.C.
Swinton, W. E. 1950. Proceed. Linnean Soc. London. 163:82–83.
Taylor, James H. 1951. Quart. Journ. Geol. Soc. London. 106:lx–lxiii.
GOTHAN, WALTHER.
Hartung, Wolfgang. 1956. Zeit. deut. geol. Gesell. 107:311–314.
Potonié, Robert. 1955. Geol. Jb. 70:xxvii–liii.
Thiergart, F. 1955. Ber. deut. bot. Gesell. 68:55–57.
GRAND'EURY, CYRILLE.
Berry, E. W. 1918. Science, n.s. 47:62–63.

Bertrand, Paul, 1919. Bull. Soc. Géol. France, ser. 4. 19:148–162.
Stockmans, F. 1972. Dict. Sci. Biogr. 5:497–498.

GROVES, JAMES.
Pugsley, H. W. 1933. Journ. Bot. Brit. and Foreign. 71:136–139.

HALLE, THORE GUSTAV.
Florin, Rudolf. 1948. Palaeontographica. 88B:vi–ix.
Lundblad, Britta. 1969. Svenskt Biogr. Lexikon. 18:15–18.
Walton, John. 1966. Year Book Roy. Soc. Edinburgh. Pp. 20–22.

HEER, OSWALD.
Anon. Nature, Oct. 25, 1883. Pp. 612–613.
Saporta, G. de. 1884. Rev. des Deux Mondes. 64:162–195. 884–915.
Scott, R. H. 1883. Geol. Mag., London. 10:575–576.
Tobein, Heinz. 1972. Dict. Sci. Biogr. 6:220–222.

HICK, THOMAS.
Cash, William. 1897. Proceed. Yorkshire Geol. and Polytech. Soc. 13:234–239.

HILDRETH. S. P.
Mathews, Albert P. 1932. Dict. Amer. Biogr. 9:21–22.

HITCHCOCK, EDWARD.
Merrill, George P. 1932. Dict. Amer. Biogr. 9:70–71.

HOLDEN, HENRY SMITH.
Harris, Tom M. 1963. Proceed. Linnean Soc. London. 175:94.
Wonnacott, F. M. 1964. Journ. Soc. Biblio. Nat. Hist. 4:230–234.

HOLDEN, RUTH.
Seward, A. C. 1917. New Phytologist. 16:154–156.

HOLLICK, ARTHUR.
Britton, N. L. 1933. Journ. N.Y. Bot. Gard. 34:121–124.
Howe, Marshall A. 1933. Bull. Torrey Bot. Club. 60:536–553.
Jeffrey, E. C. 1933. Science, n.s. 77:440–441.

HOOKE, ROBERT.
Clerge, A. M. 1908. Dict. Nat. Biogr. Brit. 9:1177–1181.
'Espinasse, Margaret. 1956. Robert Hooke. Berkeley: Univ. Calif. Press.
Waller, Richard. 1705. The posthumous works of Robert Hooke, containing his Cutlerian lectures, and other discourses. London.

HOOKER, JOSEPH D.
Boulger, G. S. 1912. Journ. Bot. Brit. and Foreign. 50:1–9, 33–43.
Scott, D. H. 1912. Proceed. Linnean Soc. London. Pp. 22–39.
Seward, A. C. 1912. Sir Joseph Hooker and Charles Darwin: the history of a forty-years' friendship. New Phytologist. 11:195–206.

HORN AF RANTZIEN, H.
Edelstam, C. 1961. Nature. 191:121–122.

HOSIUS, AUGUST.
Langer, Wolfhart. 1968. Argumenta Palaeobotanica. 2:19–26.

HOSKINS, JOHN HOBART.
Anon. 1957. Lloydia. 20:ii–vi.

HUTTON, WILLIAM.
Archbold, W. A. J. 1908. Dict. Nat. Biogr. Brit. 10:363.

JEFFREY, EDWARD C.
Wetmore, Ralph H., and Elso S. Barghoorn. 1953. Phytomorphology. 3:127–132.
JONGMANS, W. J.
Van Rummelen, F. H. 1958. Natuurhistorische Maandblad. Pp. 145–148.
Thiadens, A. A. 1957. Geologie en Mundbouw, n.w. ser. 19:417–425.
KIDSTON, ROBERT W.
Crookall, R. 1938. The Kidston Collection of fossil plants with an account of the life and work of Robert Kidston. Mem. Geol. Survey Great Brit. 34 pp.
Lang, W. H. 1925. Proceed. Roy. Soc. Obit. Notices. Pp. i–ix.
Scott, D. H. 1924. Nature. 114:321–322.
Seward, A. C. 1924. Geol. Mag. 61:477–479.
KNOWLTON, FRANK HALL.
Berry, E. W. 1927. Science. 55:7–8.
White, David. 1927. Bull. Geol. Soc. Amer. 38:52–70.
KRASSER, FRIDOLIN.
Greger, Justin. 1922. Ber. deut. bot. Gesell. 40:112–121.
Rudolph, Karl. 1922. Lotos. 70:113–140.
KRÄUSEL, RICHARD.
Banks, Harlan P. 1968. Phytomorphology. 18:178–179.
Dilcher, David L. 1967. Plant Sci. Bull. 13: no. 1.
Edwards, W. N. 1952. Palaeontographica. 92B:53–62.
Schaarschmidt, F. 1968. Ber. deut. bot. Gesell. 81:65–80.
KRYSHTOFOVICH, AFRIKAN N.
Edwards, W. N. 1956. Proceed. Geol. Soc. London. No. 1541. Pp. 127–128.
Grubov, V. I., and P. I. Dorofeev. 1934. Bot. Journ. 39:306–312 (in Russian).
LACOE, R. D.
White, David. 1901. Bull. Geol. Soc. Amer. 13:509–515.
LANG, WILLIAM HENRY.
Salisbury, E. J. 1961. Roy. Soc. London Biogr. Mem. 7:147–160.
LEIGH, CHARLES.
Sutton, C. W. 1909. Dict. Nat. Biogr. Brit. 11:872–873.
LESQUEREUX, LEO.
Lesley, J. P. 1895. Nat. Acad. Sci. Biogr. Mem. 3:189–212.
Orton, Edward. 1890. Amer. Geol. 5:284–296.
Sarton, George. 1942. Isis. 34:97–108.
LHWYD, EDWARD.
Gunther, R. T. 1945. The life and letters of Edward Lhwyd (including: Some incidents in the life of Edward Lhwyd, by Richard Ellis, p. 1–51). Vol. 14 of: Early science in Oxford. Oxford: Oxford Univ. Press.
Thomas, D. L. 1909. Dict. Nat. Biogr. Brit. 11:1096–1098.
LIGNIER, OCTAVE.
Chevalier, A. 1920. Notice biographique. Paris. 19 pp.
Jeffrey, E. C. Bot. Gazette. 62:507–508.
Stockmans, F. 1973. Dict. Sci. Biogr. 8:354.
LINDLEY, JOHN.
Anon. 1865. Journ. Bot. Brit. and Foreign. 3:384–388.

LISTER, MARTIN.
Boulger, G. S. 1909. Dict. Nat. Biogr. Brit. 11:1229–1230.
Stearns, Raymond P. 1967. A journey to Paris in the year 1698, ed. with annotations, a life of Lister, and a Lister bibliography. Urbana: Univ. Illinois Press.
LYELL, CHARLES.
Bailey, Edward. 1959. Roy. Soc. London Notes and Records. 14:121–138.
Cole, Grenville A. J. 1909. Dict. Nat. Biogr. Brit. 12:319–324.
MANTELL, GIDEON.
Bonney, T. G. 1909. Dict. Nat. Biogr. Brit. 12:984–985.
Curwen, E. C. ed. 1940. The journal of Gideon Mantell, surgeon and geologist. Oxford: Oxford Univ. Press.
MARTIN, WILLIAM.
Anon. 1811. An account of the life and writings of the late Mr. William Martin. Monthly Mag. or British Register. 32:556–565.
Woodward, B. B. 1909. Dict. Nat. Biogr. Brit. 12:1185–1186.
MARTIUS, KARL F. P. VON.
H. W. 1860. Nouv. Biogr. Gén. 33:100–102.
MORRIS, JOHN.
Anon. 1878. Geol. Mag. n.s. Dec. 2. 5:481–487.
Bonney, T. G. 1909. Dict. Nat. Biogr. Brit. 13:995.
MORTON, SAMUEL G.
Fisk, Daniel M. 1934. Dict. Amer. Biogr. 13:265–266.
NATHORST, ALFRED G.
Halle, T. G. 1921. Geol. Föreningens Förhandlingar. 43:241–311.
Holmboe, Jens. 1921. Naturen (Norway). February. Pp. 33–40.
Seward, A. C. 1921. Bot. Gazette. 71:462–465.
NĚMEJC, FRANTIŠEK.
Knobloch, Erwin. 1966. Ber. deut. Gesell. Geol. Wiss. a. Geol. Paläeont. 11:403–411.
NEUBURG, MARIA F.
Meyen, S. V. 1963. Acad. Sci. U.S.S.R. Moscow. Pp. 151–153.
NEWBERRY, JOHN S.
Britton, N. L. 1893. Bull. Torrey Bot. Club. 20:88–98.
Fairchild, Herman Le Roy. 1893. Scientific Alliance of N.Y. Proceed. 2d joint meeting, Columbia College. March 27. 39 pp.
Waller, A. E. 1943. Ohio State Archaeol. Hist. Quart. October. Pp. 324–346.
NIKITIN, PETR A.
Dorofeev. P. I. 1960. Bot. Journ. Acad. Sci. U.S.S.R. 45:619–624.
NOÉ, ADOLF C.
Croneis, Carey. 1940. Geol. Soc. Amer. Proceed. for 1939. Pp. 219–227.
NORDENSKIÖLD, ADOLPHE ERIC.
Flahault, C. 1880. Nordenskiöld, Notice sur sa vie et ses voyages. Paris: K. Nilson. 76 pp.
OLIVER, FRANCIS W.
Salisbury, E. J. 1952. Roy. Soc. London Obit. Notices, Fellows. 8:229–240.
PARKINSON, JAMES.
Bougler, G. S. 1909. Dict. Nat. Biogr. Brit. 15:314–315.

PARSONS, JAMES.
Wroth, Warwick. 1909. Dict. Nat. Biogr. Brit. 15:403–404.
PENGELLEY, WILLIAM.
Anon. 1894. Geol. Mag. n.s. Dec. 4. 1:238–239.
Anon. 1895. Quart. Journ. Geol. Soc. London. 51:liii–lvii.
PENHALLOW, DAVID P.
Barlow, Alfred E. 1910. Bull. Geol. Soc. Amer. 22:15–18.
Jeffrey, E. C. 1911. Bot. Gazette. 51:142–144.
PETRY, LOREN C.
Banks, Harlan P. 1970. Plant Sci. Bull. 16:10–11.
PIA, JULIUS.
Trauth, Friedrich. 1947. Annalen Naturhist. Museums, Vienna. 55:19–49.
PLOT, ROBERT.
Challinor, John. 1945. Dr. Plot and Staffordshire geology. Trans. North Staffordshire Field Club. 79:29–67.
Seccombe, Thomas. 1909. Dict. Nat. Biogr. Brit. 15:1310–1312.
POTONIÉ, HENRY.
Gothan, W. 1914. Jahrbuch (1913) könig. preuss. geol. Landesanstalt (Berlin). 34:535–559.
PRESL, KARL BORŽIWOG.
Anon. 1854. K. Akad. wissenschaften, Vienna, Almanach. Pp. 120–126.
RAY, JOHN.
Eyles, Joan M. 1955. Nature. 175:103–105.
Gunther, R. W. T. 1928. Two short lives of John Ray, in: Further correspondence of John Ray. London: Ray Society. 332 pp.
Raven, Charles E. 1942. John Ray, naturalist, his life and works. Cambridge: Cambridge Univ. Press.
Seward, A. C. 1937. John Ray, a biographical sketch written for the Centenary of the Cambridge Ray Club. Privately printed. 43 pp.
REID, CLEMENT.
Groves, James. 1917. Journ. Bot. 55:145–151.
G. W. L. 1917. Quart. Journ. Geol. Soc. London. 73:lxi–lxiv.
J. E. M. and E. T. N. 1917. Proceed. Roy. Soc. London. 90B:viii–x.
REID, ELEANOR.
Edwards, W. N. 1954. Proceed. Geol. Soc. London. No. 1514, pp. cxl–cxlii.
RENAULT, BERNARD.
Roche, A. 1905. Mem. Soc. Hist. Nat. Autun. Vol. 18. 159 pp.
Scott, D. H. 1906, Journ. Roy. Microscop. Soc. Pp. 129–145.
RHODE, JOHANN G.
Hoche, R. 1889. Allgemeine deut. Biogr. 28:391–392.
SAHNI, BIRBAL.
Edwards, W. N. 1950. Quart. Journ. Geol. Soc. London. 105:lxxvii–lxxix.
Halle, T. G. 1952. The Palaeobotanist. 1:22–41.
Rao, A. R. 1952. The Palaeobotanist. 1:9–16.
Thomas, H. H. 1950. Roy. Soc. London Obit. Notices, Fellows. 7:265–277.
SAPORTA, LOUIS C. J. GASTON, MARQUIS DE.
Dawson, J. W. 1895. Canadian Rec. Sci. 6:367–369.

Stockmans, F. 1975. Dict. Sci. Biogr. 12:104–105.
Williamson, W. C. 1896. Mem. and Proceed. Manchester Lit. and Phil. Soc., ser. 4. 10:210–212.
Zeiller, R. 1895. Rev. Gén. Bot. 7:353–387.

SCHENK, JOSEPH AUGUST VON.
Drude, O. 1891. Berichte deut. bot. Gesell. 9:15–26.

SCHEUCHZER, JOHANN JAKOB.
Fischer, Hans. 1973. Johann Jakob Scheuchzer, Naturforscher und Arzt. Neujahrsblatt Naturforschenden Gesell. Zürich. 168 pp.

SCHIMPER, WILHELM PHILIPP (GUILLAUME PHILIPPE).
Grad, Charles. 1880. Bull. Soc. d'hist. naturelle de Colmar. 20–21:351–392.
Richards, P. W. 1975. Dict. Sci. Biogr. 12:168–169.

SCHLOTHEIM, ERNST F. VON.
Daber, Rudolf. 1970. E. F. von Schlotheim und der Beginn der wissenschaftlichen Fragestellung in der Palaobotanik vor 150 Jahren. Wissen. Zeitschrift Humboldt-Universität Berlin, Math. Natur. Reihe. 19:249–255.
Langer, Wolfhart. 1966. Argumenta Palaeobotanica. 1:19–40.

SCOTT, DUKINFIELD HENRY.
Arber, Agnes. 1954. Nature. 174:992.
Oliver, F. W. 1935. Ann. Bot. 49:823–840.
Rendle, A. B. 1934. Journ. Bot. Pp. 83–88.
Seward, A. C. 1934. Roy. Soc. London Obit. Notices, Fellows. Pp. 137–139.

SEWARD, ALBERT CHARLES.
Harris, T. M. 1941. New Phytologist. 40:161–164.
Thomas, H. H. 1941. Roy. Soc. London Obit. Notices, Fellows. 3:867–880.

SMITH, WILLIAM.
Phillips, John. 1839. Mag. Nat. Hist. (London), n.s. 3:213–220.

SOLMS-LAUBACH, HERMANN, COUNT.
Jost. L. 1915. Ber. deut. bot. Gesell. 33:95–112 (Supplementary section).
Robinson, B. L. 1925. Proceed. Amer. Acad. Arts and Sci. 59:651–656.
Scott, D. H. 1918. Roy. Soc. London Obit. Notices. 90B:xix–xxvi.

STEINHAUER, HENRY.
Reichel, William C. 1870. A history of the rise, progress and present condition of the Moravian Seminary, 2nd ed. Philadelphia. Pp. 189–199.

STENO, NICHOLAUS.
Eyles, V. A., Axel Garboe, and Gustav Scherz. 1958. Articles in: Acta Historica Scientiarum Naturalium et Medicinalium, vol. 15. Copenhagen.
Winter, John G. 1968. Nicholaus Steno: Prodromus of a dissertation concerning a solid body. In Contributions to the history of geology, ed. George W. White. Vol. 4. New York: Hafner.

STERNBERG, CHARLES H. 1903. Life of a fossil hunter. American Inventor. 10:311–313.
——. 1931. The life of a fossil hunter. San Diego: Jensen Printing Co. 286 pp.

STERNBERG, KASPAR MARIA.
Beckinsale, Robert P., and Jan Krejčí. 1976. Dict. Sci. Biogr. 13:43–44.
Palacky, Franz, et al. 1868. Die Grafen Kaspar und Franz Sternberg, und ihr Wirken für Wissenschaft und Kunst in Böhmen. Prague. 242 pp.

STERZEL, JOHANN T.
Anon. 1915. Ber. deut. Naturwiss. Gesell. Chemnitz. 19:7–11.
STOPES, MARIE C.
Briant, Keith. 1962. Passionate Paradox: The life of Marie Stopes. New York: Norton.
Chaloner, W. G. 1959. Proceed. Geol. Assoc. 70:118–120.
Maude, Aylmer. 1933. Marie Stopes. London: Peter Davies.
Watson, D. M. S. 1959. Proceed. Geol. Soc. London. No. 1572. Pp. 152–153.
STUR, DIONYS.
Vacek, M. 1894. Jahrbuch könig̣. geol. Reichsanstalt. 44:1–12.
SZAFER, WŁADYSŁAW.
Walas, J. 1972. Acta Soc. Bot. Poloniae. 41:1–11.
THOMAS, HUGH HAMSHAW.
Harris, T. M. 1963. Roy. Soc. London Biogr. Mem. Fellows. 9:287–299.
THOMPSON, FREDERICK O.
Barghoorn, Elso S. 1953. Harvard Univ. Bot. Mus. Leaflets. 16:173–178.
UNGER, FRANZ.
Reyer, Alexander. 1871. Leben und Wirken des Naturhistorikers Dr. Franz Unger—Compiled on behalf of the Association of Physicians of Steiermark, Graz. 100 pp.
USSHER, JAMES.
Gordon, Alexander. 1909. Dict. Nat. Biogr. Brit. 20:64–72.
VELENOVSKÝ, JOSEF.
Němejc, F. 1958. Preslia. 30:277–280.
WALCH, JOHANN E. I.
Dobschütz, 1896. Allgemeine deut. Biogr. 40:652–655.
WALCOTT, CHARLES D.
Schuchert, Charles. 1928. Proceed. Amer. Acad. Arts and Sci. 62:276–285.
Yokelson, Ellis L. 1967. Nat. Acad. Sci. Biogr. Mem. 39:471–529.
WALTON, JOHN.
Hutchinson, S. A., and A. M. Young. 1971. The College Courant (Glasgow). Pp. 52–53.
WARD, LESTER F.
Cape, Emily P. 1922. Lester F. Ward: a personal sketch. New York: Putnam.
Chugerman, Samuel. 1939. Lester F. Ward, the American Aristotle. Durham, N.C.: Duke Univ. Press.
WATSON, DAVID M. S.
Parrington, F. R., and T. S. Westoll. 1974. Roy. Soc. London Biogr. Mem. Fellows. 20:483–504.
WEISS, CHRISTIAN E.
Sterzel, J. T. 1892. Jahrbuch (1890) könig̣. preuss. geol. Landesanstalt, Berlin. Pp. 1–25.
WEISS, FREDERICK E.
Thomas, H. H. 1953. Roy. Soc. London Obit. Notices, Fellows. 8:601–608.
WEYLAND, HERMANN.
Gothan, W. 1953. Palaeontographica. 95B:1–5.
Kilpper, K. 1963. Palaeontographica. 123B:1–4.

WHISTON, WILLIAM.
Stephen, Leslie. 1909. Dict. Nat. Biogr. Brit. 21:10–14.
WHITE, DAVID.
Berry, E. W. 1935. Amer. Journ. Sci. ser. 5. 29:390–391.
Mendenhall, W. C. 1935. Science. 81:244–246.
Schuchert, Charles. 1936. Nat. Acad. Sci. Biogr. Mem. 17:189–221.
Stanton, T. W. 1935. Journ. Geol. 43:778–780.
WHITE, I. C.
Fairchild, Herman L. 1928. Bull. Geol. Soc. Amer. 39:126–145.
Hayhurst, Ruth I. 1972. Newsletter West Virginia Geol. Survey. 16:12–16.
WILLIAMSON, WILLIAM C.
——. 1896. Reminiscences of a Yorkshire naturalist, ed. by A. C. Williamson. London: George Redway.
Bailey, Charles. 1886. Manchester Sci. Students' Ann. Rept. 8 pp.
Scott, D. H. 1899. Proceed. Roy. Soc. London. 60:27–32.
——. 1913. William C. Williamson, in: Makers of British botany, ed. by F. W. Oliver, pp. 243–260. Cambridge: Cambridge Univ. Press.
Solms-Laubach, H. 1895. Nature. 52:441–443.
Ward, Lester F. 1895. Saporta and Williamson and their work in paleobotany. Science, n.s. 2:141–150.
WITHAM, HENRY.
Long, Albert G. 1975. Dict. Sci. Biogr. 14:462–463.
WOODWARD, JOHN.
Clark, J. W., and T. M. Hughes. 1890. Sketch of the life and works of John Woodward, in: The life and letters of Adam Sedgwick, pp. 167–183. Cambridge: Cambridge Univ. Press.
Woodward, B. B. 1909. Dict. Nat. Biogr. Brit. 21:894–896.
ZALESSKY, MIKHAIL D.
Kryshtofovich. A. 1949. Quart. Journ. Geol. Soc. London. 104:xlii–xliii.
ZEILLER, RENÉ C.
Berry, E. W. 1916. Science, n.s. 43:201–202.
Bonnier, Gaston. 1917. Rev. Gén. Bot. 29:1–68.
Douville, H. 1917. René Zeiller, Notice nécrologique. Bull. Soc. Geol. France, ser. 4. 17:301–313.
Scott, D. H. 1916. Proceed. Linnean Soc. London. 128:74–78.

Articles and books quoted in the text or used for general information.

Allan, Thomas. 1823. Description of a vegetable impression found in the quarry of Craigleith. Trans. Roy. Soc. Edinburgh. 9:235–237.

Andrews, Henry N. 1970. Index of generic names of fossil plants, 1820–1965. U.S. Geol. Survey Bull. 1300. 354 pp.

Andrews, Henry N., et al. 1977. Early Devonian flora of the Trout Valley Formation. Rev. Palaeobot. and Palynol. 23:255–285.

Arber, E. A. N. 1921a. Devonian floras. Cambridge: Cambridge Univ. Press.
——. 1921b. A sketch of the history of palaeobotany, in: Studies in the history and method of science, ed. by Charles Singer. Oxford: Clarendon. 2:412–489.
Arnold, Chester A. 1952. A specimen of *Prototaxites* from the Kettle Point Black Shale of Ontario. Palaeontographica. 93B:45–56.
——. 1960. A lepidodendrid from Kansas. Univ. Mich. Contrib. Mus. Paleont. 15:249–267.
——. 1968. Current trends in paleobotany. Earth Sci. Rev. 4:283–309.
——. 1969. Paleobotany, in: A short history of botany in the United States, pp. 103–108, ed. by Joseph Ewan. New York and London: Hafner.
Ash, Sidney R. 1972. The search for plant fossils in the Chinle Formation. Mus. of Northern Arizona Bull. 47:45–58.
Balfour, John H. 1872. Introduction to the study of palaeontological botany. Edinburgh: Adam and Charles Black.
Banks, Harlan P. 1972. The scientific work of Suzanne Leclercq. Rev. Palaeobot. and Palynol. 14:1–5.
Beck, Charles B. 1960. The identity of *Archaeopteris* and *Callixylon*. Brittonia. 12:351–368.
Berry, Edward Wilber. 1921. Across the Andes to the Yungas. Natural History. 21:494–506.
——. 1927. Devonian floras. Amer. Journ. Sci. ser. 5. 14:109–120.
Bertrand, Paul. 1937. Notice sur les travaux scientifiques de Paul Bertrand. Lille. 64 pp.
——. 1947. Les végétaux vasculaires. Paris: Masson.
Bower, F. O. 1938. Sixty years of botany in Britain (1875–1935). London: Macmillan.
Brongniart, Adolphe. 1826. Observations on some fossil vegetables of the coal formation. Edinburgh New Phil. Journ. 3:282–289.
——. 1829. General considerations of the vegetation which has covered the surface of the earth, etc. Edinburgh New Phil. Journ. 6:349–371.
——. 1838. Reflections on the nature of the vegetables which have covered the surface of the earth, etc. Mag. Nat. Hist. 2:1–12.
——. 1850. Chronological exposition of the periods of vegetation, etc. Annals and Mag. Nat. Hist. ser. 2. 6:73–85, 192–203, 348–370.
Budantsev. L. 1969. One hundred years of the investigations of fossil floras in the Arctic. Bot. Journ. 54:486–495 (in Russian).
Bunbury, C. J. F. 1846a. Notes on the fossil plants communicated by Mr. Dawson from Nova Scotia. Quart. Journ. Geol. Soc. London. 2:136–139.
——. 1846b. On some remarkable fossil ferns from Frostburg, Maryland, collected by Mr. Lyell. Quart. Journ. Geol. Soc. London. 2:82–91.
Canright, James E. 1958. History of paleobotany in Indiana. Indiana Acad. Sci. Proceed. 67:268–273.

Challinor, J. 1948. The beginnings of scientific palaeontology in Britain. Annals of Sci. 6:46–53.

——. 1953. The early progress of British geology. I: From Leland to Woodward, 1538–1728. Annals of Sci. 9:124–253.

Chaney, Ralph W., and E. I. Sanborn. 1933. The Goshen flora of west central Oregon. Carnegie Inst. Washington. Pub. 439:1–103.

Cist, Zachariah. 1821. Account of the mines of anthracite, etc. Amer. Journ. Sci. ser. 1. 4:1–9.

Cohen, H. Hirsch. 1974. The drunkenness of Noah. University: Univ. Alabama Press.

Corsin, Paul. 1945. Les algues de l'Eodévonien de Vimy. Mém. Soc. Sci. Agri. et Arts de Lille, ser. 5. Fasc. 9, 86 pp.

Da Costa, E. M. 1758. An account of the impressions of plants on the slates of coals. Phil. Trans. Roy. Soc. London. 50:228–235.

Darrah, William C. 1936. The peel method of paleobotany. Harvard Univ. Bot. Mus. Leaflets. 4:69–83.

——. 1969. A critical review of the Upper Pennsylvanian floras of the eastern United States. Gettysburg, Pa. 220 pp.

Daugherty, Lyman H., and Howard R. Stagner. 1941. The Upper Triassic flora of Arizona. Carnegie Inst. Washington. Pub. 526. 108 pp.

Dawson, J. William. 1901. Fifty years of work in Canada, ed. by Rankine Dawson. London: Ballantyne, Hanson.

Delevoryas, T., and C. P. Person. 1975. *Mexiglossa varia* gen. et sp. nov., etc. Palaeontographica. 154B:114–120.

Dorf, Erling. 1942. Upper Cretaceous floras of the Rocky Mountain region. II. Carnegie Inst. Washington. Pub. 508. Pp. 79–159.

——. 1964. The petrified forests of Yellowstone Park. Sci. American. 210:4:107–114.

Dorofeev, P. I. 1966. The Pliocene Flora of the Matanov Garden on the River Don. Acad. Sci. U.S.S.R., Moscow. 87 pp. (in Russian with English summary).

Edwards, W. N. 1967. The early history of palaeontology. Brit. Mus. Nat. Hist. 58 pp.

Eggert, D. A. 1962. The ontogeny of Carboniferous arborescent lycopsida. Palaeontographica. 108B:43–92.

Florin, Rudolf. 1951. Evolution in cordaites and conifers. Acta Horti Bergiani. 15:285–388.

Goeppert, H. R. 1837. On the conditions of fossil plants, and on the process of petrifaction. Edinburgh New Phil. Journ. 23:73–82.

Gordon, W. T. 1908. On *Lepidophloios Scottii*, etc. Trans. Roy. Soc. Edinburgh. 46:443–453.

——. 1935. Plant life and the philosophy of geology. Brit. Assoc. Advan. Sci., Aberdeen, 1934. Pp. 49–82.

Gothan, W. 1950. Die Paleobotanik in Deutschland in den letzten 100 Jahren. Zeit. deut. geol. Gesell. 100:94–105.

Grand'Eury, C. 1905. Sur les graines trouvées attachées au *Pecopteris Pluckeneti* Schloth. Comptes Rendus Hebd. Séances Acad. Sci. 140:920–923.

Granger, Ebenezer. 1821. Notice of vegetable impressions on the rocks, etc. Amer. Journ. Sci., ser. 1. 3:5–7.

Green, J. Reynolds. 1914. A history of botany in the United Kingdom. London. 648 pp.

Harris, Tom M. 1937. The fossil flora of Scoresby Sound, East Greenland, Part 6. Meddelelser om Grönland. 112:1–114.

——. 1941. Cones of extinct Cycadales from the Jurassic rocks of Yorkshire. Phil. Trans. Roy. Soc. London. 231B:75–98.

——. 1961. The Yorkshire Jurassic flora. I. Thallophyta-Pteridophyta. Brit. Mus. Nat. Hist. 212 pp.

——. 1973. The strange Bennettitales. 19th Sir Albert Charles Seward Memorial Lecture, 1971. Birbal Sahni Inst. Palaeobot., Lucknow. 11 pp.

Hatley, Griff. 1683. A letter concerning some formed stones found at Hunton in Kent. Phil. Trans. Roy. Soc. London. 13:463–465.

Heer. O. 1869. The last discoveries in the extreme North. Ann. Mag. Nat. Hist. 4:81–101.

——. 1876. The primaeval world of Switzerland. Trans. by James Heywood. 2 vols. London: Longmans, Green.

Hewitson, W. H. 1962. Comparative morphology of the Osmundaceae. Ann. Missouri Bot. Gard. 49:57–93.

Hildreth, S. P. 1836. Observations on the bituminous coal deposits of the valley of the Ohio, etc. Amer. Journ. Sci. ser. 1. 29:1–148 (with appendix by Samuel Morton, p. 149–154).

——. 1837. Miscellaneous observations made during a tour in May, 1835, etc. Amer. Journ. Sci. ser. 1. 31:1–84.

Hirmer, Max. 1927. Handbuch der Paläobotanik, Vol. 1. Berlin: von Oldenbourg.

Høeg, Ove Arbo. 1942. The Downtonian and Devonian flora of Spitsbergen. Norges Svalbard-og Ishavs-Undersøkelser, No. 83. Oslo. 228 pp.

Hollick, Arthur, 1930. The Upper Cretaceous floras of Alaska. U. S. Geol. Survey Prof. Paper 159. 123 pp.

Holmes, W. H. 1879. Fossil forests of the volcanic Tertiary formations of the Yellowstone National Park. U.S. Geol. Survey Bull. 5:125–132.

Hooke, Robert. 1665. Micrographia. Royal Soc. London. 273 pp. Reprint ed., New York: Dover, 1961.

Hooker, J. D. 1848a. On the vegetation of the Carboniferous period. Mem. Geol. Survey Great Brit. 2:387–430.

——. 1848b. Remarks on the structure and affinities of Lepidostrobi. Mem. Geol. Survey Great Brit. 2:440–456.

——. 1855. On some minute seed-vessels from the Eocene beds of Lewisham. Quart. Journ. Geol. Soc. London. 11:562–570.

Hooker, J. D., and E. W. Binney. 1855. On the structure of certain limestone nodules enclosed in seams of bituminous coal, etc. Phil. Trans. Roy. Soc. London. 145:149–156.

Hutton, James. 1788. Theory of the Earth. Trans. Roy. Soc. Edinburgh. 1:208–304.

Jahn Melvin E., and Daniel J. Woolf. 1963. The lying stones of Dr. Johann Bartholomew Adam Beringer, being his Lithographiae Wirceburgensis. Berkeley: Univ. California Press.

Jussieu, Antoine de. 1718. Des causes des impressions des plantes marquées sur certaines pierres des environs de Saint-Chaumont. Mem. Roy. Sci. Paris. Pp. 287–297.

Kidston, Robert. 1904. On the fructification of *Neuropteris heterophylla* Brongniart. Phil. Trans. Roy. Soc. London. 197B:1–5.

——. 1923–1925. Fossil plants of the Carboniferous rocks of Great Britain. Mem. Geol. Survey Great Brit. Palaeontology. Vol. 4, parts 1–6. London.

Kidston, R., and W. H. Lang. 1917. On Old Red Sandstone plants showing structure, etc. Part I. Trans Roy. Soc. Edinburgh. 51:761–784.

Kirchheimer, Franz. 1936. Beiträge zur Kenntnis der Tertiär flora: Fruchte und Samen aus dem deutschen Tertiär. Palaeontographica. 82B:73–141.

Knowlton, F. H. 1899. The fossil flora of the Yellowstone National Park. U.S. Geol. Survey Mon. 32:651–881.

Kryshtofovich, A. N. 1956. History of paleobotany in the U.S.S.R. Acad. Sci. U.S.S.R., Moscow. 108 pp. (in Russian).

Lacey, William S. 1963. Palaeobotanical techniques, in: Viewpoints in biology, ed. by J. D. Carthy and C. L. Duddington. London: Butterworth.

Leisman, Gilbert A. 1968. A century of progress in paleobotany in Kansas. Trans. Kansas Acad. Sci. 71:301–308.

Lindley, John, and William Hutton. 1831–1837. The fossil flora of Great Britain. 3 vols. London: James Ridgway.

Lister, Martin. 1671. A letter of Mr. Martin Lister, etc. Phil. Trans. Roy. Soc. London. 6:2281–2284.

Long, Albert G. 1944. On the prothallus of *Lagenostoma ovoides* Will. Ann. Bot., n.s. 8:105–117.

——. 1959–1960. The fossil plants of Berwickshire: a review of past work. Part 1, 1959; part 2, 1960. History Berwick. Nat. Club. 34:248–273; 35:26–47.

——. 1961. Some pteridosperm seeds from the Calciferous Sandstone Series of Berwickshire. Trans. Roy. Soc. Edinburgh. 44:401–419.

——. 1966. Some Lower Carboniferous fructifications from Berwickshire. Trans. Roy. Soc. Edinburgh. 66:345–375.

——. 1976. Palaeobotanical reminiscences. History Berwick. Nat. Club. 40:179–189.

Lyell, Charles. 1845. Travels in North America. 2 vols. London.

——. 1850. A second visit to the United States of North America. 2 vols. New York.

——. 1872. Principles of Geology. 11th ed. Vol. 1. London.

McCartney, Eugene S. 1924. Fossil lore in Greek and Latin literature. Papers Mich. Acad. Sci. Arts and Letters. 3:23– 38.

MacGinitie, H. D. 1953. Fossil plants of the Florissant beds, Colorado. Carnegie Inst. Washington. Pub. 599. 198 pp.

——. 1969. The Eocene Green River flora of northwestern Colorado and northeastern Utah. Univ. California Press, Pub. Geol. Sci. Vol. 83. 140 pp.

Mamay, Sergius H. 1976. Paleozoic origin of the cycads. U.S. Geol. Survey Prof. Paper No.934. 48 pp.

Martin, William. 1809. Outlines of an attempt to establish a knowledge of extraneous fossils on scientific principles. Macclesfield: J. Wilson.

Martius, K. F. P. de. 1825. On certain antediluvian plants, etc. Edinburgh New Phil. Journ. 12:47–56.

Metcalfe, C. R. 1976. History of the Jodrell Laboratory, Royal Botanic Gardens, Kew, Surrey, England. 35 pp.

Miller, Charles N., Jr. 1971. Evolution of the fern family Osmundaceae based on anatomical studies. Contrib. Mus. Paleont. Univ. Michigan. 23:105–169.

Murray, Peter. 1829. Account of a deposit of fossil plants, etc. Edinburgh New Phil. Journ. 6:311–317.

Nathorst, A. G. 1899. The Swedish Arctic expedition of 1898. Geog. Journ., London. 14:51–76.

Nicol, William. 1834. Observations on the structure of recent and fossil conifers. Edinburgh New Phil. Journ. 16:137–158.

Nordenskiöld, Adolphe E. 1882. The voyage of the Vega round Asia and Europe. Trans. by Alexander Leslie. New York: Macmillan.

North, F. J. 1931. From Giraldus Cambrensis to the geological map. Trans. Cardiff Naturalists Soc. 64:20–97.

Oliver, F. W. 1906. The seed, a chapter in evolution. Brit. Assoc. Adv. Sci. at York, Trans. Pp. 725–738.

——, ed. 1913. Makers of British Botany. Cambridge: Cambridge Univ. Press.

Oliver, F. W., and D. H. Scott, 1903. On *Lagenostoma lomaxi*, the seed of *Lyginodendron*. Proceed. Roy. Soc. London. 71:477–481.

——. 1904. On the structure of the Palaeozoic seed *Lagenostoma lomaxi*. Phil. Trans. Roy. Soc. London. 197B:193–247.

Pant, D. D. 1977. The plant of *Glossopteris*. Journ. Indian Bot. Soc. 56:1–23.

Parkinson, James. 1811. Observations on some of the strata in the neighborhood of London, etc. Trans. Geol. Soc. 1:324–354.

——. 1820. Organic remains of a former world. Vol. 1. London: Sherwood, Neely and Jones.

Parsons, James. 1757. An account of some fossil fruits, and other bodies, found in the island of Shepey. Phil. Trans. Roy. Soc. London. 50:396–407.

Phillips, Tom L., H. W. Pfefferkorn, and R. A. Peppers. 1973. Development of paleobotany in the Illinois Basin. Illinois State Geol. Survey Circular 480. 86 pp.

Phillips, Tom L., M. J. Avcin, and D. Berggren. 1976. Fossil peat of the Illinois Basin. Illinois Geol. Survey Educational Series 11. 39 pp.

Pichi-Sermolli, Rodolfo E. G. 1959. Pteridophyta (pp. 421–493), in: Vistas in Botany, ed. by W. B. Turrill. New York: Pergamon Press.

Plot, Robert. 1677. The natural history of Oxfordshire. London. 358 pp.

Pulteney, Richard. 1790. Historical and biographical sketches of the progress of botany in England from its origin. 2 vols. London.

Ray, John. 1673. Observations topographical, moral & physiological, made on a journey through part of the Low-Countries, Germany, Italy, and France. London: John Martyn.

Reid, Clement. 1899. The origin of the British flora. London: Dulau.

Reid, Clement, and Eleanor M. Reid. 1910. The lignite of Bovey Tracey. Phil. Trans. Roy. Soc. London. 201B:161–178.

Reid, Eleanor M., and Majorie E. J. Chandler. 1933. The London Clay flora. London: Brit. Mus. Nat. Hist. 561 pp.

Riggs, Elmer S. 1926. Fossil hunting in Patagonia. Journ. Amer. Mus. (Nat. Hist.). 26:537–544.

Rudwick, Martin J. S. 1972. The meaning of fossils. London: Macdonald.

Sahni, Birbal. 1920. On the structure and affinities of *Acmopyle pancheri* Pilger. Phil. Trans. Roy. Soc. London. 210B:253–310.

——. 1938. Recent advances in Indian palaeobotany. Lucknow Univ. Studies, No. 2. 100 pp.

——. 1948. The Pentoxyleae: a new group of Jurassic gymnosperms from the Rajmahal Hills of India. Bot. Gazette. 110:47–80.

Scheuchzer, Johann Jakob. 1723. Herbarium Diluvianum. 119 pp.

Schimper, W. P. 1869–74. Traité de paléontologie végétale ou la flore du monde primitif. 3 vols. Paris: J. B. Baillière.

Schlotheim, E. F. 1820. Die Petrefactenkunde auf ihrem jetzigen Standpunkte, etc. Gotha. 436 pp.

Schneer, Cecil. 1954. The rise of historical geology in the seventeenth century. Isis. 45:256–268.

Scott, D. H. 1905a. What were the Carboniferous ferns? Journ. Roy. Microscop. Soc. Pp. 137–149.

——. 1905b. The early history of seed-bearing plants, as recorded in the Carboniferous flora. Mem. and Proceed. Manchester Lit. and Phil. Soc. 49:(12)1–32.

——. 1908. The present position of Paleozoic botany. Smithsonian Ann. Rept. for 1907. Pp. 371–405.

——. 1911. Presidential address. Proceed. Linnean Soc. London. 123:17–29.

——. 1912. Presidential address. Proceed. Linnean Soc. London. 124:26–39.

——. 1925. German reminiscences of the early 'eighties. New Phytologist. 24:9–16.

——. "Letters." A collection of letters written to Scott and preserved in the library of the British Museum (Natural History).

Seward, A. C. 1922. A summer in Greenland. Cambridge: Cambridge Univ. Press.

Sheppard, T. 1922. William Smith: his maps and memoirs. Proceed. Yorkshire Geol. Soc., n.s. 19:75–213.

Stafleu, Franz A. 1966. Brongniart's Histoire des Végétaux Fossiles. Taxon. 15:320–324.

Steinhauer, Henry. 1818. On fossil reliquia of unknown vegetables in the coal strata. Trans. Amer. Phil. Soc., n.s. 1:265–297.

Stopes, M. C. 1914. Palaeobotany: its past and its future. Knowledge. 37:15–24.

Stur, Dionys R. J. 1875–1877. Beiträge zur Kenntnis der Flora der Vorwelt. Die Culm Flora. Kgl. geol. Reichsanst. Part 1, 1875, 8:106 pp.; Part 2, 1877, 8:366 pp.

Szafer, Władysław. 1961. Miocene flora from Stare Gliwice in Upper Silesia. Inst. Geol. Prace 33:205 pp. (English pp. 162–195).

Thomas, H. H. 1925. The Caytoniales, a new group of angiospermous plants from the Jurassic rocks of Yorkshire. Phil. Trans. Roy. Soc. London. 213B:299–363.

——. 1947. The rise of geology and its influence on contemporary thought. Annals of Sci. 5:325–341.

Traverse, Alfred. 1974. 25 years of botany: paleopalynology, 1947–1972. Ann. Missouri Bot. Gard. 61:203–236.

Unger, F. 1865. The sunken island of Atlantis. Journ. Bot. Brit. and Foreign. 3:12–26.

Walton, Johh. 1959. Palaeobotany in Great Britain, in: Vistas in botany, ed. by W. B. Turrill. London: Pergamon Press.

——. 1965. A great friendship, its origin and consequences. 12th Sir Albert Charles Seward Memorial Lecture, December 12, 1964. Birbal Sahni Inst. Paleobot., Lucknow. 10 pp.

Ward, John. 1740. The lives of the professors of Gresham College. London.

Ward, Lester F. 1885. Sketch of paleobotany. 5th Ann. Rept. U.S. Geol. Survey, Washington. Pp. 363–469.

——. 1889. The geographical distribution of fossil plants. 8th Ann. Rept. U.S. Geol. Survey, Washington. Pp. 663–960.

——. 1892. Principles and methods of geologic correlation by means of fossil plants. Amer. Geol. 9:34–47.

——. 1900. Status of the Mesozoic floras of the United States—the older Mesozoic. 20th Ann. Rept. U.S. Geol. Survey, Washington. Pt. 2, pp. 213–430.

White, David. 1904. The seeds of *Aneimites*. Smithsonian Misc. Collections. 47:322–331.

Wieland, G. R. 1906, 1916. American fossil cycads. Vol. 1, 1906, 295 pp. Vol. 2, 1916, 277 pp. Carnegie Inst. Washington. Pub. 34.

——. 1913. The Liassic flora of the Mixteca Alta of Mexico. Amer. Jour. Sci. 36:251–281.

——. 1935. The Cerro Cuadrado petrified forest. Carnegie Inst. Washington. Pub. 449. 180 pp.

Woodward, Horace B. 1911. History of geology. London.
Zeiller, R. 1905. Une nouvelle classe de gymnosperms: les ptéridospermées. Rev. Gén. Sci. 16:718–727.
Zittel, Karl von. 1901. History of geology and paleontology to the end of thc nineteenth century. Trans. by Maria M. Ogilvie-Gordon. London: Walter Scott.

Index

Page numbers in italic indicate illustrations.

Library of Congress Cataloging in Publication Data

Andrews, Henry Nathaniel, 1910–
The fossil hunters.

Bibliography: P.
Includes index.
1. Paleobotany—History. 2. Paleobotanists—
Biography. I. Title.
QE904.A1A5 561′.09 79–24101
ISBN 0-8014-1248-X